Basic Thermodynamics

Basic Thermodynamics

Arjun Sharma

A.J. Books
Ghaziabad (U.P.)

Edition : 2015

ISBN : 978–93–85085–06–2

Published by :

A.J. BOOKS

B-46-A, Nandgram,

Ghaziabad (U.P.)

Printed in India.

Preface

Thermodynamics is a branch of physics concerned with heat and temperature and their relation to energy and work. It defines macroscopic variables, such as internal energy, entropy, and pressure, that partly describe a body of matter or radiation. It states that the behaviour of those variables is subject to general constraints, that are common to all materials, not the peculiar properties of particular materials. These general constraints are expressed in the four laws of thermodynamics. Thermodynamics describes the bulk behaviour of the body, not the microscopic behaviours of the very large numbers of its microscopic constituents, such as molecules. Its laws are explained by statistical mechanics, in terms of the microscopic constituents. Thermodynamics applies to a wide variety of topics in science and engineering. Historically, thermodynamics developed out of a desire to increase the efficiency and power output of early steam engines, particularly through the work of French physicist Nicolas Léonard Sadi Carnot (1824) who believed that the efficiency of heat engines was the key that could help France win the Napoleonic Wars. Initially, thermodynamics, as applied to heat engines, was concerned with the thermal properties of their 'working materials' such as steam, in an effort to increase the efficiency and power output of engines. Thermodynamics later expanded to the study of energy transfers in chemical processes, for example to the investigation, published in 1840, of the heats of chemical reactions by Germain Hess, which was not originally explicitly concerned with the relation between energy exchanges by heat and work. From this evolved the study of Chemical thermodynamics and the role of entropy in chemical reactions.

Thermodynamic systems are theoretical constructions used to model physical systems that exchange matter and energy in terms of the laws of thermodynamics. The study of thermodynamical systems has developed into several related branches, each using a different fundamental model as a theoretical or experimental basis, or applying the principles to varying types of systems. There are three fundamental kinds of entity in thermodynamics, states of a system, processes of a system, and thermodynamic operations. This allows three fundamental approaches to thermodynamic reasoning, that in terms of states of thermodynamic equilibrium of a system, and that in terms of time-invariant processes of a system, and that in terms of cyclic processes of a system. Thermodynamic potentials are different quantitative measures of the stored energy in a system. Potentials are used to measure energy changes in systems as they evolve from an initial state to a final state. The potential used depends on the constraints of the system, such as constant temperature or pressure. For example, the Helmholtz and Gibbs energies are the energies available in a system to do useful work when the temperature and volume or the pressure and temperature are fixed, respectively. Originally thermodynamics concerned material and radiative phenomena that are experimentally reproducible. Gradually, the laws of thermodynamics came to be used to explain phenomena that occur outside the experimental laboratory. For example, phenomena on the scale of the earth's atmosphere cannot be reproduced in a laboratory experiment. But processes in the atmosphere can be modelled by use of thermodynamic ideas, extended well beyond the scope of laboratory equilibrium thermodynamics.

The book presents a clear and simple exposition of thermodynamic principles to enable beginners to penetrate its fundamental ideas buried under a haze of abstractness and to appreciate the logical development of thermodynamic reasoning. It will be an important introductory resource guide for graduate engineering students.

Book writing-whether original or compilation is a laborious complicated job requiring academic labour.

The present work is purely a research compilation based on authoritative publications and researches authored by eminent engineers and subject experts.

Assistance has also been sought from Websites through Internet.

The author, humbly acknowledges the contributions of all those eminent engineers and subject experts alongwith their respective publishers and Websites in preparation of this book.

—Arjun Sharma

Contents

Chapter

1

Introduction

Basic Concepts

The universe is made up of bits of matter, and the bits of matter have mass and energy. Neither mass nor energy may be created or destroyed, but in some special cases interconversions between mass and energy may occur. The things that happen in nature occur because matter is trying to gather to itself as much mass as it can (gravity) and to release as much of its energy as it can (the Sun, for example). This gathering of mass and loss of energy requires a price to be paid - the price of freedom. It seems that mass, energy, and freedom are the coins of the universe. Mass and energy must be conserved: if one part of the universe gathers more "wealth" as mass or energy, those must come from somewhere else in the universe, so there is always competition in the trading of these commodities. Freedom, on the other hand, is not conserved. However, there are some strange restraints on freedom. The freedom of the universe can take many forms, and appears to be able to increase without limit. However, the total freedom of the universe is not allowed to decrease.

The energy or the mass of a part of the universe may increase or decrease, but only if there is a corresponding decrease or increase somewhere else in the universe. The freedom in that part of the universe may increase with no change in the freedom of the rest of the universe. There might be decreases in freedom in the rest of the universe, but the sum of the increase and decrease must result in a net increase. There can be a decrease in the freedom in one part of the universe, but ONLY if there is an equal or greater increase in the rest of the universe.

Most of us have a general idea of what mass and energy are, and we may have a fair understanding of how we can quantify them, or to say how much of them we have. Freedom is a more complicated concept. The freedom within a part of the universe may take two major forms: the freedom of the mass and the freedom of the energy. The amount of freedom is related to the number of different ways the mass or the energy in that part of the universe may be arranged while not gaining or losing any mass or energy. We will concentrate on a specific part of the universe, perhaps within a closed container. If the mass within the container is distributed into a lot of tiny little balls (atoms) flying blindly about, running into each other and anything else (like walls) that may be in their way, there is a huge number of different ways the atoms could be arranged at any one time. Each atom could at different times occupy any place within the container that was not already occupied by another atom, but on average the atoms will be uniformly distributed throughout the container. If we can mathematically estimate the number of different ways the atoms may be arranged, we can quantify the freedom of the mass. If somehow we increase the size of the container, each atom can move around in a greater amount of space, and the number of ways the mass may be arranged will increase.

Now let us turn to the freedom of the energy within the container. If the mass is in the form of atoms flying around, energy is only in the form of the kinetic energy of these atoms, and the energy of the electrons in the atoms moving around their nucleus (and sometimes we have to consider the neutrons, protons, and other stuff moving around in the nucleus). An atom's kinetic energy is related to its mass and its velocity. The energy in the container is the sum of the kinetic energies of all of the atoms, but the velocity is not the same for each atom, and the atoms are continually exchanging this energy through collisions with each other and through collisions with the walls. In the same way that the mass may have freedom in the number of ways the atoms may be arranged in space, the energy may have freedom in the number of ways that the velocities and directions of the atoms may be arranged. The velocities of the molecules are closely related to the temperature.

The energy and freedom of gaseous atoms appear only in velocities and directions, and is called translational energy and translational freedom. In the case of molecules (a molecule is a group of two or more atoms held together by chemical bonds) there are additional freedoms and additional forms of energy. Bonds between atoms act as springs allowing the atoms to vibrate within the molecule, so that the molecules may contain different levels of vibrational energy and vibrational freedom. Additionally, the entire molecule may rotate on different axes, allowing different levels of rotational energy and rotational freedom.

While the freedom of mass is related to the volume in which the mass is distributed, the freedom of energy is related to the temperature. An increase in the temperature of a gas leads directly to an increase in energy, and this can only occur if there is a decrease in energy somewhere else in the universe. We say that energy is transferred to the gas and the container (we will call that the "system")) from somewhere in the remainder of the universe (we will call that the "surroundings"). The increase in energy is accompanied by an increase in the energetic freedom of the system.

The thermodynamic term for quantifying freedom is entropy, and it is given the symbol S. Like freedom, the entropy of a system increases with the temperature and with volume. The effect of volume is more easily seem in terms of concentration, especially in the case of mixtures. For a certain number of atoms or molecules, an increase in volume results in a decrease in concentration. Therefore, the entropy of a system increases as the concentrations of the components decrease. The part of entropy which is determined by energetic freedom is called thermal entropy, and the part that is determined by concentration is called configurational entropy.

Thermodynamic System

A thermodynamic system is a precisely specified macroscopic region of the universe, defined by boundaries or walls of particular natures, together with the physical surroundings of that region, which determine processes that are allowed to affect the interior of the region, studied using the principles of thermodynamics.

All space in the universe outside the thermodynamic system is known as the *surroundings*, the *environment*, or a *reservoir*. A system is separated from its surroundings by a *boundary*, which may be notional or real but, by convention, delimits a finite volume. Exchanges of work, heat, or matter and energy between the system and the surroundings may take place across this boundary. A thermodynamic system is classified by the nature of the exchanges that are allowed to occur across its boundary.

A thermodynamic system has a characteristic set of thermodynamic parametres, or state variables. They are experimentally measurable macroscopic properties, such as volume, pressure, temperature, electric field, and others. The set of values of the thermodynamic parametres necessary to uniquely define a state of a system is called the thermodynamic state of a system. The state variables of a system are normally related by one or more functional relationships, the equations of state.

An isolated system is in thermodynamic equilibrium when its state does not change with time. Systems that are not isolated can be in thermodynamic equilibrium, or in a state constant or precisely cyclically changing in time - a steady state - that is far from equilibrium, or they can be in a changing state that is not in thermodynamic equilibrium. Classical thermodynamics considers only states of thermodynamic equilibrium or states constant or precisely cycling in time. Originally, in 1824, Sadi Carnot described a thermodynamic system as the working substance of a heat engine under study.

Overview

Thermodynamics describes the physics of matter using the concept of the *thermodynamic system*, a region of the universe that is under study. All quantities, such as pressure or mechanical work, in an equation refer to the system unless labelled otherwise. As thermodynamics is fundamentally concerned with the flow and balance of energy and matter, systems are distinguished depending on the kinds of interaction they undergo and the types of energy they exchange with the surrounding environment.

Interactions of thermodynamic systems			
Type of system	***Mass flow***	***Work***	***Heat***
Open	✓	✓	✓
Closed	✗	✓	✓
Thermally isolated	✗	✓	✗
Mechanically isolated	✗	✗	✓
Isolated	✗	✗	✗

Isolated systems are completely isolated from their environment. They do not exchange heat, work or matter with their environment.

The only truly isolated system there could be is the universe, but even that is up for debate if the Big Bang is considered. *Closed systems* are able to exchange energy (heat and work) but not matter with their environment. A greenhouse is an example of a closed system exchanging heat but not work with its environment. Whether a system exchanges heat, work or both is usually thought of as a property of its boundary. *Open systems* may exchange any form of energy as well as matter with their environment. A boundary allowing matter exchange is called permeable. The ocean would be an example of an open system.

In practice, a system can never be absolutely isolated from its environment, because there is always at least some slight coupling, such as gravitational attraction. In analyzing a system in steady-state, the energy into the system is equal to the energy leaving the system. An example system is the system of hot liquid water and solid table salt in a sealed, insulated test tube held in a vacuum (the surroundings). The test tube constantly loses heat in the form of black-body radiation, but the heat loss progresses very slowly. If there is another process going on in the test tube, for example the dissolution of the salt crystals, it probably occurs so quickly that any heat lost to the test tube during that time can be neglected. Thermodynamics in general does not measure time, but it does sometimes accept limitations on the time frame of a process.

History

The first to develop the concept of a thermodynamic system was the French physicist Sadi Carnot whose 1824 *Reflections on the Motive Power of Fire* studied what he called the *working substance*, e.g., typically a body of water vapour, in steam engines, in regards to the system's ability to do work when heat is applied to it. The working substance could be put in contact with either a heat reservoir (a boiler), a cold reservoir (a stream of cold water), or a piston (to which the working body could do work by pushing on it). In 1850, the German physicist Rudolf Clausius generalized this picture to include the concept of the surroundings, and began referring to the system as a "working body." In his 1850 manuscript *On the Motive Power of Fire*, Clausius wrote:

> *"With every change of volume (to the working body) a certain amount work must be done by the gas or upon it, since by its expansion it overcomes an external pressure, and since its compression can be brought about only by an exertion of external pressure. To this excess*

of work done by the gas or upon it there must correspond, by our principle, a proportional excess of heat consumed or produced, and the gas cannot give up to the "surrounding medium" the same amount of heat as it receives."

The article Carnot heat engine shows the original piston-and-cylinder diagram used by Carnot in discussing his ideal engine; below, we see the Carnot engine as is typically modelled in current use:

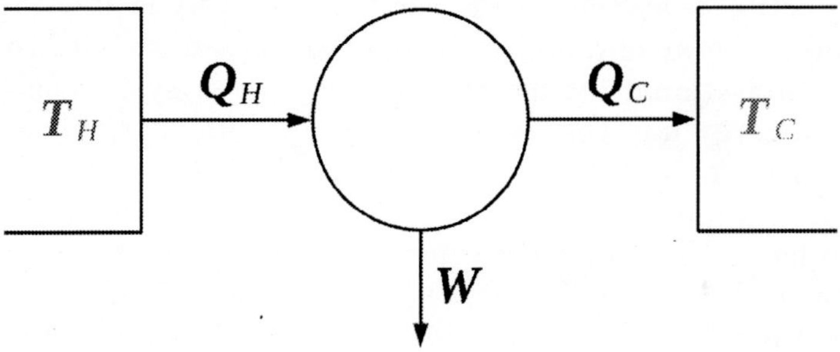

Figure: *Carnot engine diagram (modern) - where heat flows from a high temperature* T_H *furnace through the fluid of the "working body" (working substance) and into the cold sink* T_C*, thus forcing the working substance to do mechanical work*W *on the surroundings, via cycles of contractions and expansions.*

In the diagram shown, the "working body" (system), a term introduced by Clausius in 1850, can be any fluid or vapour body through which heat$Q$ can be introduced or transmitted through to produce work. In 1824, Sadi Carnot, in his famous paper *Reflections on the Motive Power of Fire*, had postulated that the fluid body could be any substance capable of expansion, such as vapour of water, vapour of alcohol, vapour of mercury, a permanent gas, or air, etc. Though, in these early years, engines came in a number of configurations, typically Q_H was supplied by a boiler, wherein water boiled over a furnace; Q_C was typically a stream of cold flowing water in the form of a condenser located on a separate part of the engine. The output work W was the movement of the piston as it turned a crank-arm, which typically turned a pulley to lift water out of flooded salt mines. Carnot defined work as "weight lifted through a height."

Boundary

A system boundary is a real or imaginary two-dimensional closed surface that encloses or demarcates the volume or region that a thermodynamic system occupies, across which quantities such as heat, mass, or work can flow. In short, a thermodynamic boundary is a geometrical division between a system and its surroundings. Topologically, it is usually considered nearly or piecewise smoothly homeomorphic with a two-sphere, because a system is usually considered simply connected.

Boundaries can also be fixed (e.g. a constant volume reactor) or moveable (e.g. a piston). For example, in a reciprocating engine, a fixed boundary means the piston is locked at its position; as such, a constant volume process occurs. In that same engine, a moveable boundary allows the piston to move in and out. Boundaries may be real or imaginary. For closed systems, boundaries are real while for open system boundaries are often imaginary. For theoretical purposes, a boundary may be declared adiabatic, isothermal, diathermal, insulating, permeable, or semipermeable—but actual physical materials that provide such idealized properties are not always readily available.

Anything that passes across the boundary that effects a change in the internal energy must be accounted for in the energy balance equation. The volume can be the region surrounding a single atom resonating energy, such as Max Planck defined in 1900; it can be a body of steam or air in a steam engine, such as Sadi Carnot defined in 1824; it can be the body of a tropical cyclone, such as Kerry Emanuel theorized in 1986 in the field of atmospheric thermodynamics; it could also be just one nuclide (i.e. a system of quarks) as hypothesized in quantum thermodynamics.

Surroundings

The system is the part of the universe being studied, while the *surroundings* is the remainder of the universe that lies outside the boundaries of the system. It is also known as the *environment*, and the *reservoir*. Depending on the type of system, it may interact with the system by exchanging mass, energy (including heat and work), momentum, electric charge, or other conserved properties. The environment is ignored in analysis of the system, except in regards to these interactions.

Open System

In open systems, matter may flow in and out of the system boundaries. The first law of thermodynamics for open systems states: *the increase in the internal energy of a system is equal to the amount of energy added to the system by matter flowing in and by heating, minus the amount lost by matter flowing out and in the form of work done by the system.*

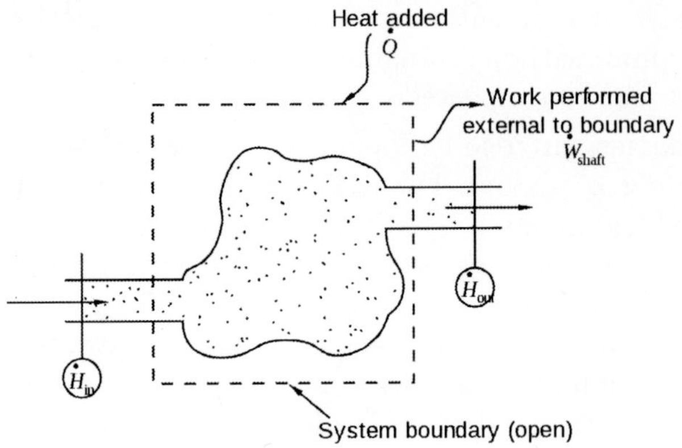

Figure: *During steady, continuous operation, an energy balance applied to an open system equates shaft work performed by the system to heat added plus net enthalpy added.*

The first law for open systems is given by:

$$dU = dU_{in} + \delta Q - dU_{out} - \delta W$$

where U_{in} is the average internal energy entering the system and U_{out} is the average internal energy leaving the system.

The region of space enclosed by open system boundaries is usually called a control volume, and it may or may not correspond to physical walls. If we choose the shape of the control volume such that all flow in or out occurs perpendicular to its surface, then the flow of matter into the system performs work as if it were a piston of fluid pushing mass into the system, and the system performs work on the flow of matter out as if it were driving a piston of fluid. There are then, two types of work performed: *flow work* described above, which is performed on the fluid (this is also often called *PV work*) and *shaft work*, which may be performed on some mechanical device. These two types of work are expressed in the equation:

$$\delta W = d(P_{out}V_{out}) - d(P_{in}V_{in}) + \delta W_{shaft}$$

Substitution into the equation above for the control volume *cv* yields:

$$dU_{cv} = dU_{in} + d(P_{in}V_{in}) - dU_{out} - d(P_{out}V_{out}) + \delta Q - \delta W_{shaft}$$

The definition of enthalpy, *H*, permits us to use this thermodynamic potential to account for both internal energy and PV work in fluids for open systems:

$$dU_{cv} = dH_{in} - dH_{out} + \delta Q - \delta W_{shaft}$$

During steady-state operation of a device, any system property within the control volume is independent of time. Therefore, the internal energy of the system enclosed by the control volume remains constant, which implies that dU_{cv} in the expression above may be set equal to zero. This yields a useful expression for the power generation or requirement for these devices in the absence of chemical reactions:

$$\frac{\delta W_{shaft}}{dt} = \frac{dH_{in}}{dt} - \frac{dH_{out}}{dt} + \frac{\delta Q}{dt}$$

This expression is described by the diagram above.

Closed System

In a closed system, no mass may be transferred in or out of the system boundaries. The system always contains the same amount of matter, but heat and work can be exchanged across the boundary of the system. Whether a system can exchange heat, work, or both is dependent on the property of its boundary.

- Adiabatic boundary – not allowing any heat exchange: A thermally isolated system
- Rigid boundary – not allowing exchange of work: A mechanically isolated system

One example is fluid being compressed by a piston in a cylinder. Another example of a closed system is a bomb calorimetre, a type of constant-volume calorimetre used in measuring the heat of combustion of a particular reaction. Electrical energy travels across the boundary to produce a spark between the electrodes and initiates combustion. Heat transfer occurs across the boundary after combustion but no mass transfer takes place either way.

Beginning with the first law of thermodynamics for an open system, this is expressed as:

$$\Delta U = Q - W + m_i(h + \frac{1}{2}v^2 + gz)_i - m_e(h + \frac{1}{2}v^2 + gz)_e$$

where U is internal energy, Q is the heat added to the system, W is the work done by the system, and since no mass is transferred in or out of the system, both expressions involving mass flow are zero and the first law of thermodynamics for a closed system is derived. The first law of thermodynamics for a closed system states that the increase of internal energy of the system equals the amount of heat added to the system minus the work done by the system. For infinitesimal changes the first law for closed systems is stated by:

$$dU = \delta Q - \delta W.$$

If the work is due to a volume expansion by dV at a pressure P than:

$$\delta W = PdV.$$

For a homogeneous system, in which only reversible processes can take place, the second law of thermodynamics reads:

$$\delta Q = TdS$$

where T is the absolute temperature and S is the entropy of the system. With these relations the fundamental thermodynamic relationship, used to compute changes in internal energy, is expressed as:

$$\delta U = TdS - PdV.$$

For a simple system, with only one type of particle (atom or molecule), a closed system amounts to a constant number of particles. However, for systems undergoing a chemical reaction, there may be all sorts of molecules being generated and destroyed by the reaction process. In this case, the fact that the system is closed is expressed by stating that the total number of each elemental atom is conserved, no matter what kind of molecule it may be a part of. Mathematically:

$$\sum_{j=1}^{m} a_{ij} N_j = b_i^0$$

where N_j is the number of j-type molecules, a_{ij} is the number of atoms of element i in molecule j and b_i^0 is the total number of atoms of element i in the system, which remains constant, since the system is closed. There is one such equation for each element in the system.

Isolated System

An isolated system is more restrictive than a closed system as it does not interact with its surroundings in any way. Mass and energy remains constant within the system, and no energy or mass transfer takes place across the boundary. As time passes in an isolated system, internal differences in the system tend to even out and pressures and temperatures tend to equalize, as do density differences. A system in which all equalizing processes have gone practically to completion is in a state of thermodynamic equilibrium.

Truly isolated physical systems do not exist in reality (except perhaps for the universe as a whole), because, for example, there is always gravity between a system with mass and masses elsewhere. However, real systems may behave nearly as an isolated system for finite (possibly very long) times. The concept of an isolated system can serve as a useful model approximating many real-world situations. It is an acceptable idealization used in constructing mathematical models of certain natural phenomena.

In the attempt to justify the postulate of entropy increase in the second law of thermodynamics, Boltzmann's H-theorem used equations, which assumed that a system (for example, a gas) was isolated. That is all the mechanical degrees of freedom could be specified, treating the walls simply as mirrorboundary conditions. This inevitably led to Loschmidt's paradox. However, if the stochastic behaviour of the molecules in actual walls is considered, along with the randomizing effect of the ambient, background thermal radiation, Boltzmann's assumption of molecular chaos can be justified.

The second law of thermodynamics for isolated systems states that the entropy of an isolated system not in equilibrium tends to increase over time, approaching maximum value at equilibrium. Overall, in an isolated system, the internal energy is constant and the entropy can never decrease. A *closed* system's entropy can decrease e.g. when heat is extracted from the system. It is important to note that isolated systems are not equivalent to closed systems. Closed systems cannot exchange matter with the surroundings, but can exchange energy. Isolated systems can exchange neither matter nor energy with their surroundings, and as such are only theoretical and do not exist in reality (except, possibly, the entire universe). It is worth noting that 'closed system' is often used in thermodynamics discussions when 'isolated system' would be correct - i.e. there is an assumption that energy does not enter or leave the system.

Mechanically Isolated System

A mechanically isolated system can exchange no work energy with its environment, but may exchange heat energy and/or mass with its environment. The internal energy of a mechanically isolated system may therefore change due to the exchange of heat energy and mass. For a simple system, mechanical isolation is equivalent to constant volume and any process which occurs in such a simple system is said to be isochoric.

Systems in Equilibrium

At thermodynamic equilibrium, a system's properties are, by definition, unchanging in time. Systems in equilibrium are much simpler and easier to understand than systems not in equilibrium. Often, when analyzing a thermodynamic process, it can be assumed that each intermediate state in the process is at equilibrium. This also considerably simplifies the analysis.

In isolated systems it is consistently observed that as time goes on internal rearrangements diminish and stable conditions are approached. Pressures and temperatures tend to equalize, and matter arranges itself into one or a few relatively homogeneous phases. A system in which all processes of change have gone practically to completion is considered in a state of thermodynamic equilibrium. The thermodynamic properties of a system in equilibrium are unchanging in time. Equilibrium system states are much easier to describe in a deterministic manner than non-equilibrium states.

In thermodynamic processes, large departures from equilibrium during intermediate steps are associated with increases in entropy and increases in the production of heat rather than useful work. It can be shown that for a process to be reversible, each step in the process must be reversible. For a step in a process to be reversible, the system must be in equilibrium throughout the step. That ideal cannot be accomplished in practice because no step can be taken without perturbing the system from equilibrium, but the ideal can be approached by making changes slowly.

Control Volume

In fluid mechanics and thermodynamics, a control volume is a mathematical abstraction employed in the process of creating mathematical models of physical processes. In an inertial frame of reference, it is a volume fixed in space or moving with constant velocity through which the fluid (gas or liquid) flows. The surface enclosing the control volume is referred to as the control surface.

At steady state, a control volume can be thought of as an arbitrary volume in which the mass of the fluid remains constant. As fluid moves through the control volume, the mass entering the control volume is equal to the mass leaving the control volume. At steady state, and in the absence of work and heat transfer, the energy within the control volume remains constant. It is analogous to the classical mechanics concept of the free body diagram.

Overview

Typically, to understand how a given physical law applies to the system under consideration, one first begins by considering how it applies to a small, control volume, or “representative volume”. There is nothing special about a particular control volume, it simply represents a small part of the system to which physical laws can be easily applied. This gives rise to what is termed a volumetric, or volume-wise formulation of the mathematical model. One can then argue that since the physical laws behave in a certain way on a particular control volume, they behave the same way on all such volumes, since that particular control volume was not special in any way. In this way, the corresponding point-wise formulation of the mathematical model can be developed so it can describe the physical behaviour of an entire (and maybe more complex) system.

In fluid mechanics the conservation equations (for instance, the Navier-Stokes equations) are in integral form. They therefore apply on volumes. Finding forms of the equation that are *independent* of the control volumes allows simplification of the integral signs.

Substantive Derivative

Computations in fluid mechanics often require that the regular time derivation operator d/dt is replaced by the substantive derivative operator D/Dt. This can be seen as follows. Consider a bug that is moving through a volume where there is some scalar, e.g. pressure, that varies with time and position: $p = p(t,x,y,z)$.

If the bug during the time interval from t to $t+dt$ moves from (x,y,z) to $(x+dx, y+dy, z+dz)$, then the bug experiences a change dp in the scalar value,

$$dp = \frac{\partial p}{\partial t}dt + \frac{\partial p}{\partial x}dx + \frac{\partial p}{\partial y}dy + \frac{\partial p}{\partial z}dz$$

(the total differential). If the bug is moving with velocity $\mathrm{v} = (v_x, v_y, v_z)$, the change in position is $\mathrm{v}dt = (v_x dt, v_y dt, v_z dt)$, and we may write

$$\begin{aligned} dp &= \frac{\partial p}{\partial t}dt + \frac{\partial p}{\partial x}v_x dt + \frac{\partial p}{\partial y}v_y dt + \frac{\partial p}{\partial z}v_z dt \\ &= \left(\frac{\partial p}{\partial t} + \frac{\partial p}{\partial x}v_x + \frac{\partial p}{\partial y}v_y + \frac{\partial p}{\partial z}v_z\right)dt \\ &= \left(\frac{\partial p}{\partial t} + \mathrm{v}\cdot\nabla p\right)dt. \end{aligned}$$

where ∇p is the gradient of the scalar field p. If the bug is just a fluid particle moving with the fluid’s velocity field, the same formula applies, but now the velocity vector is that of the fluid. The last parenthesized expression is the substantive derivative of the scalar pressure. Since the pressure p in this computation is an arbitrary scalar field, we may abstract it and write the substantive derivative operator as

$$\frac{D}{Dt} = \frac{\partial}{\partial t} + \mathrm{v}\cdot\nabla. \tag{1}$$

Microscopic and Macroscopic Descriptions of a System

The microscopic description of a system is the complete description of each particle in this system. In the above example, the microscopic description of the gas would be the list of the state of each molecule: position and velocity in this problem. It would require a great deal of data for this description; there are roughly 10^{19} molecules in a cube of air one centimetre on a side at room temperature and pressure. The macroscopic description, which is in terms of a few (two!) properties is thus far more accessible and useable for engineering applications, although it is restricted to equilibrium states.

To address the description of entropy on a microscopic level, we need to state some results concerning microscopic systems. These results and the computations and arguments below are taken almost entirely from the excellent discussion of Engineering Thermodynamics by Reynolds and Perkins.

For a given macroscopic system, there are many microscopic states. A key idea from quantum mechanics is that the states of atoms, molecules, and entire systems are discretely quantized. This means that a system of particles under certain constraints, like being in a box of a specified size, or having a fixed total energy, can exist in a finite number of allowed microscopic states. This number can be very big, but it is finite. The microstates of the system keep changing with time from one quantum state to another as molecules move and collide with one another. The probability for the system to be in a particular quantum state is defined by its quantum-state probability p_i. The set of the p_i is called the distribution of probability. The sum of the probabilities of all the allowed quantum states must be unity, hence for any time t,

$$\sum_i p_i = 1. \tag{2}$$

When the system reaches equilibrium, the individual molecules still change from one quantum state to another. In equilibrium, however, the system state does not change with time; so the probabilities for the different quantum states are independent of time.

This distribution is then called the equilibrium distribution, and the probability p_i can be viewed as the fraction of time a system spends in the ith quantum state. In what follows, we limit consideration to equilibrium states.

We can get back to macroscopic quantities from the microscopic description using the probability distribution. For instance, the macroscopic energy of the system would be the weighted average of the successive energies of the system (the energies of the quantum states); the energies are weighted by the relative time the system spends in the corresponding microstates. In terms of probabilities, the average energy, $\langle E \rangle$, is

$$\langle E \rangle = \sum_i p_i \varepsilon_i, \tag{3}$$

where ε_i is the energy of a quantum state.

The probability distribution provides information on the randomness of the equilibrium quantum states. For example, suppose the system can only exist in three states (1, 2 and 3). If the distribution probability is

$$p_1 = 1, \quad p_2 = 0, \quad p_3 = 0,$$

the system is in quantum state 1 and there is no randomness. If we were asked what quantum state the system is in, we would be able to say it is always in state 1. If the distribution were

$$p_1 = 0.3, \; p_2 = 0.2, \; p_3 = 0.5,$$

or

$$p_1 = 0.5, \; p_2 = 0.2, \; p_3 = 0.3,$$

the randomness would not be zero and would be equal in both cases. We would be more uncertain about the instantaneous quantum state than in the first situation.

Maximum randomness corresponds to the case where the three states are equally probable:

$$p_1 = 1/3, \; p_2 = 1/3, \; p_3 = 1/3.$$

In this case, we can only guess the instantaneous state with 33% probability.

Concept of Continuum

Matter exists in molecules. Solids have a great cohesive forece and hence they take any shape. Liquids also have good cohesive force between molecules but not as much as solids. They take the shape of the container. Gases have very less cohesive force between molecles and hence they move randomly in the given volume. In the Macroscopic point of view, the volumes considerd are large compared to the size of the molecules. It is assumed that the volume under consideration will have enough number of molecules present in it, so that the definition of density, mass etc. will not alter eventhough there is movement of molecules in and out.

Let us take a volume comparable to the volume of molecule. Due to the random movement of the molecules, the volume under consideration may have molecules at an instant and may not have molecules at another instant. In that case, the definition of density and other properties will not have any meaning. Hence it is considered that, always, the volume under consideration will have many molecules so that the definition of the properties will have meaning.

Thermodynamic Equilibrium

In thermodynamics, a thermodynamic system is in thermodynamic equilibrium when it is in thermal equilibrium, mechanical equilibrium, radiative equilibrium, and chemical equilibrium. Equilibrium means a state of balance. In a state of thermodynamic equilibrium, there are no net flows of matter or of energy, no phase changes, and no unbalanced potentials (or driving forces), within the system. A system that is in thermodynamic equilibrium experiences no changes when it is isolated from its surroundings.

In non-equilibrium systems there are net flows of matter or energy, or phase changes are occurring; if such changes can be triggered to occur in a system in which they are not already occurring, it is said to be in a metastable equilibrium.

When a body of material starts from a non-equilibrium state of inhomogeneity or chemical non-equilibrium, and is then isolated, it spontaneously evolves towards its own internal state of thermodynamic equilibrium. It is not necessary that all aspects of internal thermodynamic equilibrium be reached simultaneously; some can be established before others.

Overview

Classical thermodynamics deals with states of dynamic equilibrium. The state of a system at thermodynamic equilibrium is the one for which some thermodynamic potential is minimized, or for which the entropy (S) is maximized, for specified conditions. One such potential is the Helmholtz free energy (A), for a system with surroundings at controlled constant temperature and volume:

$$A = U - TS.$$

Another potential, the Gibbs free energy (G), is minimized at thermodynamic equilibrium in a system with surroundings at controlled constant temperature and pressure:

$$G = U - TS + PV.$$

where T denotes the absolute thermodynamic temperature, P the pressure, S the entropy, V the volume, and U the internal energy of the system.

Thermodynamic equilibrium is the unique stable stationary state that is approached or eventually reached as the system interacts with its surroundings over a long time. The above-mentioned potentials are mathematically constructed to be the thermodynamic quantities that are minimized under the particular conditions in the specified surroundings.

Conditions for Thermodynamic Equilibrium

- For a completely isolated system, S is maximum at thermodynamic equilibrium.
- For a system with controlled constant temperature and volume, A is minimum at thermodynamic equilibrium.
- For a system with controlled constant temperature and pressure, G is minimum at thermodynamic equilibrium.

The various types of equilibriums are achieved as follows:

- Two systems are in *thermal equilibrium* when their temperatures are the same.
- Two systems are in *mechanical equilibrium* when their pressures are the same.
- Two systems are in *diffusive equilibrium* when their chemical potentials are the same.
- All forces are balanced.

Relation of Exchange Equilibrium Between Systems

Often the surroundings of a thermodynamic system may also be regarded as another thermodynamic system. In this view, one may consider the system and its surroundings as two systems in mutual contact, with long-range forces also linking them. The enclosure of the system is the surface of contiguity or boundary between the two systems. In the thermodynamic formalism, that surface is regarded as having specific properties of permeability. For example, the surface of contiguity may be supposed to be permeable only to heat, allowing energy to transfer only as heat. Then the two systems are said to be in thermal equilibrium when the long-range forces are unchanging in time and the time rate of transfer of energy as heat between them has reached an absolutely stable stationary state; this is an example of a contact equilibrium. Other kinds of contact equilibrium are defined by other kinds of specific permeability. Contact equilibrium is an example of a relation of exchange equilibrium between systems. When two systems are in contact equilibrium with respect to a particular kind of permeability, they have common values of the intensive variables that belong to that particular kind of permeability. Examples of such intensive variables are temperature, pressure, chemical potential.

A contact equilibrium may be regarded also as an exchange equilibrium. There is a zero balance of transfer of some quantity between the two systems in contact equilibrium. For example, for a wall permeable only to heat, the rates of diffusion of internal energy as heat between the two systems are equal and opposite. For transfer across an adiabatic wall between the two systems, the "permeability" is to energy as work; the rates of transfer of energy as work between them are equal and opposite. If the wall is a simple wall, then the rates of transfer of volume across it are also equal and opposite; and the pressures on either side of it are equal. If the adiabatic wall is more complicated, with a sort of leverage, having an area-ratio, then the pressures of the two systems in exchange equilibrium are in the inverse ratio of the volume exchange ratio; this keeps the zero balance of rates of transfer as work.

Thermodynamic State of Internal Equilibrium of a System

A collection of matter may be entirely isolated from its surroundings. Then if left undisturbed for an indefinitely long time, it reaches a state in which no changes occur within it, and there are no flows within it. This is a thermodynamic state of internal equilibrium. Such states are a principal concern in what is known as classical or equilibrium thermodynamics, for they are the only states of the system that are regarded as well defined in that subject. A system in contact equilibrium with another system can by a thermodynamic operation be isolated, and upon the event of isolation, no change occurs in it. A system in a relation of contact equilibrium with another system may thus also be regarded as being in its own state of internal thermodynamic equilibrium.

Multiple Contact Equilibrium

The thermodynamic formalism allows that a system may have contact with several other systems at once, which may or may not also have mutual contact, the contacts having respectively different permeabilities. If these systems are all jointly isolated from the rest of the world, those of them that are in contact then reach respective contact equilibria with one another. If several systems are free of adiabatic walls between each other, but are jointly isolated from the rest of the world, then they reach a state of multiple contact equilibrium, and they have a common temperature, and a joint total internal energy, and a joint total entropy.

In the formalism of equilibrium thermodynamics, one of the systems may be designated as the system of interest (here now referred to as the system), while the others (here now referred to as the surrounding subsystems) may then be regarded as its surroundings, all of them jointly being isolated from the rest of the world, and still no surrounding subsystem in contact with the system being separated from it by an adiabatic wall. Immovable adiabatic walls are then allowed between the surrounding subsystems, and initial thermodynamic equilibrium is not required across those walls. Still, in order that the system be in contact equilibrium with its surroundings, the system, and each of the surrounding subsystems that is in contact with the system, and each of the surrounding subsystems that is not separated from other surrounding subsystems by an adiabatic wall, must share the common temperature.

It is then allowed that thermodynamic operations occur as events within a definitely limited time in the surroundings, though directly affecting neither the contact of the system of interest with its surroundings, nor its interior; for example, an immovable adiabatic wall may be placed or removed within the surroundings. Consequently, the system is for a time driven away from its own initial internal state of thermodynamic equilibrium, and then the whole is allowed to reach a new eventual and final mutual equilibrium. This train of events is called a thermodynamic process, and is allowed in equilibrium thermodynamics just because the initial and final states are of thermodynamic equilibrium, even though there is transient intermediate departure from thermodynamic equilibrium, when neither the system nor its surroundings are in well defined states of internal equilibrium. Such processes, more or less describable within the thermodynamic formalism, do not exhaust the class of permissible thermodynamic processes; engineered machines and artificial devices are permitted within the surroundings; further account of this is beyond the scope of the present article. The allowance of such operations and devices in the surroundings but not in the system is the reason why Kelvin in one of his statements of the second law of thermodynamics spoke of "inanimate" agency; a system in thermodynamic equilibrium is inanimate.

It is often convenient to suppose that some of the surrounding subsystems are so much larger than the system that the process can affect the intensive variables only of the surrounding subsystems, and they are then called reservoirs for relevant intensive variables.

Local and Global Equilibrium

It is useful to distinguish between global and local thermodynamic equilibrium. In thermodynamics, exchanges within a system and between the system and the outside are controlled by intensive parametres. As an example, temperature controls heat exchanges. *Global thermodynamic equilibrium* (GTE) means that those intensive parametres are homogeneous throughout the whole system, while *local thermodynamic equilibrium* (LTE) means that those intensive parametres are varying in space and time, but are varying so slowly that, for any point, one can assume thermodynamic equilibrium in some neighbourhood about that point.

If the description of the system requires variations in the intensive parametres that are too large, the very assumptions upon which the definitions of these intensive parametres are based will break down, and the system will be in neither global nor local equilibrium. For example, it takes a certain number of collisions for a particle to equilibrate to its surroundings. If the average distance it has moved during these collisions removes it from the neighbourhood it is equilibrating to, it will never equilibrate, and there will be no LTE. Temperature is, by definition, proportional to the average internal energy of an equilibrated neighbourhood. Since there is no equilibrated neighbourhood, the concept of temperature breaks down, and the temperature becomes undefined.

It is important to note that this local equilibrium may apply only to a certain subset of particles in the system. For example, LTE is usually applied only to massive particles. In a radiating gas, the photons being emitted and absorbed by the gas need not be in thermodynamic equilibrium with each other or with the massive particles of the gas in order for LTE to exist. In some cases, it is not considered necessary for free electrons to be in equilibrium with the much more massive atoms or molecules for LTE to exist.

As an example, LTE will exist in a glass of water that contains a melting ice cube. The temperature inside the glass can be defined at any point, but it is colder near the ice cube than far away from it. If energies of the molecules located near a given point are observed, they will be distributed according to the Maxwell-Boltzmann distribution for a certain temperature. If the energies of the molecules located near another point are observed, they will be distributed according to the Maxwell-Boltzmann distribution for another temperature.

Local thermodynamic equilibrium does not require either local or global stationarity. In other words, each small locality need not have a constant temperature. However, it does require that each small locality change slowly enough to practically sustain its local Maxwell-Boltzmann distribution of molecular velocities. A global non-equilibrium state can be stably stationary only if it is maintained by exchanges between the system and the outside. For example, a globally stably stationary state could be maintained inside the glass of water by continuously adding finely powdered ice into it in order to compensate for the melting, and continuously draining off the meltwater. Transport phenomena are processes that lead a system from local to global thermodynamic equilibrium. Going back to our example, the diffusion of heat will lead our glass of water towards global thermodynamic equilibrium, a state in which the temperature of the glass is completely homogeneous.

Reservations

Careful and well informed writers about thermodynamics, in their accounts of thermodynamic equilibrium, often enough make provisos or reservations to their statements. Some writers leave such reservations merely implied or more or less unstated.

For example, one widely cited writer, H. B. Callen writes in this context: "In actuality, few systems are in absolute and true equilibrium." He refers to radioactive processes and remarks that they may take "cosmic times to complete, [and] generally can be ignored". He adds "In practice, the criterion for equilibrium is circular. *Operationally, a system is in an equilibrium state if its properties are consistently described by thermodynamic theory!*"

J.A. Beattie and I. Oppenheim write: "Insistence on a strict interpretation of the definition of equilibrium would rule out the application of thermodynamics to practically all states of real systems."

Another author, cited by Callen as giving a "scholarly and rigorous treatment", and cited by Adkins as having written a "classic text", A.B. Pippard writes in that text: "Given long enough a supercooled vapour will eventually condense,.... The time involved may be so enormous, however, perhaps 10^{100} years or more,.... For most purposes, provided the rapid change is not artificially stimulated, the systems may be regarded as being in equilibrium."

Another author, A. Münster, writes in this context. He observes that thermonuclear processes often occur so slowly that they can be ignored in thermodynamics. He comments: "The concept 'absolute equilibrium' or 'equilibrium with respect to all imaginable processes', has therefore, no physical significance." He therefore states that: "... we can consider an equilibrium only with respect to specified processes and defined experimental conditions."

According to L. Tisza: "... in the discussion of phenomena near absolute zero. The absolute predictions of the classical theory become particularly vague because the occurrence of frozen-in nonequilibrium states is very common."

Definitions

The most general kind of thermodynamic equilibrium of a system is through contact with the surroundings that allows simultaneous passages of all chemical substances and all kinds of energy. A system in thermodynamic equilibrium may move with uniform acceleration through space but must not change its shape or size while doing so; thus it is defined by a rigid volume in space. It may lie within external fields of force, determined by external factors of far greater extent than the system itself, so that events within the system cannot in an appreciable amount affect the external fields of force. The system can be in thermodynamic equilibrium only if the external force fields are uniform, and are determining its uniform acceleration, or if it lies in a non-uniform force field but is held stationary there by local forces, such as mechanical pressures, on its surface.

Thermodynamic equilibrium is a primitive notion of the theory of thermodynamics. According to P.M. Morse: "It should be emphasized that the fact that there are thermodynamic states,..., and the fact that there are thermodynamic variables which are uniquely specified by the equilibrium state... are *not* conclusions deduced logically from some philosophical first principles. They are conclusions ineluctably drawn from more than two centuries of experiments." This means that thermodynamic equilibrium is not to be defined solely in terms of other theoretical concepts of thermodynamics. M. Bailyn proposes a fundamental law of thermodynamics that defines and postulates the existence of states of thermodynamic equilibrium.

Textbook definitions of thermodynamic equilibrium are often stated carefully, with some reservation or other.

For example, A. Münster writes: "An isolated system is in thermodynamic equilibrium when, in the system, no changes of state are occurring at a measurable rate." There are two reservations stated here; the system is isolated; any changes of state are immeasurably slow. He discusses the second proviso by giving an account of a mixture oxygen and hydrogen at room temperature in the absence of a catalyst. Münster points out that a thermodynamic equilibrium state is described by fewer macroscopic variables than is any other state of a given system. This is partly, but not entirely, because all flows within and through the system are zero.

R. Haase's presentation of thermodynamics does not start with a restriction to thermodynamic equilibrium because he intends to allow for non-equilibrium thermodynamics. He considers an arbitrary system with time invariant properties. He tests it for thermodynamic equilibrium by cutting it off from all external influences, except external force fields. If after insulation, nothing changes, he says that the system was in *equilibrium*.

In a section headed "Thermodynamic equilibrium", H.B. Callen defines equilbrium states in a paragraph. He points out that they "are determined by intrinsic factors" within the system. They are "terminal states", towards which the systems evolve, over time, which may occur with "glacial slowness". This statement does not explicitly say that for thermodynamic equilibrium, the system must be isolated; Callen does not spell out what he means by the words "intrinsic factors".

Another textbook writer, C.J. Adkins, explicitly allows thermodynamic equilibrium to occur in a system which is not isolated. His system is, however, closed with respect to transfer of matter. He writes: "In general, the approach to thermodynamic equilibrium will involve both thermal and work-like interactions with the surroundings." He distinguishes such thermodynamic equilibrium from thermal equilibrium, in which only thermal contact is mediating transfer of energy.

Another textbook author, J.R. Partington, writes: "(i) *An equilibrium state is one which is independent of time.*" But, referring to systems "which are only apparently in equilibrium", he adds : "Such systems are in states of 3 false equilibrium. " Partington's statement does not explicitly state that the equilibrium refers to an isolated system. Like Münster, Partington also refers to the mixture of oxygen and hydrogen. He adds a proviso that "In a true equilibrium state, the smallest change of any external condition which influences the state will produce a small change of state..." This proviso means that thermodynamic equilibrium must be stable against small perturbations; this requirement is essential for the strict meaning of thermodynamic equilibrium.

A student textbook by F.H. Crawford has a section headed "Thermodynamic Equilibrium". It distinguishes several drivers of flows, and then says: "These are examples of the apparently universal tendency of isolated systems towards a state of complete mechanical, thermal, chemical, and electrical—or, in a single word, *thermodynamic—equilibrium.*"

A monograph on classical thermodynamics by H.A. Buchdahl considers the "equilibrium of a thermodynamic system", without actually writing the phrase "thermodynamic equilibrium". Referring to systems closed to exchange of matter, Buchdahl writes: "If a system is in a terminal condition which is properly static, it will be said to be in *equilibrium.*" Buchdahl's monograph also discusses amorphous glass, for the purposes of thermodynamic description. It states: "More precisely, the glass may be regarded as being *in equilibrium* so long as experimental tests show that 'slow' transitions are in effect reversible." It is not customary to make this proviso part of the definition of thermodynamic equilibrium, but the converse is usually assumed: that if a body in thermodynamic equilibrium is subject to a sufficiently slow process, that process may be considered to be sufficiently nearly reversible, and the body remains sufficiently nearly in thermodynamic equilibrium during the process.

A. Münster carefully extends his definition of thermodynamic equilibrium for isolated systems by introducing a concept of *contact equilibrium*. This specifies particular processes that are allowed when considering thermodynamic equilibrium for non-isolated systems, with special concern for open systems, which may gain or lose matter from or to their surroundings. A contact equilibrium is between the system of interest and a system in the surroundings, brought into contact with the system of interest, the contact being through a special kind of wall; for the rest, the whole joint system is isolated. Walls of this special kind were also considered by C. Carathéodory, and are mentioned by other writers also. They are selectively permeable. They may be permeable only to mechanical work, or only to heat, or only to some particular chemical substance. Each contact equilibrium defines an intensive parametre; for example, a wall permeable only to heat defines an empirical temperature. A contact equilibrium can exist for each chemical constituent of the system of interest. In a contact equilibrium, despite the possible exchange through the selectively permeable wall, the system of interest is changeless, as if it were in isolated thermodynamic equilibrium. This scheme follows the general rule that "... we can consider an equilibrium only with respect to specified processes and defined experimental conditions." Thermodynamic equilibrium for an open system means that, with respect to every relevant kind of selectively permeable wall, contact equilibrium exists when the respective intensive parametres of the system and surroundings are equal. This definition does not consider the most general kind of thermodynamic equilibrium, which is through unselective contacts. This definition does not simply state that no current of matter or energy exists in the interior or at the boundaries; but it is compatible with the following definition, which does so state.

M. Zemansky also distinguishes mechanical, chemical, and thermal equilibrium. He then writes: "When the conditions for all three types of equilibrium are satisfied, the system is said to be in a state of thermodynamic equilibrium".

P.M. Morse writes that thermodynamics is concerned with "*states of thermodynamic equilibrium*". He also uses the phrase "thermal equilibrium" while discussing transfer of energy as heat between a body and a heat reservoir in its surroundings, though not explicitly defining a special term 'thermal equilibrium'.

J.R. Waldram writes of "a definite thermodynamic state". He defines the term "thermal equilibrium" for a system "when its observables have ceased to change over time". But shortly below that definition he writes of a piece of glass that has not yet reached its "*full* thermodynamic equilibrium state".

Considering equilibrium states, M. Bailyn writes: "Each intensive variable has its own type of equilibrium." He then defines thermal equilibrium, mechanical equilibrium, and material equilibrium. Accordingly, he writes: "If all the intensive variables become uniform, *thermodynamic equilibrium* is said to exist." He is not here considering the presence of an external force field.

J.G. Kirkwood and I. Oppenheim define thermodynamic equilibrium as follows: "A system is in a state of *thermodynamic equilibrium* if, during the time period allotted for experimentation, (a) its intensive properties are independent of time and (b) no current of matter or energy exists in its interior or at its boundaries with the surroundings." It is evident that they are not restricting the definition to isolated or to closed systems. They do not discuss the possibility of changes that occur with "glacial slowness", and proceed beyond the time period allotted for experimentation. They note that for two systems in contact, there exists a small subclass of intensive properties such that if all those of that small subclass are respectively equal, then all respective intensive properties are equal. States of thermodynamic equilibrium may be defined by this subclass, provided some other conditions are satisfied.

Characteristics of a State of Internal Thermodynamic Equilibrium

Homogeneity in the Absence of External Forces

A thermodynamic system consisting of a single phase in the absence of external forces, in its own internal thermodynamic equilibrium, is homogeneous. This means that the material in any small volume element of the system can be interchanged with the material of any other geometrically congruent volume element of the system, and the effect is to leave the system thermodynamically unchanged. In general, a strong external force field makes a system of a single phase in its own internal thermodynamic equilibrium inhomogeneous with respect to some intensive variables. For example, a relatively dense component of a mixture can be concentrated by centrifugation.

Uniform Temperature

Such equilibrium inhomogeneity, induced by external forces, does not occur for the intensive variable temperature. According to E.A. Guggenheim, "The most important conception of thermodynamics is temperature." Planck introduces his treatise with a brief account of heat and temperature and thermal equilibrium, and then announces: "In the following we shall deal chiefly with homogeneous, isotropic bodies of any form, possessing throughout their substance the same temperature and density, and subject to a uniform pressure acting everywhere perpendicular to the surface." As did Carathéodory, Planck was setting aside surface effects and external fields and anisotropic crystals. Though referring to temperature, Planck did not there explicitly refer to the concept of thermodynamic equilibrium. In contrast, Carathéodory's scheme of presentation of classical thermodynamics for closed systems postulates the concept of an "equilibrium state" following Gibbs (Gibbs speaks routinely of a "thermodynamic state"), though not explicitly using the phrase 'thermodynamic equilibrium', nor explicitly postulating the existence of a temperature to define it.

The temperature within a system in thermodynamic equilibrium is uniform in space as well as in time. This is so in all cases, including those of non-uniform external force fields. In order that a system may be in its own internal state of thermodynamic equilibrium, it is of course necessary, but not sufficient, that it be in its own internal state of thermal equilibrium; it is possible for a system to reach internal mechanical equilibrium before it reaches internal thermal equilibrium.

Number of Real Variables Needed for Specification

In his exposition of his scheme of closed system equilibrium thermodynamics, C. Carathéodory initially postulates that experiment reveals that a definite number of real variables define the states that are the points of the manifold of equilibria. In the opinion of Prigogine and Defay (1945): "It is a matter of experience that when we have specified a certain number of macroscopic properties of a system, then all the other properties are fixed." This opinion of Prigogine and Defay is in precise agreement with this postulate of Carathéodory. As noted above, according to A. Münster, the number of variables needed to define a thermodynamic equilibrium is the least for any state of a given isolated system. As noted above, J.G. Kirkwood and I. Oppenheim point out that a state of thermodynamic equilibrium may be defined by a special subclass of intensive variables, with a definite number of members in that subclass.

If the thermodynamic equilibrium lies in an external force field, it is only the temperature that can in general be expected to be spatially uniform. Intensive variables other than temperature will in general be non-uniform if the external force field is non-zero. In such a case, in general, additional variables are needed to describe the spatial non-uniformity.

Stability Against Small Perturbations

As noted above, J.R. Partington points out that a state of thermodynamic equilibrium is stable against small perturbations.

Without this condition, in general, experiments intended to study systems in thermodynamic equilibrium are in severe difficulties.

Approach to Thermodynamic Equilibrium Within an Isolated System

When a body of material starts from a non-equilibrium state of inhomogeneity or chemical non-equilibrium, and is then isolated, it spontaneously evolves towards its own internal state of thermodynamic equilibrium. It is not necessary that all aspects of internal thermodynamic equilibrium be reached simultaneously; some can be established before others. For example, in many cases of such evolution, internal mechanical equilibrium is established much more rapidly than the other aspects of the eventual thermodynamic equilibrium. Another example is that, in many cases of such evolution, thermal equilibrium is reached much more rapidly than chemical equilibrium.

Thermal Equilibrium

An explicit distinction between 'thermal equilibrium' and 'thermodynamic equilibrium' is made by B. C. Eu. He considers two systems in thermal contact, one a thermometre, the other a system in which there are occurring several irreversible processes, entailing non-zero fluxes; the two systems are separated by a wall permeable only to heat. He considers the case in which, over the time scale of interest, it happens that both the thermometre reading and the irreversible processes are steady. Then there is thermal equilibrium without thermodynamic equilibrium. Eu proposes consequently that the zeroth law of thermodynamics can be considered to apply even when thermodynamic equilibrium is not present; also he proposes that if changes are occurring so fast that a steady temperature cannot be defined, then "it is no longer possible to describe the process by means of a thermodynamic formalism. In other words, thermodynamics has no meaning for such a process." This illustrates the importance for thermodynamics of the concept of temperature.

Thermal equilibrium is achieved when two systems in thermal contact with each other cease to have a net exchange of energy. It follows that if two systems are in thermal equilibrium, then their temperatures are the same.

Thermal equilibrium occurs when a system's macroscopic thermal observables have ceased to change with time. For example, an ideal gas whose distribution function has stabilised to a specific Maxwell-Boltzmann distribution would be in thermal equilibrium. This outcome allows a single temperature and pressure to be attributed to the whole system. For an isolated body, it is quite possible for mechanical equilibrium to be reached before thermal equilibrium is reached, but eventually, all aspects of equilibrium, including thermal equilibrium, are necessary for thermodynamic equilibrium.

Non-equilibrium

A system's internal state of thermodynamic equilibrium should be distinguished from a "stationary state" in which thermodynamic parametres are unchanging in time but the system is not isolated, so that there are, into and out of the system, non-zero macroscopic fluxes which are constant in time. Non-equilibrium thermodynamics is a branch of thermodynamics that deals with systems that are not in thermodynamic equilibrium. Most systems found in nature are not in thermodynamic equilibrium because they are changing or can be triggered to change over time, and are continuously and discontinuously subject to flux of matter and energy to and from other systems. The thermodynamic study of non-equilibrium systems requires more general concepts than are dealt with by equilibrium thermodynamics. Many natural systems still today remain beyond the scope of currently known macroscopic thermodynamic methods.

Exact and Inexact Differentials

Exact Differential

In multivariate calculus, a differential is said to be exact (or perfect), as contrasted with an inexact differential, if it is of the form dQ, for some differentiable function Q.

Overview

Definition: We work in three dimensions, with similar definitions holding in any other number of dimensions. In three dimensions, a form of the type

$$A(x, y, z)dx + B(x, y, z)dy + C(x, y, z)dz$$

is called a differential form. This form is called *exact* on a domain $D \subset \mathbb{R}^3$ in space if there exists some scalar function $Q = Q(x, y, z)$ defined on D such that

$$dQ \equiv \left(\frac{\partial Q}{\partial x}\right)_{y,z} dx + \left(\frac{\partial Q}{\partial y}\right)_{z,x} dy + \left(\frac{\partial Q}{\partial z}\right)_{x,y} dz = Adx + Bdy + Cdz$$

throughout D. This is equivalent to saying that the vector field (A, B, C) is a conservative vector field, with corresponding potential Q.

One Dimension

In one dimension, a differential form

$$A(x)dx$$

is exact as long as A has an antiderivative; in this case let Q be the antiderivative of A. Otherwise, if A does *not* have an antiderivative, we cannot write $dQ = A(x)dx$ and so the differential form is inexact.

Two and Three Dimensions

By symmetry of second derivatives, for any "nice" (non-pathological) function Q we have

$$\frac{\partial^2 Q}{\partial x \partial y} = \frac{\partial^2 Q}{\partial y \partial x}$$

Hence, it follows that in a simply-connected region *R* of the *xy*-plane, a differential

$$A(x, y)dx + B(x, y)dy$$

is an exact differential if and only if the following holds:

$$\left(\frac{\partial A}{\partial y}\right)_x = \left(\frac{\partial B}{\partial x}\right)_y$$

For three dimensions, a differential

$$dQ = A(x, y, z)dx + B(x, y, z)dy + C(x, y, z)dz$$

is an exact differential in a simply-connected region *R* of the *xyz*-coordinate system if between the functions *A*, *B* and *C* there exist the relations:

$$\left(\frac{\partial A}{\partial y}\right)_{x,z} = \left(\frac{\partial B}{\partial x}\right)_{y,z} ; \left(\frac{\partial A}{\partial z}\right)_{x,y} = \left(\frac{\partial C}{\partial x}\right)_{y,z} ; \left(\frac{\partial B}{\partial z}\right)_{x,y} = \left(\frac{\partial C}{\partial y}\right)_{x,z}$$

Note: The subscripts outside the parenthesis indicate which variables are being held constant during differentiation. Due to the definition of the partial derivative, these subscripts are not required, but they are included as a reminder.

These conditions are equivalent to the following one: If *G* is the graph of this vector valued function then for all tangent vectors *X*,Y of the *surface G* then $s(X, Y) = 0$ with *s* the symplectic form.

These conditions, which are easy to generalize, arise from the independence of the order of differentiations in the calculation of the second derivatives. So, in order for a differential *dQ*, that is a function of four variables to be an exact differential, there are six conditions to satisfy.

In summary, when a differential dQ is exact:

- the function Q exists;
- $\int_i^f dQ = Q(f) - Q(i)$, independent of the path followed.

In thermodynamics, when dQ is exact, the function Q is a state function of the system. The thermodynamic functions U, S, H, A and G are state functions. Generally, neither work nor heat is a state function. An *exact differential* is sometimes also called a 'total differential', or a 'full differential', or, in the study of differential geometry, it is termed an exact form.

Partial Differential Relations

If three variables, x, $y$and z are bound by the condition $F(x, y, z) = \text{constant}$ for some differentiable function $F(x, y, z)$, then the following total differentials exist

$$dx = \left(\frac{\partial x}{\partial y}\right)_z dy + \left(\frac{\partial x}{\partial z}\right)_y dz$$

$$dz = \left(\frac{\partial z}{\partial x}\right)_y dx + \left(\frac{\partial z}{\partial y}\right)_x dy.$$

Substituting the first equation into the second and rearranging, we obtain

$$dz = \left(\frac{\partial z}{\partial x}\right)_y \left[\left(\frac{\partial x}{\partial y}\right)_z dy + \left(\frac{\partial x}{\partial z}\right)_y dz\right] + \left(\frac{\partial z}{\partial y}\right)_x dy,$$

$$dz = \left[\left(\frac{\partial z}{\partial x}\right)_y \left(\frac{\partial x}{\partial y}\right)_z + \left(\frac{\partial z}{\partial y}\right)_x\right] dy + \left(\frac{\partial z}{\partial x}\right)_y \left(\frac{\partial x}{\partial z}\right)_y dz,$$

$$\left[1 - \left(\frac{\partial z}{\partial x}\right)_y \left(\frac{\partial x}{\partial z}\right)_y\right] dz = \left[\left(\frac{\partial z}{\partial x}\right)_y \left(\frac{\partial x}{\partial y}\right)_z + \left(\frac{\partial z}{\partial y}\right)_x\right] dy.$$

Since y and z are independent variables, dy and dz may be chosen without restriction. For this last equation to hold in general, the bracketed terms must be equal to zero.

Reciprocity Relation

Setting the first term in brackets equal to zero yields

$$\left(\frac{\partial z}{\partial x}\right)_y \left(\frac{\partial x}{\partial z}\right)_y = 1.$$

A slight rearrangement gives a reciprocity relation,

$$\left(\frac{\partial z}{\partial x}\right)_y = \frac{1}{\left(\frac{\partial x}{\partial z}\right)_y}.$$

There are two more permutations of the foregoing derivation that give a total of three reciprocity relations between x, $y$and z. Reciprocity relations show that the inverse of a partial derivative is equal to its reciprocal.

Cyclic Relation

The cyclic relation is also known as the cyclic rule or the Triple product rule. Setting the second term in brackets equal to zero yields

$$\left(\frac{\partial z}{\partial x}\right)_y \left(\frac{\partial x}{\partial y}\right)_z = -\left(\frac{\partial z}{\partial y}\right)_x.$$

Using a reciprocity relation for $\frac{\partial z}{\partial y}$ on this equation and reordering gives a cyclic relation (the triple product rule),

$$\left(\frac{\partial x}{\partial y}\right)_z \left(\frac{\partial y}{\partial z}\right)_x \left(\frac{\partial z}{\partial x}\right)_y = -1.$$

If, *instead*, a reciprocity relation for $\frac{\partial x}{\partial y}$ is used with subsequent rearrangement, a standard form for implicit differentiation is obtained:

$$\left(\frac{\partial y}{\partial x}\right)_z = -\frac{\left(\frac{\partial z}{\partial x}\right)_y}{\left(\frac{\partial z}{\partial y}\right)_x}.$$

Some Useful Equations Derived from Exact Differentials in Two Dimensions

Suppose we have five state functions z, x, y, u, and v. Suppose that the state space is two dimensional and any of the five quantities are exact differentials. Then by the chain rule

$$dz = \left(\frac{\partial z}{\partial x}\right)_y dx + \left(\frac{\partial z}{\partial y}\right)_x dy = \left(\frac{\partial z}{\partial u}\right)_v du + \left(\frac{\partial z}{\partial v}\right)_u dv \quad (1)$$

but also by the chain rule:

$$dx = \left(\frac{\partial x}{\partial u}\right)_v du + \left(\frac{\partial x}{\partial v}\right)_u dv \tag{2}$$

and

$$dy = \left(\frac{\partial y}{\partial u}\right)_v du + \left(\frac{\partial y}{\partial v}\right)_u dv \tag{3}$$

so that:

$$dz = \left[\left(\frac{\partial z}{\partial x}\right)_y \left(\frac{\partial x}{\partial u}\right)_v + \left(\frac{\partial z}{\partial y}\right)_x \left(\frac{\partial y}{\partial u}\right)_v\right] du \tag{4}$$

$$+\left[\left(\frac{\partial z}{\partial x}\right)_y \left(\frac{\partial x}{\partial v}\right)_u + \left(\frac{\partial z}{\partial y}\right)_x \left(\frac{\partial y}{\partial v}\right)_u\right] dv$$

which implies that:

$$\left(\frac{\partial z}{\partial u}\right)_v = \left(\frac{\partial z}{\partial x}\right)_y \left(\frac{\partial x}{\partial u}\right)_v + \left(\frac{\partial z}{\partial y}\right)_x \left(\frac{\partial y}{\partial u}\right)_v \tag{5}$$

Letting $v = y$ gives:

$$\left(\frac{\partial z}{\partial u}\right)_y = \left(\frac{\partial z}{\partial x}\right)_y \left(\frac{\partial x}{\partial u}\right)_y \tag{6}$$

Letting $u = y$ gives:

$$\left(\frac{\partial z}{\partial y}\right)_v = \left(\frac{\partial z}{\partial y}\right)_x + \left(\frac{\partial z}{\partial x}\right)_y \left(\frac{\partial x}{\partial y}\right)_v \tag{7}$$

Letting $u = y$, $v = z$ gives:

$$\left(\frac{\partial z}{\partial y}\right)_x = -\left(\frac{\partial z}{\partial x}\right)_y \left(\frac{\partial x}{\partial y}\right)_z \tag{8}$$

using $\partial a / \partial b)_c = 1 / (\partial b / \partial a)_c$ (gives the triple product rule:

$$\left(\frac{\partial z}{\partial x}\right)_y \left(\frac{\partial x}{\partial y}\right)_z \left(\frac{\partial y}{\partial z}\right)_x = -1 \tag{9}$$

Inexact Differential

An inexact differential or imperfect differential is a specific type of differential used in thermodynamics to express the path dependence of a particular differential. It is contrasted with the concept of the exact differential in calculus, which can be expressed as the gradient of another function and is therefore path independent. Consequently, an inexact differential cannot be expressed in terms of its antiderivative for the purpose of integral calculations i.e. its value cannot be inferred just by looking at the initial and final states of a given system. It is primarily used in calculations involving heat and work because they are not state functions.

Definition

An inexact differential is commonly defined as a differential form dx where there is no corresponding function x such that: $x = \int dx$. More precisely, an inexact differential is a differential form that cannot be expressed as the differential of a function. In the language of calculus, for a given vector field F, $\delta F = F\,dr$ is an inexact differential if there is no function f such that

$$F = \nabla f$$

The fundamental theorem of calculus for line integrals requires path independence in order to express the values of a given vector field in terms of the partial derivatives of another function that is the multivariate analogue of the antiderivative. This is because there can be no unique representation of an antiderivative for inexact differentials since their variation is inconsistent along different paths. This stipulation of path independence is a necessary addendum to the fundamental theorem of calculus because in one-dimensional calculus there is only one path in between two points defined by a function.

First Law of Thermodynamics

Inexact differentials are known especially for their presence in the first law of thermodynamics:

$$\mathrm{d}U = \delta Q - \delta W$$

The symbol δ instead of the plain d, which originated from the work of German mathematician Carl Gottfried Neumann indicates that Q and W are path-dependent.

Internal energy U is a *state function*, meaning its change can be inferred just by comparing two different states of the system (not its transition path), which we can therefore indicate with U_1 and U_2. Since we can go from state U_1 to state U_2 either by providing heat $Q = U_2 - U_1$ or work $W = U_2 - U_1$, such a change of state does not identify uniquely the values of provided W and Q, but only the change in internal energy ΔU.

Examples

Although difficult to express mathematically, the inexact differential is very simple conceptually. There are many everyday examples that are much more relevant to inexact differentials in the actual context in which it is used.

The easiest example is the difference between net distance and total distance. For example, in walking from Point *A* to Point *B* one covers a net distance *B-A* that is equal to the total distance. If one then returns to Point *A*, however, net distance is now *0* while total distance covered is *2*(B-A)*. This example captures the essential idea behind the inexact differential in one dimension.

Precisely, the differential of *net distance* is simply the exact one form $df = 1dx$ with corresponding function $\Delta f = x_B - x_A$. It is exact because 1 has antiderivative x everywhere on the real line. On the other hand, the differential of *total distance* is the inexact one form $\delta g = \mathrm{sgn}(dx)dx$ (i.e. the sign function). It is inexact because sgn(x) has antiderivative |x| which is not differentiable at x = 0. Therefore $\Delta g \neq x_B - x_A$ and instead we must look at the path dependence. In our example, in the first leg of the journey, sgn(dx) is 1 since x is increasing. In the second leg, sgn(dx) is –1 since x is decreasing. We can then evaluate the total distance as:

$$\Delta g = \int_A^B dx + \int_B^A (-dx) = 2\int_A^B dx = 2(B-A)$$

It is known that, with some skill, it is possible to start a fire only using friction and tinder. This is a way to convert mechanical energy (work, *W*) into an increase of internal energy, ΔU, which finally results into an increase in the local temperature of wood, its gasification and combustion, thereby creating a fire.

It is also possible to start a fire by adding heat using a lighter. This is a way to convert heat (*Q*) into an increase of internal energy, ΔU, but it has the same result as in the example involving work.

Both friction and heat transfer increase the internal energy of the system, since work and heat are both form of energy transform. Therefore, *the sum* of exchanged heat and work is an exact differential (*dU*), but since they are equivalent and the lack of one can be compensated by the presence of the other, singularly they are inexact differentials. In other words, specifying the change in internal energy alone is insufficient to determine the heat evolved or the work done because there is no differentiation between the two forms of energy.

Integrating Factors

It is sometimes possible to convert an inexact differential into an exact one by means of an integrating factor. The most common example of this in thermodynamics is the definition of entropy:

$$dS = \frac{\delta Q_{\text{rev}}}{T}$$

In this case, δQ is an inexact differential, because its effect on the state of the system can be compensated by δW. However, when divided by the absolute temperature*and* when the exchange occurs at reversible conditions (therefore the $_{\text{rev}}$ subscript), it produces an exact differential: the entropy S is also a state function.

Thermodynamic Cycle

A thermodynamic cycle consists of a collection of thermodynamic processes transferring heat and work, while varying pressure, temperature, and other state variables, eventually returning a system to its initial state. In the process of going through this cycle, the system may perform work on its surroundings, therefore acting as a heat engine.

State quantities depend only on the thermodynamic state, and cumulative variation of such properties adds up to zero during a cycle. Process quantities (or path quantities), such as heat and work are process dependent, and cumulative heat and work are non-zero. Since the change in internal energy (a state quantity) over the course of a cycle is zero, the first law of thermodynamics dictates that the net heat input is equal to the net work output over a cycle. The repeating nature of the process path allows for continuous operation, making the cycle an important concept in thermodynamics. Thermodynamic cycles are often represented mathematically as quasistatic processes to model the workings of an actual device.

Heat and Work

Two primary classes of thermodynamic cycles are power cycles and heat pump cycles. Power cycles are cycles which convert some heat input into a mechanical work output, while heat pump cycles transfer heat from low to high temperatures using mechanical work input. Cycles composed entirely of quasistatic processes can operate as power or heat pump cycles by controlling the process direction. On a pressure volume diagram or temperature entropy diagram, the clockwise and counterclockwise directions indicate power and heat pump cycles, respectively.

Relationship to Work

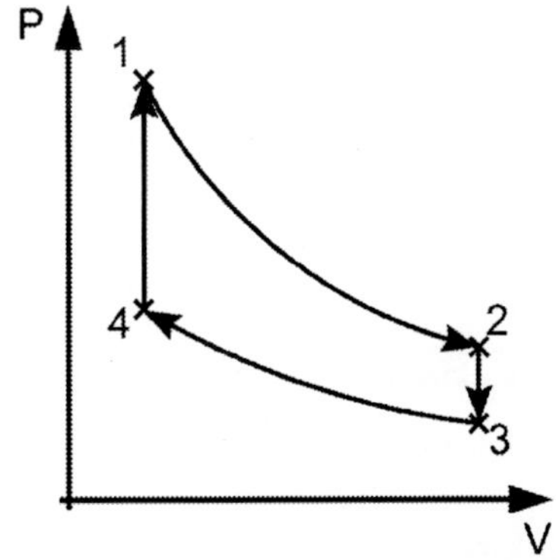

Figure: *Example of P-V diagram of a thermodynamic cycle.*

Because the net variation in state properties during a thermodynamic cycle is zero, it forms a closed loop on a PV diagram. A PV diagram's *Y* axis shows pressure (*P*) and *X* axis shows volume (*V*). The area enclosed by the loop is the work (*W*) done by the process:

$$W = \oint P\, dV \tag{1}$$

This work is equal to the balance of heat (Q) transferred into the system:

$$W = Q = Q_{in} - Q_{out} \tag{2}$$

Equation (2) makes a cyclic process similar to an isothermal process: even though the internal energy changes during the course of the cyclic process, when the cyclic process finishes the system's energy is the same as the energy it had when the process began.

If the cyclic process moves clockwise around the loop, then W will be positive, and it represents a heat engine. If it moves counterclockwise, then W will be negative, and it represents a heat pump.

Each Point in the Cycle

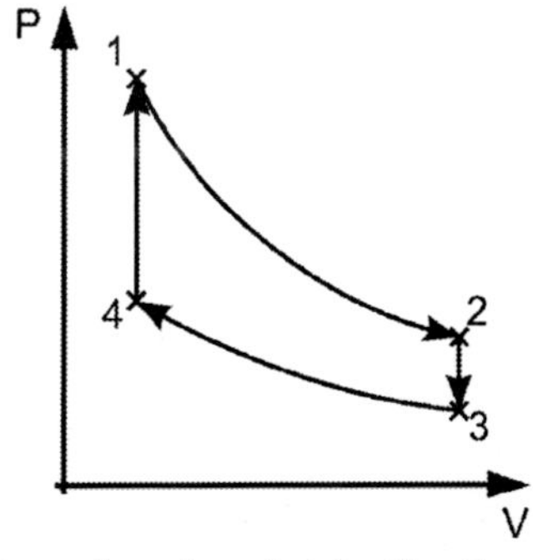

Figure: *Description of each point in the thermodynamic cycles.*

1→2: Isentropic Expansion: Constant entropy (s), Decrease in pressure (P), Increase in volume (v), Decrease in temperature (T)

2→3: Isochoric Cooling: Constant volume(v), Decrease in pressure (P), Decrease in entropy (S), Decrease in temperature (T)

3→4: Isentropic Compression: Constant entropy (s), Increase in pressure (P), Decrease in volume (v), Increase in temperature (T)

4→1: Isochoric Heating: Constant volume (v), Increase in pressure (P), Increase in entropy (S), Increase in temperature (T)

Some Other Processes:

Adiabatic: Heat transfer (Q) during the cycle would amount to 0 (enthalpy (h) = 0 kj/kg). Q= Δh

Isothermal: Temperature in the cycle remains constant (T=0)

Isobaric: Pressure in a cycle will remain constant. This would be attainable by a piston cylinder device.

Power Cycles

Thermodynamic power cycles are the basis for the operation of heat engines, which supply most of the world's electric power and run the vast majority of motor vehicles. Power cycles can be divided according to the type of heat engine they seek to model. The most common cycles used to model internal combustion engines are the Otto cycle, which models gasoline engines, and the Diesel cycle, which models diesel engines. Cycles that model external combustion engines include the Brayton cycle, which models gas turbines, and the Rankine cycle, which models steam turbines.

For example the pressure-volume mechanical work done in the heat engine cycle, consisting of 4 thermodynamic processes, is:

$$W = W_{1\to2} + W_{2\to3} + W_{3\to4} + W_{4\to1} \tag{3}$$

$$W_{1\to2} = \int_{V_1}^{V_2} PdV, \text{negative, work done on system}$$

$$W_{2\to3} = \int_{V_2}^{V_3} PdV, \text{zero work if V2 equal V3}$$

$$W_{3\to4} = \int_{V_3}^{V_4} PdV, \text{positive, work done on system}$$

$$W_{4\to1} = \int_{V_4}^{V_1} PdV, \text{zero work if V4 equal V1}$$

If no volume change happens in process 4->1 and 2->3, equation (3) simplifies to:

$$W = W_{1\to2} + W_{3\to4} \tag{4}$$

Heat Pump Cycles

Thermodynamic heat pump cycles are the models for heat pumps and refrigerators. The difference between the two is that heat pumps are intended to add heat to part of the environment, while refrigerators are designed to remove heat. The most common refrigeration cycle is the vapour compression cycle, which models systems using refrigerants that change phase. The absorption refrigeration cycle is an alternative that absorbs the refrigerant in a liquid solution rather than evaporating it. Gas refrigeration cycles include the reversed Brayton cycle and the Hampson-Linde cycle. Regeneration in gas refrigeration allows for the liquefaction of gases.

Modelling Real Systems

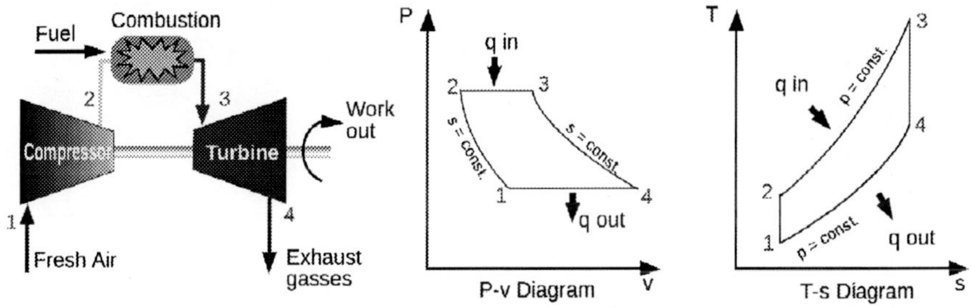

Figure: *Example of a real system modelled by an idealized process: PV and TS diagrams of a Brayton cycle mapped to actual processes of a gas turbine engine*

Thermodynamic cycles may be used to model real devices and systems, typically by making a series of assumptions. Simplifying assumptions are often necessary to reduce the problem to a more manageable form. For example, as shown in the figure, devices such a gas turbine or jet engine can be modelled as a Brayton cycle. The actual device is made up of a series of stages, each of which is itself modelled as an idealized thermodynamic process. Although each stage which acts on the working fluid is a complex real device, they may be modelled as idealized processes which approximate their real behaviour. A further assumption is that the exhaust gases would be passed back through the inlet with a corresponding loss of heat, thus completing the idealized cycle.

The difference between an idealized cycle and actual performance may be significant. For example, the following images illustrate the differences in work output predicted by an ideal Stirling cycle and the actual performance of a Stirling engine:

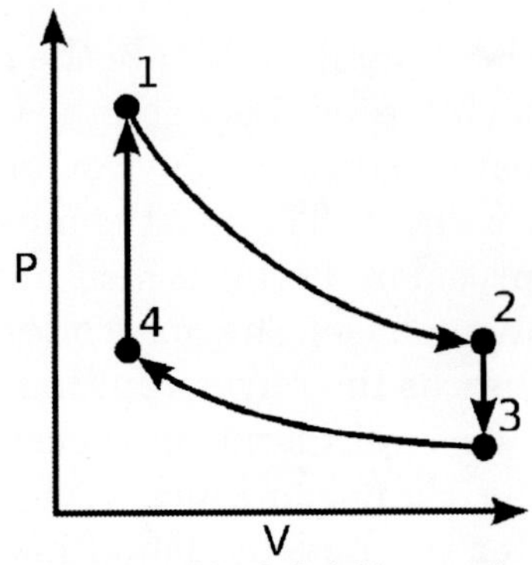

Figure: *Ideal Stirling cycle*

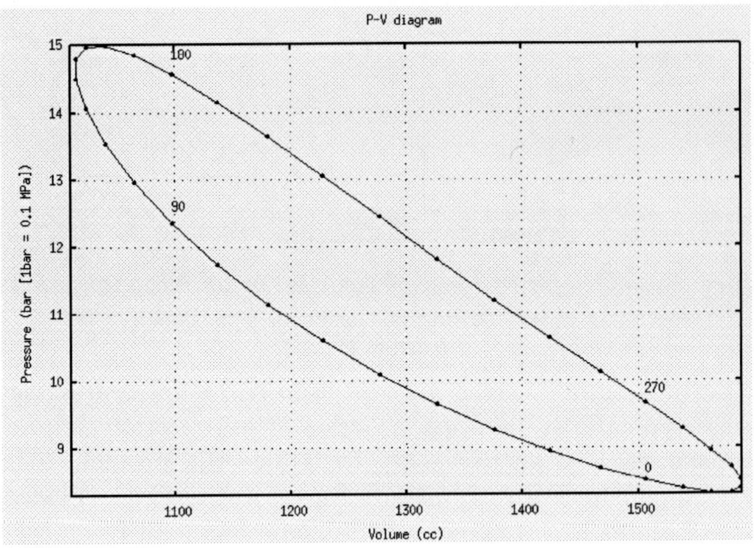

Figure: *Actual performance*

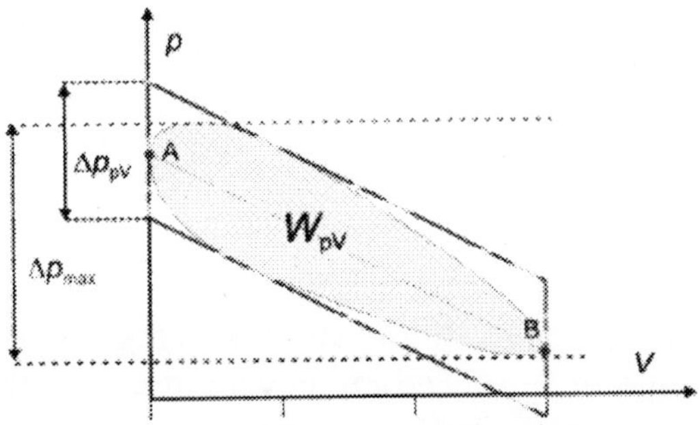

Figure: *Actual and ideal overlaid, showing difference in work output*

As work output is represented by the interior of the cycle, there is a significant difference between the predicted work output of the ideal cycle and the actual work output shown by a real engine. It may also be observed that the real individual processes diverge from their idealized counterparts; e.g., isochoric expansion (process 1-2) occurs with some actual volume change.

Well-known Thermodynamic Cycles

In practice, simple idealized thermodynamic cycles are usually made out of four thermodynamic processes. Any thermodynamic processes may be used. However, when idealized cycles are modelled, often processes where one state variable is kept constant are used, such as an isothermal process (constant temperature), isobaric process (constant pressure), isochoric process (constant volume), isentropic process (constant entropy), or an isenthalpic process (constant enthalpy). Often adiabatic processes are also used, where no heat is exchanged.

Some example thermodynamic cycles and their constituent processes are as follows:

Cycle	*Process 1-2 (Compression)*	*Process 2-3 (Heat Addition)*	*Process 3-4 (Expansion)*	*Process 4-1 (Heat Rejection)*	*Notes*
Power cycles normally with external combustion - or heat pump cycles:					
Bell Coleman	adiabatic	isobaric	adiabatic	isobaric	A reversed Brayton cycle
Carnot	isentropic	isothermal	isentropic	isothermal	
Ericsson	isothermal	isobaric	isothermal	isobaric	the second Ericsson cycle from 1853
Rankine	adiabatic	isobaric	adiabatic	isobaric	Steam engine
Scuderi	adiabatic	variable pressure and volume	adiabatic	isochoric	
Stirling	isothermal	isochoric	isothermal	isochoric	
Stoddard	adiabatic	isobaric	adiabatic	isobaric	
Power cycles normally with internal combustion:					
Brayton	adiabatic	isobaric	adiabatic	isobaric	Jet engines the external combustion version of this cycle is known as first Ericsson cycle from 1833
Diesel	adiabatic	isobaric	adiabatic	isochoric	
Lenoir	isobaric	isochoric	adiabatic		Pulse jets (Note: Process 1-2 accomplishes both the heat rejection and the compression)
Otto	adiabatic	isochoric	adiabatic	isochoric	Gasoline / petrol engines

Ideal Cycle

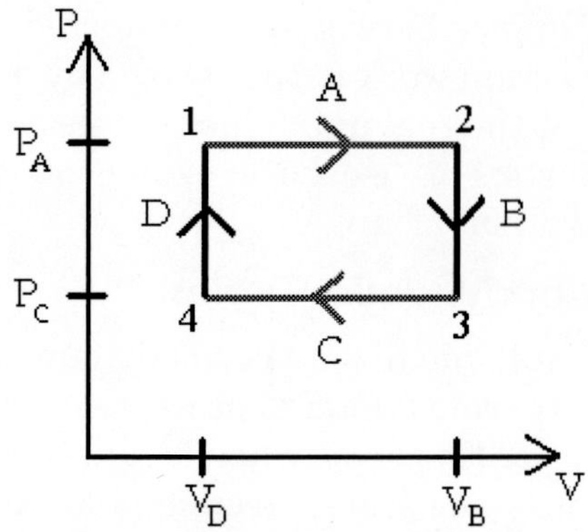

Figure: *An illustration of an ideal cycle heat engine (arrows clockwise).*

An ideal cycle is constructed out of:

1. TOP and BOTTOM of the loop: a pair of parallel isobaric processes
2. LEFT and RIGHT of the loop: a pair of parallel isochoric processes

Carnot Cycle

The Carnot cycle is a cycle composed of the totally reversible processes of isentropic compression and expansion and isothermal heat addition and rejection. The thermal efficiency of a Carnot cycle depends only on the absolute temperatures of the two reservoirs in which heat transfer takes place, and for a power cycle is:

$$\eta = 1 - \frac{T_L}{T_H}$$

where T_L is the lowest cycle temperature and T_H the highest. For Carnot power cycles the coefficient of performance for a heat pump is:

$$COP = 1 + \frac{T_L}{T_H - T_L}$$

and for a refrigerator the coefficient of performance is:

$$COP = \frac{T_L}{T_H - T_L}$$

The second law of thermodynamics limits the efficiency and COP for all cyclic devices to levels at or below the Carnot efficiency. The Stirling cycle and Ericsson cycle are two other reversible cycles that use regeneration to obtain isothermal heat transfer.

Stirling Cycle

A Stirling cycle is like an Otto cycle, except that the adiabats are replaced by isotherms. It is also the same as an Ericsson cycle with the isobaric processes substituted for constant volume processes.

1. TOP and BOTTOM of the loop: a pair of quasi-parallel isothermal processes
2. LEFT and RIGHT sides of the loop: a pair of parallel isochoric processes

Heat flows into the loop through the top isotherm and the left isochore, and some of this heat flows back out through the bottom isotherm and the right isochore, but most of the heat flow is through the pair of isotherms. This makes sense since all the work done by the cycle is done by the pair of isothermal processes, which are described by $Q=W$.

This suggests that all the net heat comes in through the top isotherm. In fact, all of the heat which comes in through the left isochore comes out through the right isochore: since the top isotherm is all at the same warmer temperature T_H and the bottom isotherm is all at the same cooler temperature T_C, and since change in energy for an isochore is proportional to change in temperature, then all of the heat coming in through the left isochore is cancelled out exactly by the heat going out the right isochore.

State Functions and Entropy

If Z is a state function then the balance of Z remains unchanged during a cyclic process:

$$\oint dZ = 0 .$$

Entropy is a state function and is defined as

$$S = \frac{Q}{T}$$

so that

$$\Delta S = \frac{\Delta Q}{T} ,$$

then it is clear that for any cyclic process,

$$\oint dS = \oint \frac{dQ}{T} = 0$$

meaning that the net entropy change over a cycle is 0.

Reversible Process (Thermodynamics)

In thermodynamics, a reversible process, or *reversible cycle* if the process is cyclic, is a process that can be "reversed" by means of infinitesimal changes in some property of the system without entropy production (i.e. dissipation of energy). Due to these infinitesimal changes, the system is in thermodynamic equilibrium throughout the entire process. Since it would take an infinite amount of time for the reversible process to finish, perfectly reversible processes are impossible. However, if the system undergoing the changes responds much faster than the applied change, the deviation from reversibility may be negligible. In a reversible cycle, the system and its surroundings will be exactly the same after each cycle.

An alternative definition of a *reversible process* is a process that, after it has taken place, can be reversed and causes no change in either the system or its surroundings. In thermodynamic terms, a process "taking place" would refer to its transition from its initial state to its final state.

Irreversibility

In an irreversible process, finite changes are made; therefore the system is not at equilibrium throughout the process. At the same point in an irreversible cycle, the system will be in the same state, but the surroundings are permanently changed after each cycle.

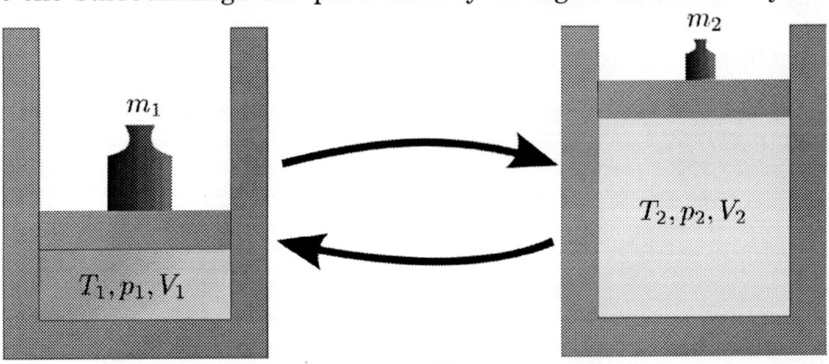

***Figure:** Reversible adiabatic process: The state on the left can be reached from the state on the right as well as vice versa without exchanging heat with the environment.*

Boundaries and States

A reversible process changes the state of a system in such a way that the net change in the combined entropy of the system and its surroundings is zero. Reversible processes define the boundaries of how efficient heat engines can be in thermodynamics and engineering: a reversible process is one where no heat is lost from the system as "waste", and the machine is thus as efficient as it can possibly be.

In some cases, it is important to distinguish between reversible and quasistatic processes. Reversible processes are always quasistatic, but the converse is not always true. For example, an infinitesimal compression of a gas in a cylinder where there exists friction between the piston and the cylinder is a quasistatic, but not reversible process. Although the system has been driven from its equilibrium state by only an infinitesimal amount, heat has been irreversibly lost due to friction, and cannot be recovered by simply moving the piston infinitesimally in the opposite direction.

Engineering Archaisms

Historically, the term *Tesla principle* was used to describe (amongst other things) certain reversible processes invented by Nikola Tesla. However, this phrase is no longer in conventional use. The principle was that some systems could be reversed and operated in a complementary manner. It was developed during Tesla's research in alternating currents where the current's magnitude and direction varied cyclically. During a demonstration of the Tesla turbine, the disks revolved and machinery fastened to the shaft was operated by the engine. If the turbine's operation was reversed, the disks acted as a pump.

Quasistatic Process

In thermodynamics, a quasistatic process is a thermodynamic process that happens infinitely slowly. No real process is quasistatic, but such processes can be approximated by performing them very slowly.

Quasistatic and Reversibility

A quasistatic process ensures that the system will go through a sequence of states that are infinitesimally close to equilibrium (so the system remains in quasistatic equilibrium), in which case the process is typically reversible. Any reversible process is necessarily a quasistatic one. However, some quasistatic processes are irreversible, if there is heat flowing (in or out of the system) or if entropy is being created in some other way. An example of a quasistatic process that is not reversible is a compression against a system with a piston subject to friction — Although the system is always in thermal equilibrium, the friction ensures the generation of dissipative entropy, which directly goes against the definition of reversible. A notable example of a process that is not even quasistatic is the slow heat exchange between two bodies at two finitely different temperatures, where the heat exchange rate is controlled by an approximately adiabatic partition between the two bodies — in this case, no matter how slowly the process takes place, the states of the two bodies are never infinitesimally close to equilibrium, since thermal equilibrium requires that the two bodies be at precisely the same temperature.

Some ambiguity exists in the literature concerning the distinction between quasistatic and reversible processes, as these are sometimes taken as synonyms. The reason is precisely because of the proven theorem that any reversible process is also a quasistatic one, even though we have also shown that the converse is not true. It is practically not useful to differentiate between the two because any engineer would remember to include friction when calculating the dissipative entropy generation. The above definition is closer to the intuitive understanding of the word "quasi-" (almost) "static", while remaining technically different from reversible processes.

PV-Work in Various Quasi-static Processes

1. Constant pressure: Isobaric processes,

   $$W_{1-2} = \int PdV = P(V_2 - V_1)$$

2. Constant volume: Isochoric processes,

   $$W_{1-2} = \int PdV = 0$$

3. Constant temperature: Isothermal processes,

   $$W_{1-2} = \int PdV, \quad PV = P_1V_1 = C$$

   $$W_{1-2} = P_1V_1 \ln\frac{V_2}{V_1}$$

Irreversible Process

In science, a process that is not reversible is called irreversible. This concept arises most frequently in thermodynamics, as applied to processes.

In thermodynamics, a change in the thermodynamic state of a system and all of its surroundings cannot be precisely restored to its initial state by infinitesimal changes in some property of the system without expenditure of energy. A system that undergoes an irreversible process may still be capable of returning to its initial state; however, the impossibility occurs in restoring the environment to its own initial conditions. An irreversible process increases the entropy of the universe. However, because entropy is a state function, the change in entropy of a system is the same whether the process is reversible or irreversible. The second law of thermodynamics can be used to determine whether a process is reversible or not.

All complex natural processes are irreversible. The phenomenon of irreversibility results from the fact that if a thermodynamic system, which is any system of sufficient complexity, of interacting molecules is brought from one thermodynamic state to another, the configuration or arrangement of the atoms and molecules in the system will change in a way that is not easily predictable. A certain amount of "transformation energy" will be used as the molecules of the "working body" do work on each other when they change from one state to another. During this transformation, there will be a certain amount of heat energy loss or dissipation due to intermolecular friction and collisions; energy that will not be recoverable if the process is reversed.

Many biological processes that were once thought to be reversible have been found to actually be a pairing of two irreversible processes. Whereas a single enzyme was once believed to catalyze both the forward and reverse chemical changes, research has found that two separate enzymes of similar structure are typically needed to perform what results in a pair of thermodynamically irreversible processes.

Absolute Versus Statistical Reversibility

Thermodynamics defines the statistical behaviour of large numbers of entities, whose exact behaviour is given by more specific laws. Since the fundamental theoretical laws of physics are all time-reversible, however experimentally, probability of real reversibility is low, former presuppositions can be fulfilled and/or former state recovered only to higher or lower degree. The irreversibility of thermodynamics must be statistical in nature; that is, that it must be merely highly unlikely, but not impossible, that a system will lower in entropy.

History

The German physicist Rudolf Clausius, in the 1850s, was the first to mathematically quantify the discovery of irreversibility in nature through his introduction of the concept of entropy. In his 1854 memoir "On a Modified Form of the Second Fundamental Theorem in the Mechanical Theory of Heat" Clausius states:

It may, moreover, happen that instead of a descending transmission of heat accompanying, in the one and the same process, the ascending transmission, another permanent change may occur which has the peculiarity of *not being reversible* without either becoming replaced by a new permanent change of a similar kind, or producing a descending transmission of heat.

Simply, Clausius states that it is impossible for a system to transfer heat from a cooler body to a hotter body. For example, a cup of hot coffee placed in an area of room temperature (~72 °F) will transfer heat to its surroundings and thereby cool down with the temperature of the room slightly increasing (~72.3 °F). However, that same initial cup of coffee will never absorb heat from its surroundings causing it to grow even hotter with the temperature of the room decreasing (~71.7 °F). Therefore, the process of the coffee cooling down is irreversible unless extra energy is added to the system.

However, a paradox arose when attempting to reconcile microanalysis of a system with observations of its macrostate. Many processes are mathematically reversible in their microstate when analyzed using classical Newtonian mechanics. From 1872 to 1875, Ludwig Boltzmann reinforced the statistical explanation of this paradox in the form of Boltzmann's entropy formula stating that as the number of possible microstates a system might be in increases, the entropy of the system increases and it becomes less likely that the system will return to an earlier state. His formulas quantified the work done by William Thomson, 1st Baron Kelvin who had argued that:

The equations of motion in abstract dynamics are perfectly reversible; any solution of these equations remains valid when the time variable t is replaced by –t. Physical processes, on the other hand, are irreversible: for example, the friction of solids, conduction of heat, and diffusion. Nevertheless, the principle of dissipation of energy is compatible with a molecular theory in which each particle is subject to the laws of abstract dynamics.

Another explanation of irreversible systems was presented by French mathematician Henri Poincaré. In 1890, he published his first explanation of nonlinear dynamics, also called chaos theory. Applying the chaos theory to the second law of thermodynamics, the paradox of irreversibility can be explained in the errors associated with scaling from microstates to macrostates and the degrees of freedom used when making experimental observations. Sensitivity to initial conditions relating to the system and its environment at the microstate compounds into an exhibition of irreversible characteristics within the observable, physical realm.

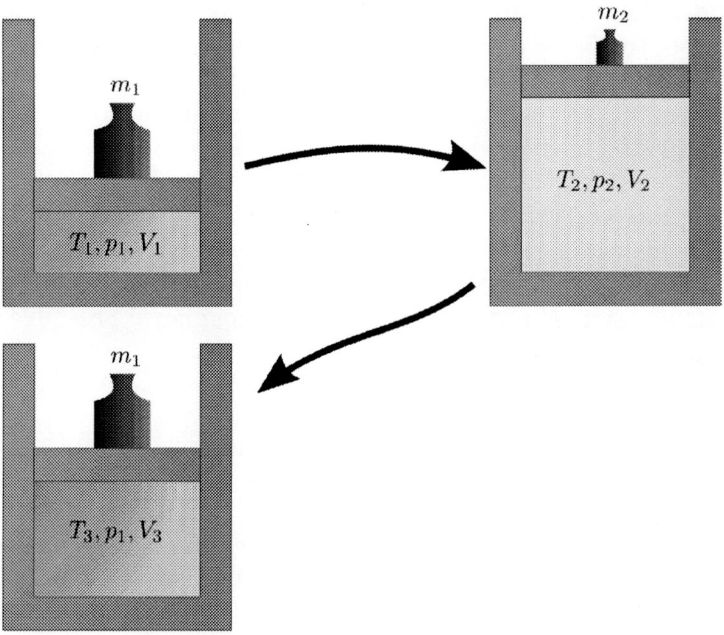

Figure: *Irreversible adiabatic process: If the cylinder is a perfect insulator, the initial top-left state cannot be reached anymore after it is changed to the one on the top-right. Instead, the state on the bottom left is assumed when going back to the original pressure because energy is converted into heat.*

Examples of Irreversible Processes

In the physical realm, many irreversible processes are present to which the inability to achieve 100% efficiency in energy transfer can be attributed. The following is a list of spontaneous events which contribute to the irreversibility of processes.

- Heat transfer through a finite temperature difference
- Friction
- Plastic deformation
- Flow of electric current through a resistance
- Magnetization or polarization with a hysteresis
- Unrestrained expansion of fluids
- Spontaneous chemical reactions
- Spontaneous mixing of matter of varying composition/states

A Joule expansion is an example of classical thermodynamics, as it is easy to work out the resulting increase in entropy. It occurs where a volume of gas is kept in one side of a thermally isolated container (via a small partition), with the other side of the container being evacuated; the partition between the two parts of the container is then opened, and the gas fills the whole container. The internal energy of the gas remains the same, while the volume increases. The original state cannot be recovered by simply compressing the gas to its original volume, since the internal energy will be increased by this compression. The original state can only be recovered by then cooling the re-compressed system, and thereby irreversibly heating the environment. The diagram to the right applies only if the first expansion is "free" (Joule expansion). i.e. there can be no atmospheric pressure outside the cylinder and no weight lifted.

Complex Systems

The difference between reversible and irreversible events has particular explanatory value in complex systems (such as living organisms, or ecosystems). According to the biologists Humberto Maturana and Francisco Varela, living organisms are characterized by autopoiesis, which enables their continued existence. More primitive forms of self-organizing systems have been described by the physicist and chemist Ilya Prigogine. In the context of complex systems, events which lead to the end of certain self-organising processes, like death, extinction of a species or the collapse of a meteorological system can be considered as irreversible. Even if a clone with the same organizational principle (e.g. identical DNA-structure) could be developed, this would not mean that the former distinct system comes back into being. Events to which the self-organizing capacities of organisms, species or other complex systems can adapt, like minor injuries or changes in the physical environment are reversible. However, adaptation depends on import of negentropy into the organism, thereby increasing irreversible processes in its environment. Ecological principles, like those of sustainability and the precautionary principle can be defined with reference to the concept of reversibility.

Work (Thermodynamics)

In thermodynamics, work performed by a system is the energy transferred by the system to another that is accounted for by changes in the external generalized mechanical constraints on the system. As such, thermodynamic work is a generalization of the concept of mechanical work in physics.

The external generalized mechanical constraints may be chemical, electromagnetic, (including radiative), gravitational or pressure/volume or other simply mechanical constraints, including momental, as in radiative transfer. Thermodynamic work is defined to be measurable solely from knowledge of such external macroscopic constraint variables. These macroscopic variables always occur in conjugate pairs, for example pressure and volume, magnetic flux density and magnetization, mole fraction and chemical potential. In the SI system of measurement, work is measured in joules (symbol: J). The rate at which work is performed is power.

It is customary to calculate amount of energy transferred as work through quantities external to the system of interest, and thus belonging to its surroundings. Nevertheless, for historical reasons, the customary sign convention is to consider work done by the system on its surroundings as positive. Although all real physical processes entail some dissipation of kinetic energy, it is matter of principle that the dissipation that results from transfer of energy as work occurs only inside the system; energy dissipated outside the system, in the process of transfer of energy, is not counted as thermodynamic work. Thermodynamic work does not account for any energy transferred between systems as heat.

Mechanical thermodynamic work is performed by actions such as compression, and including shaft work, stirring, and rubbing. In the simplest case, for example, there are work of change of volume against a resisting pressure, and work without change of volume, known as isochoric work. An example of isochoric work is when an outside agency, in the surrounds of the system, drives a frictional action on the surface of the system. In this case the dissipation is not necessarily actually confined to the system, and the quantity of energy so transferred as work must be estimated through the overall change of state of the system as measured by both its mechanically and externally measurable deformation variables (such as its volume), and its non-deformation variable (usually internal to the system, for example its empirical temperature, regarded not as a temperature but simply as a mechanically measurable variable). In a process of transfer of energy by work, the internal energy of the final state of the system is then measured by the amount of adiabatic work of change of volume that would have been necessary to reach it from the initial state, such adiabatic work being measurable only through the externally measurable mechanical or deformation variables of the system, but including also full information about the forces exerted by the surroundings on the system during the process. In the case of some of Joule's measurements, the process was so arranged that heat produced outside the system by the frictional process was practically entirely transferred into the system during the process, so that the quantity of work done by the surrounds on the system could be calculated as shaft work, an external mechanical variable. For closed systems, internal energy changes in a system other than as work transfer are as heat.

History

1824:

Work, i.e. "weight *lifted* through a height", was originally defined in 1824 by Sadi Carnot in his famous paper *Reflections on the Motive Power of Fire*. Specifically, according to Carnot:

We use here motive power (work) to express the useful effect that a motor is capable of producing. This effect can always be likened to the elevation of a weight to a certain height. It has, as we know, as a measure, the product of the weight multiplied by the height to which it is raised.

1845:

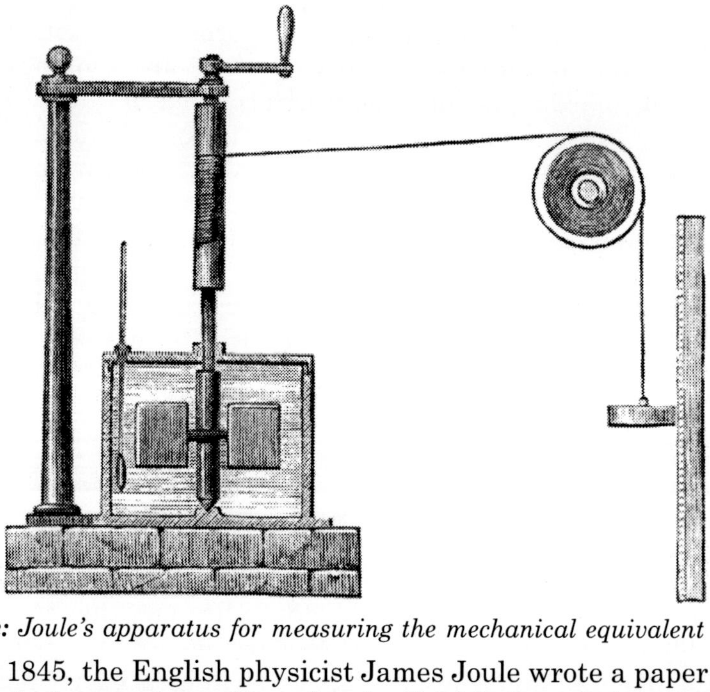

Figure: *Joule's apparatus for measuring the mechanical equivalent of heat.*

In 1845, the English physicist James Joule wrote a paper *On the mechanical equivalent of heat* for the British Association meeting in Cambridge. In this work, he reported his best-known experiment, in which the work released through the action of a "weight *falling* through a height" was used to turn a paddle-wheel in an insulated barrel of water.

In this experiment, the friction and agitation of the paddle-wheel on the body of water caused heat to be generated which, in turn, increased the temperature of water. Both the temperature change ΔT of the water and the height of the fall Δh of the weight mg were recorded. Using these values, Joule was able to determine the mechanical equivalent of heat. Joule estimated a mechanical equivalent of heat to be 819 ft•lbf/Btu (4.41 J/cal). The modern day definitions of heat, work, temperature, and energy all have connection to this experiment.

Overview

The first law of thermodynamics relates changes in the internal energy of a closed thermodynamic system to two forms of energy transfer, as heat and as work. Essential to the thermodynamic concept of work is that the energy transfer in fictive principle be able to occur at a finite rate without any of it necessarily being dissipated by friction or chemical degradation, which are necessarily dissipative. A thermodynamic dissipative process is one in which energy, internal, bulk flow kinetic, or system potential, is transduced from some initial form to some final form, the capacity to do mechanical work of the final form being less that that of the initial form. For example, transfer of energy as heat is dissipative because it is a transfer of internal energy from a body at one temperature to a body at a lower temperature. The second law of thermodynamics implies that this reduces the capacity of that internal energy to do mechanical work.

The concept of thermodynamic work is more general than that of simple mechanical work because it includes other types of energy transfers as well. Thermodynamic work is strictly and fully defined by its external generalized mechanical variables. The other form of energy transfer between closed systems is as heat. Heat is measured by change of temperature of a known quantity of calorimetric material substance; it is of the essence of heat transfer that it is not mediated by the external generalized mechanical variables that define work. This distinction between work and heat is essential to thermodynamics.

Work refers to forms of energy transfer between closed systems that can be accounted for in terms of changes in the external *macroscopic* physical constraints on the system, for example energy which goes into expanding the volume of a system against an external pressure, by driving a piston-head out of a cylinder against an external force. The electrical work required to move a charge against an external electrical field can be measured.

This is in contrast to heat, which is primarily the energy that is transported or transduced as the *microscopic* thermal motions of particles and their associated inter-molecular potential energies, or by thermal radiation. There are two forms of macroscopic heat transfer by direct contact between a closed system and its surroundings: conduction, and thermal radiation. There are several forms of dissipative transduction of energy that can occur internally within a system at a microscopic level, such as friction including bulk and shear viscosity, chemical reaction, unconstrained expansion as in Joule expansion and in diffusion, and phase change; these are not transfers of heat between systems. Convection of internal energy is a form a transport of energy but is in general not, as sometimes mistakenly supposed (a relic of the caloric theory of heat), a form of transfer of energy as heat, because convection is not in itself a microscopic motion of microscopic particles or their intermolecular potential energies, or photons; nor is it a of transfer of energy as work. Nevertheless, if the wall between the system and its surroundings is thick and contains fluid, in the presence of a gravitational field, convective circulation within the wall can be considered as indirectly mediating transfer of energy as heat between the system and its surroundings, though they are not in direct contact.

Formal Definition

In thermodynamics, the quantity of work done by a closed system on its surroundings is defined by factors strictly confined to the interface of the surroundings with the system and to the surroundings of the system, for example an extended gravitational field in which the system sits, that is to say, to things external to the system. There are a few especially important kinds of thermodynamic work.

A simple example of one of those important kinds is pressure-volume work. The pressure of concern is that exerted by the surroundings on the surface of the system, and the volume of interest is the negative of the increment of volume gained by the system from the surroundings. It is usually arranged that the pressure exerted by the surroundings on the surface of the system is well defined and equal to the pressure exerted by the system on the surroundings. This arrangement for transfer of energy as work can be varied in a particular way that depends on the strictly mechanical nature of pressure-volume work. The variation consists in letting the coupling between the system and surroundings be through a rigid rod that links pistons of different areas for the system and surroundings. Then for a given amount of work transferred, the exchange of volumes involves different pressures, inversely with the piston areas, for mechanical equilibrium. This cannot be done for the transfer of energy as heat because of its non-mechanical nature.

Another important kind of work is isochoric work, that is to say work that involves no eventual overall change of volume of the system between the initial and the final states of the process. Examples are friction on the surface of the system as in Rumford's experiment; shaft work such as in Joule's experiments; and slow vibrational action on the system that leaves its eventual volume unchanged, but involves friction within the system. Isochoric work for a body in its own state of internal thermodynamic equilibrium is done only by the surroundings on the body, not by the body on the surroundings, so that the sign of isochoric work with the present sign convention is always negative.

When work is done by a closed system that cannot pass heat in or out because it is adiabatically isolated, the work is referred to as being adiabatic in character. Adiabatic work can be of the pressure-volume kind or of the isochoric kind, or both.

According to the first law of thermodynamics for a closed system, any net increase in the internal energy U must be fully accounted for, in terms of heat δQ entering the system and the work δW done by the system:

$$dU = \delta Q - \delta W.$$

The letter d indicates an exact differential, expressing that internal energy U is a property of the state of the system; they depend only on the original state and the final state, and not upon the path taken. In contrast, the Greek deltas (δ's) in this equation reflect the fact that the heat transfer and the work transfer are *not* properties of the final state of the system. Given only the initial state and the final state of the system, one can only say what the total change in internal energy was, not how much of the energy went out as heat, and how much as work. This can be summarized by saying that heat and work are not state functions of the system.

The minus sign in front of δW indicates that a positive amount of work done *by* the system leads to energy being lost from the system. This is the sign convention for work in many textbooks on physics. This sign convention entails that a non-zero quantity of isochoric work always has a negative sign, because of the second law of thermodynamics.

(An alternate sign convention is to consider the work performed *on* the system by its surroundings as positive. This leads to a change in sign of the work, so that $dU = \delta Q + \delta W$. This is the convention adopted by many modern textbooks of physical chemistry.)

Pressure-volume Work

Pressure-volume work, (or *PV* work) occurs when the volume, *V* of a system changes. *PV* work is often measured in units of litre-atmospheres, where $1\text{L}\cdot\text{atm} = 101.325\text{J}$. However the litre-atmosphere is not a recognised unit in the SI system of units, which measures P in Pascal (Pa), V in m^3 and PV in Joule (J), where 1 J = 1 Pa-m^3. *PV* work is an important topic in chemical thermodynamics.

For a reversible process in a closed system, PV work is represented by the following differential equation:

$$\delta W = PdV$$

where

δW denotes an infinitesimal increment of work done *by* the system;

*P*denotes the pressure inside the system and outside the system, against which the system expands; the two pressures are practically equal for a reversible process;

dV denotes the infinitesimal increment of the volume of the system.

Moreover,

$$W = \int_{V_i}^{V_f} PdV.$$

where

W denotes the work done *by* the system during the whole of the reversible process.

The first law of thermodynamics can then be expressed as

$$dU = \delta Q - PdV.$$

(In the alternate sign convention where *W* = work done *on* the system, $\delta W = -PdV$, but $dU = \delta Q - PdV$ is unchanged.)

Path Dependence

As for all kinds of work, in general PV work is path-dependent and is therefore a thermodynamic process function. The statement that a process is reversible and adiabatic serves as a specification of the path, but does not determine the path uniquely, because the path can include several slow goings backward and forward in volume, as long as there is no transfer of energy as heat. The first law of thermodynamics states $dU = \delta Q - \delta W$. For an adiabatic process, $\delta Q = 0$ and thus the integral amount work done is equal to the change in internal energy. For a reversible adiabatic process, the integral amount of work done during the process depends only on the initial and final states of the process, and is the one and the same for every intermediate path.

If the process took a path other than an adiabatic path, the work would be different. This would only be possible if heat flowed into/ out of the system. In a non-adiabatic process, there are indefinitely many paths between the initial and final states.

In the current methematical notation, the differential δW is an inexact differential.

In another notation, δW is written πW (with a line through the d). This notation indicates that πW is not an exactone-form. The line-through is merely a flag to warn us there is actually no function (0-form) W which is the potential of πW. If there were, indeed, this function W, we should be able to just use Stokes Theorem to evaluate this putative function, the potential of πW, at the boundary of the path, that is, the initial and final points, and therefore the work would be a state function. This impossibility is consistent with the fact that it does not make sense to refer to *the work on a point* in the PV diagram; work presupposes a path.

Free Energy and Exergy

The amount of useful work which may be extracted from a thermodynamic system is determined by the second law of thermodynamics. Under many practical situations this can be represented by the thermodynamic availability, or Exergy, function. Two important cases are: in thermodynamic systems where the temperature and volume are held constant, the measure of useful work attainable is the Helmholtz free energy function; and in systems where the temperature and pressure are held constant, the measure of useful work attainable is the Gibbs free energy.

Non-mechanical Forms of Work

Non-mechanical work in thermodynamics is work determined by long-range forces established by events in the surroundings of the system, but penetrating into the system as force fields. The long-range forces are forces in the ordinary physical sense of the word, not the so-called 'thermodynamic forces' of non-equilibrium thermodynamic terminology.

The non-mechanical work of long-range forces can have either positive or negative sign, work being done by the system on the surroundings, or *vice versa*. Work done by long-range forces can be done indefinitely slowly, so as to approach the fictive reversible quasi-static ideal, in which entropy is not created in the system by the process.

In thermodynamics, non-mechanical work is to be contrasted with work that is done by forces in immediate contact between the system and its surroundings. If the putative 'work' of a process cannot be defined as either long-range work or else as contact work, then sometimes it cannot be described by the thermodynamic formalism as work at all. Nevertheless, the thermodynamic formalism allows that energy can be transferred between an open system and its surroundings by processes for which work is not defined. An example is when the wall between the system and its surrounds is not considered as idealized and vanishingly thin, so that processes can occur within the wall, such as friction affecting the transfer of matter across the wall; in this case, the forces of transfer are neither strictly long-range nor strictly due to contact between the system and its surrounds; the transfer of energy can then be considered as by convection, and assessed in sum just as transfer of internal energy. This is conceptually different from transfer of energy as heat through a thick fluid-filled wall in the presence of a gravitational field, between a closed system and its surroundings; in this case there may convective circulation within the wall but the process may still be considered as transfer of energy as heat between the system and its surroundings; if the whole wall is moved by the application of force from the surroundings, without change of volume of the wall, so as to change the volume of the system, then it is also at the same time transferring energy as work. A chemical reaction within a system can lead to electrical long-range forces and to electric current flow, which transfer energy as work between system and surroundings, though the system's chemical reactions themselves (except for the special limiting case in which in they are driven through devices in the surroundings so as to occur along a line of thermodynamic equilibrium) are always irreversible and do not directly interact with the surroundings of the system.

Non-mechanical work contrasts with pressure-volume work which may be regarded as a kind of contact work, because it occurs through direct material contact with the matter at the boundary of the system. Such work can often be accurately described by changes in state variables of the system, such as the time courses of changes in the pressure and volume of the system; the volume of the system is classified as a "deformation variable". Pressure-volume work can have either positive or negative sign. Pressure-volume work can be made to approach the fictive reversible quasi-static ideal.

Non-mechanical work also contrasts with shaft work which however is not so readily described by changes in the state variables of the system. This is because it can be isochoric, that is to say, it can be done without a change in the volume of the system between the initial and final states of the process. It is nevertheless mechanical. A system that is initially in a state of thermodynamic equilibrium cannot do shaft work on the surroundings, though the thermodynamic formalism allows that the surroundings can do shaft work on the system. This explains the curious use of the phrase "inanimate material agency" by Kelvin in one of his statements of the second law of thermodynamics; the surroundings are considered to be able to create elaborate changes such as indefinitely prolonged, varied, or ceased rotation of a driving shaft, while a system that starts in a state of thermodynamic equilibrium is inanimate and cannot spontaneously do that. Thus the sign of shaft work is always negative, work being done on the system by the surroundings. Shaft work can hardly be done indefinitely slowly; consequently it always produces entropy within the system, because it relies on friction or viscosity within the system for its transfer. The foregoing comments about shaft work apply only when one ignores that the system can store angular momentum and its related energy.

Examples of non-mechanical work modes include

- Electrical work – where the force is defined by the surroundings' *voltage* (the electrical potential) and the generalized displacement is change of spatial distribution of *electrical charge*
- Magnetic work – where the force is defined by the surroundings' *magnetic field strength* and the generalized displacement is change of total *magnetic dipole moment*
- Electrical polarization work – where the force is defined by the surroundings' *electric field strength* and the generalized displacement is change of the *polarization of the medium* (the sum of the electric dipole moments of the molecules)
- Gravitational work – where the force is defined by the surroundings' gravitational field and the generalized displacement is change of the spatial distribution of the matter within the system.

Path Function and Point Function

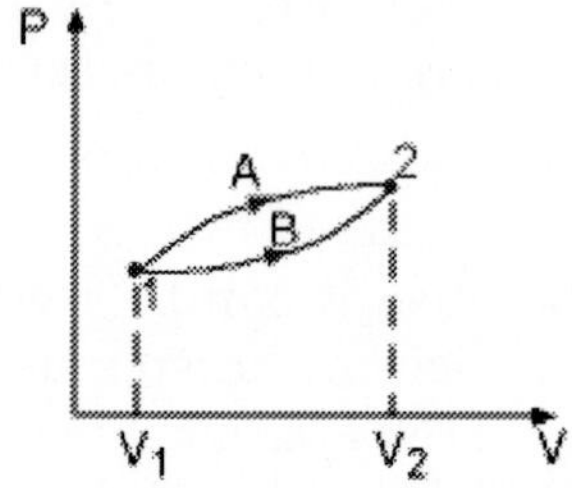

Figure: *Path Function and Point Function*

Path function and Point function are introduced to identify the variables of thermodynamics.

- *Path function:* Their magnitudes depend on the path followed during a process as well as the end states. Work (W), heat (Q) are path functions.

  Process A: W_A = 10 kJ

  Process b: W_B = 7 kJ

- *Point Function:* They depend on the state only, and not on how a system reaches that state. All properties are point functions.

  Process A: $V_2 - V_1 = 3\ m^3$

  Process B: $V_2 - V_1 = 3\ m^3$

Zeroth Law of Thermodynamics

The zeroth law of thermodynamics states that if two systems are each in thermal equilibrium with a third system, they are also in thermal equilibrium with each other.

Two systems are said to be in the relation of thermal equilibrium if they are linked by a wall permeable only to heat, and do not change over time. (As a convenience of language, systems are sometimes also said to be in a relation of thermal equilibrium if they are not linked so as to be able to transfer heat to each other, but would not do so if they were connected by a wall permeable only to heat.)

The physical meaning of the law was expressed by Maxwell in the words: “All heat is of the same kind.” For this reason, another statement of the law is “All diathermal walls are equivalent.”

The law is important for the mathematical formulation of thermodynamics, which needs the assertion that the relation of thermal equilibrium is an equivalence relation. This information is needed for the mathematical definition of temperature that will agree with the physical existence of valid thermometres.

Zeroth Law as Equivalence Relation

A system is said to be in thermal equilibrium when it experiences no net change of its observable state over time. The most precise statement of the zeroth law is that thermal equilibrium constitutes an equivalence relation on pairs of thermodynamic systems. In other words, the set of all equilibrated thermodynamic systems may be divided into subsets in which every system belongs to one and only one subset, and is in thermal equilibrium with every other member of that subset, and is not in thermal equilibrium with a member of any other subset. This means that a unique "tag" can be assigned to every system, and if the "tags" of two systems are the same, they are in thermal equilibrium with each other, and if they are not, they are not. Ultimately, this property is used to justify the use of thermodynamic temperature as a tagging system. Thermodynamic temperature provides further properties of thermally equilibrated systems, such as order and continuity with regard to "hotness" or "coldness", but these properties are not implied by the standard statement of the zeroth law.

If it is specified that a system is in thermal equilibrium with itself (i.e., thermal equilibrium is reflexive), then the zeroth law may be stated as follows: If *A*, *B*, and *C* are distinct thermodynamic systems:

If *A* and *C* are each in thermal equilibrium with *B*, *A* is also in equilibrium with *C*.

This statement asserts that thermal equilibrium is a Euclidean relation between thermodynamic systems. If we also grant that all thermodynamic systems are in thermal equilibrium with themselves, then thermal equilibrium is also a reflexive relation. Relations that are both reflexive and Euclidean are equivalence relations. One consequence of this reasoning is that thermal equilibrium is a transitive relationship: If *A* is in thermal equilibrium with *B* and *B* is in thermal equilibrium with *C*, then *A* is in thermal equilibrium with *C*. Another consequence is that the equilibrium relationship is symmetric: If *A* is in thermal equilibrium with *B*, then *B* is in thermal equilibrium with *A*. Thus we may say that two systems are in thermal equilibrium with each other, or that they are in mutual equilibrium. Implicitly assuming both reflexivity and symmetry, the zeroth law is therefore often expressed as:

If two systems are in thermal equilibrium with a third system, then they are in thermal equilibrium with each other.

Again, implicitly assuming both reflexivity and symmetry, the zeroth law is occasionally expressed as the transitive relationship:

If *A* is in thermal equilibrium with *B* and if *B* is in thermal equilibrium with *C*, then *A* is in thermal equilibrium with *C*.

Foundation of Temperature

The zeroth law establishes thermal equilibrium as an equivalence relationship. An equivalence relationship on a set (such as the set of thermally equilibrated systems) divides that set into a collection of distinct subsets ("disjoint subsets") where any member of the set is a member of one and only one such subset. In the case of the zeroth law, these subsets consist of systems which are in mutual equilibrium. This partitioning allows any member of the subset to be uniquely "tagged" with a label identifying the subset to which it belongs. Although the labelling may be quite arbitrary, temperature is just such a labelling process which uses the real number system for tagging. The zeroth law justifies the use of suitable thermodynamic systems as thermometres to provide such a labelling, which yield any number of possible empirical temperature scales, and justifies the use of the second law of thermodynamics to provide an absolute, or thermodynamic temperature scale. Such temperature scales bring additional continuity and ordering (i.e., "hot" and "cold") properties to the concept of temperature.

In the space of thermodynamic parametres, zones of constant temperature form a surface, that provides a natural order of nearby surfaces. One may therefore construct a global temperature function that provides a continuous ordering of states. The dimensionality of a surface of constant temperature is one less than the number of thermodynamic parametres, thus, for an ideal gas described with three thermodynamic parametres *P*, *V* and *n*, it is a two-dimensional surface.

For example, if two systems of ideal gases are in equilibrium, then $P_1V_1/N_1 = P_2V_2/N_2$ where P_i is the pressure in the *i*th system, V_i is the volume, and N_i is the amount (in moles, or simply the number of atoms) of gas.

The surface $PV/N = const$ defines surfaces of equal thermodynamic temperature, and one may label defining *T* so that $PV/N = RT$, where *R* is some constant. These systems can now be used as a thermometre to calibrate other systems. Such systems are known as "ideal gas thermometres".

History

According to Arnold Sommerfeld, Ralph H. Fowler invented the title 'the zeroth law of thermodynamics' when he was discussing the 1935 text of Saha and Srivastava. They write on page 1 that "every physical quantity must be measurable in numerical terms". They presume that temperature is a physical quantity and then deduce the statement "If a body A is in temperature equilibrium with two bodies B and C, then B and C themselves will be in temperature equilibrium with each other." They then in a self-standing paragraph italicize as if to state their basic postulate: "*Any of the physical properties of A which change with the application of heat may be observed and utilised for the measurement of temperature.*" They do not themselves here use the term 'zeroth law of thermodynamics'. There are very many statements of these physical ideas in the physics literature long before this text, in very similar language. What was new here was just the label 'zeroth law of thermodynamics'. Fowler, with co-author Edward A. Guggenheim, wrote of the zeroth law as follows:

...we introduce the postulate: *If two assemblies are each in thermal equilibrium with a third assembly, they are in thermal equilibrium with each other.*

They then proposed that "it may be shown to follow that the condition for thermal equilibrium between several assemblies is the equality of a certain single-valued function of the thermodynamic states of the assemblies, which may be called the temperature t, any one of the assemblies being used as a "thermometre" reading the temperature t on a suitable scale. This postulate of the "*Existence of temperature*" could with advantage be known as *the zeroth law of thermodynamics*." The first sentence of this present article is a version of this statement.

Thermodynamic Temperature

Thermodynamic temperature is the absolute measure of temperature and it is one of the principal parametres of thermodynamics.

Thermodynamic temperature is defined by the second law of thermodynamics in which the theoretically lowest temperature is the null or zero point. At this point, called absolute zero, the particle constituents of matter have minimal motion and can become no colder. In the quantum-mechanical description, matter at absolute zero is in its ground state, which is its state of lowest energy. Thermodynamic temperature is therefore often also called absolute temperature.

The International System of Units specifies a particular scale for thermodynamic temperature. It uses the Kelvin scale for measurement and selects the triple point of water at 273.16K as the fundamental fixing point. Other scales have been in use historically. The Rankine scale, using the degree Fahrenheit as its unit interval, is still in use as part of the English Engineering Units in the United States in some engineering fields. ITS-90 gives a practical means of estimating the thermodynamic temperature to a very high degree of accuracy.

Roughly, the temperature of a body at rest is a measure of the mean of the energy of the translational, vibrational and rotational motions of matter's particle constituents, such as molecules, atoms, and subatomic particles. The full variety of these kinetic motions, along with potential energies of particles, and also occasionally certain other types of particle energy in equilibrium with these, contribute the total internal energy of a substance. Internal energy is loosely called the heat energy or thermal energy in conditions when no work is done upon the substance by its surroundings, or by the substance upon the surroundings. Internal energy may be stored in a number of ways within a substance, each way constituting a "degree of freedom". At equilibrium, each degree of freedom will have on average the same energy: $k_B T / 2$ where k_B is the Boltzmann constant, unless that degree of freedom is in the quantum regime. The internal degrees of freedom (rotation, vibration, etc.) may be in the quantum regime at room temperature, but the translational degrees of freedom will be in the classical regime except at extremely low temperatures (fractions of Kelvins) and it may be said that, for most situations, the thermodynamic temperature is specified by the average translational kinetic energy of the particles.

Overview

Temperature is a measure of the random submicroscopic motions and vibrations of the particle constituents of matter. These motions comprise the internal energy of a substance. More specifically, the thermodynamic temperature of any bulk quantity of matter is the measure of the average kinetic energy per classical (i.e., non-quantum) degree of freedom of its constituent particles. "Translational motions" are almost always in the classical regime. Translational motions are ordinary, whole-body movements in three-dimensional space in which particles move about and exchange energy in collisions. Figure shows translational motion in gases; Figure shows translational motion in solids. Thermodynamic temperature's null point, absolute zero, is the temperature at which the particle constituents of matter are as close as possible to complete rest; that is, they have minimal motion, retaining only quantum mechanical motion. Zero kinetic energy remains in a substance at absolute zero.

Throughout the scientific world where measurements are made in SI units, thermodynamic temperature is measured in kelvins (symbol: K). Many engineering fields in the U.S. however, measure thermodynamic temperature using the Rankine scale.

By international agreement, the unit *kelvin* and its scale are defined by two points: absolute zero, and the triple point of Vienna Standard Mean Ocean Water (water with a specified blend of hydrogen and oxygen isotopes). Absolute zero, the lowest possible temperature, is defined as being precisely 0 K *and* –273.15 °C. The triple point of water is defined as being precisely 273.16 K *and* 0.01 °C. This definition does three things:

1. It fixes the magnitude of the kelvin unit as being precisely 1 part in 273.16 parts the difference between absolute zero and the triple point of water;
2. It establishes that one kelvin has precisely the same magnitude as a one-degree increment on the Celsius scale; and
3. It establishes the difference between the two scales' null points as being precisely 273.15 kelvins (0 K =–273.15 °C and 273.16 K = 0.01 °C).

Temperatures expressed in kelvins are converted to degrees Rankine simply by multiplying by 1.8 as follows: $T_{°R} = 1.8T_K$, where T_K and $T_{°R}$ are temperatures in kelvin and degrees Rankine respectively. Temperatures expressed in degrees Rankine are converted to kelvins by *dividing* by 1.8 as follows: $T_K = {}^{T_{°R}}/_{1.8}$.

Practical Realization

Although the Kelvin and Celsius scales are defined using absolute zero (0 K) and the triple point of water (273.16 K and 0.01 °C), it is impractical to use this definition at temperatures that are very different from the triple point of water. ITS-90 is then designed to represent the thermodynamic temperature as closely as possible throughout its range. Many different thermometre designs are required to cover the entire range. These include helium vapour pressure thermometres, helium gas thermometres, standard platinum resistance thermometres (known as SPRTs, PRTs or Platium RTDs) and monochromatic radiation thermometres.

For some types of thermometre the relationship between the property observed (e.g., length of a mercury column) and temperature, is close to linear, so for most purposes a linear scale is sufficient, without point-by-point calibration. For others a calibration curve or equation is required. The mercury thermometre, invented before the thermodynamic temperature was understood, originally *defined* the temperature scale; its linearity made readings correlate well with true temperature, i.e. the "mercury" temperature scale was a close fit to the true scale.

The Relationship of Temperature, Motions, Conduction, and Thermal Energy

The Nature of Kinetic Energy, Translational Motion, and Temperature: The thermodynamic temperature is a measure of the average energy of the translational, vibrational, and rotational motions of matter's particle constituents (molecules, atoms, and subatomic particles). The full variety of these kinetic motions, along with potential energies of particles, and also occasionally certain other types of particle energy in equilibrium with these, contribute the total internal energy (loosely, the thermal energy) of a substance. Thus, internal energy may be stored in a number of ways (degrees of freedom) within a substance. When the degrees of freedom are in the classical regime ("unfrozen") the temperature is very simply related to the average energy of those degrees of freedom at equilibrium. The three translational degrees of freedom are unfrozen except for the very lowest temperatures, and their kinetic energy is simply related to the thermodynamic temperature over the widest range. The heat capacity, which relates heat input and temperature change, is discussed below.

The relationship of kinetic energy, mass, and velocity is given by the formula $E_k = {}^1/_2 mv^2$. Accordingly, particles with one unit of mass moving at one unit of velocity have precisely the same kinetic energy, and precisely the same temperature, as those with four times the mass but half the velocity.

Except in the quantum regime at extremely low temperatures, the thermodynamic temperature of any *bulk quantity* of a substance (a statistically significant quantity of particles) is directly proportional to the mean average kinetic energy of a specific kind of particle motion known as *translational motion.* These simple movements in the three x, y, and z–axis dimensions of space means the particles move in the three spatial *degrees of freedom*. The temperature derived from this translational kinetic energy is sometimes referred to as ***kinetic temperature*** and is equal to the thermodynamic temperature over a very wide range of temperatures. Since there are three translational degrees of freedom (e.g., motion along the x, y, and z axes), the translational kinetic energy is related to the kinetic temperature by:

$$\bar{E}=\frac{3}{2}k_B T_k$$

where:

- $\bar{E}$ is the mean kinetic energy in joules (J) and is pronounced "E bar"
- $k_B = 1.3806504(24)\times10^{-23}$ J/K is the Boltzmann constant and is pronounced "Kay sub bee"
- T_k is the kinetic temperature in kelvins (K) and is pronounced "Tee"

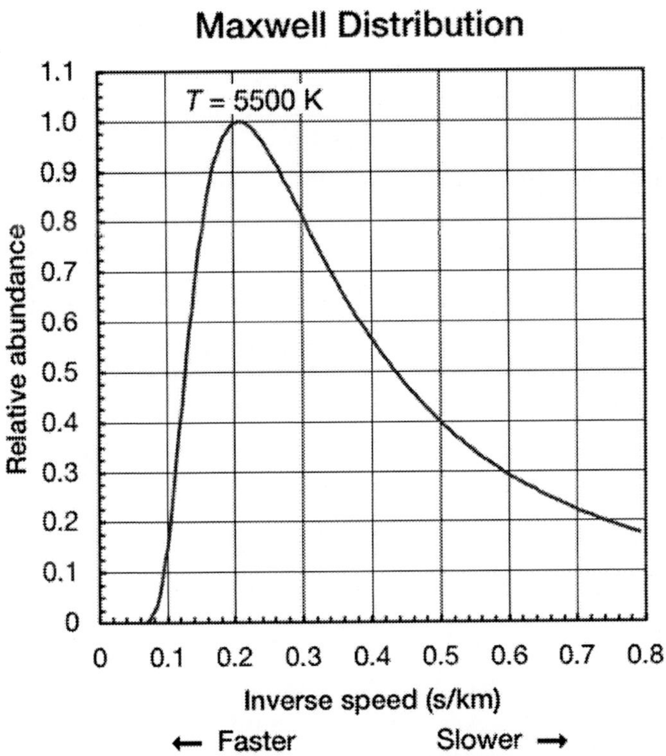

***Figure:** The translational motions of helium atoms occur across a range of speeds.*

While the Boltzmann constant is useful for finding the mean kinetic energy of a particle, it's important to note that even when a substance is isolated and in thermodynamic equilibrium (all parts are at a uniform temperature and no heat is going into or out of it), the translational motions of individual atoms and molecules occurs across a wide range of speeds. At any one instant, the proportion of particles moving at a given speed within this range is determined by probability as described by the Maxwell–Boltzmann distribution. The graph shown here in *Fig. 2* shows the speed distribution of 5500 K helium atoms. They have a *most probable* speed of 4.780 km/s. However, a certain proportion of atoms at any given instant are moving faster while others are moving relatively slowly; some are momentarily at a virtual standstill (off the *x*–axis to the right). This graph uses *inverse speed* for its *x*–axis so the shape of the curve can easily be compared to the curves in figure. In both graphs, zero on the *x*–axis represents infinite temperature. Additionally, the *x* and *y*–axis on both graphs are scaled proportionally.

The High Speeds of Translational Motion

Although very specialized laboratory equipment is required to directly detect translational motions, the resultant collisions by atoms or molecules with small particles suspended in a fluid produces Brownian motion that can be seen with an ordinary microscope. The translational motions of elementary particles are *very* fast and temperatures close to absolute zero are required to directly observe them. For instance, when scientists at the NIST achieved a record-setting cold temperature of 700 nK (billionths of a kelvin) in 1994, they used optical lattice laser equipment to adiabatically cool caesium atoms. They then turned off the entrapment lasers and directly measured atom velocities of 7 mm per second in order to calculate their temperature. Formulas for calculating the velocity and speed of translational motion are given in the following footnote.

The Internal Motions of Molecules and Specific Heat

There are other forms of internal energy besides the kinetic energy of translational motion. As can be seen in the animation at right, molecules are complex objects; they are a population of atoms and thermal agitation can strain their internal chemical bonds in three different ways: via rotation, bond length, and bond angle movements. These are all types of *internal degrees of freedom*. This makes molecules distinct from *monatomic* substances (consisting of individual atoms) like the noble gaseshelium and argon, which have only the three translational degrees of freedom. Kinetic energy is stored in molecules' internal degrees of freedom, which gives them an *internal temperature*. Even though these motions are called *internal*, the external portions of molecules still move—rather like the jiggling of a stationary water balloon. This permits the two-way exchange of kinetic energy between internal motions and translational motions with each molecular collision. Accordingly, as energy is removed from molecules, both their kinetic temperature (the temperature derived from the kinetic energy of translational motion) and their internal temperature simultaneously diminish in equal proportions. This phenomenon is described by the equipartition theorem, which states that for any bulk quantity of a substance in equilibrium, the kinetic energy of particle motion is evenly distributed among all the active (i.e. unfrozen) degrees of freedom available to the particles. Since the internal temperature of molecules are usually equal to their kinetic temperature, the distinction is usually of interest only in the detailed study of non-local thermodynamic equilibrium (LTE) phenomena such as combustion, the sublimation of solids, and the diffusion of hot gases in a partial vacuum.

The kinetic energy stored internally in molecules causes substances to contain more internal energy at any given temperature and to absorb additional internal energy for a given temperature increase. This is because any kinetic energy that is, at a given instant, bound in internal motions is not at that same instant contributing to the molecules' translational motions. This extra thermal energy simply increases the amount of energy a substance absorbs for a given temperature rise. This property is known as a substance's specific heat capacity.

Different molecules absorb different amounts of thermal energy for each incremental increase in temperature; that is, they have different specific heat capacities. High specific heat capacity arises, in part, because certain substances' molecules possess more internal degrees of freedom than others do. For instance, nitrogen, which is a diatomic molecule, has *five* active degrees of freedom at room temperature: the three comprising translational motion plus two rotational degrees of freedom internally. Since the two internal degrees of freedom are essentially unfrozen, in accordance with the equipartition theorem nitrogen has five-thirds the specific heat capacity per mole (a specific number of molecules) as do the monatomic gases. Another example is gasoline. Gasoline can absorb a large amount of thermal energy per mole with only a modest temperature change because each molecule comprises an average of 21 atoms and therefore has many internal degrees of freedom. Even larger, more complex molecules can have dozens of internal degrees of freedom.

The Diffusion of Thermal Energy: Entropy, Phonons, and Mobile Conduction Electrons

Heat conduction is the diffusion of thermal energy from hot parts of a system to cold. A system can be either a single bulk entity or a plurality of discrete bulk entities. The term *bulk* in this context means a statistically significant quantity of particles (which can be a microscopic amount). Whenever thermal energy diffuses within an isolated system, temperature differences within the system decrease (and entropy increases).

One particular heat conduction mechanism occurs when translational motion, the particle motion underlying temperature, transfers momentum from particle to particle in collisions. In gases, these translational motions are of the nature shown above in *Fig. 1.* As can be seen in that animation, not only does momentum (heat) diffuse throughout the volume of the gas through serial collisions, but entire molecules or atoms can move forward into new territory, bringing their kinetic energy with them. Consequently, temperature differences equalize throughout gases very quickly—especially for light atoms or molecules; convection speeds this process even more.

Translational motion in *solids*, however, takes the form of *phonons.* Phonons are constrained, quantized wave packets travelling at the speed of sound for a given substance. The manner in which phonons interact within a solid determines a variety of its properties, including its thermal conductivity. In electrically insulating solids, phonon-based heat conduction is *usually* inefficient and such solids are considered *thermal insulators* (such as glass, plastic, rubber, ceramic, and rock). This is because in solids, atoms and molecules are locked into place relative to their neighbours and are not free to roam.

Metals however, are not restricted to only phonon-based heat conduction. Thermal energy conducts through metals extraordinarily quickly because instead of direct molecule-to-molecule collisions, the vast majority of thermal energy is mediated via very light, mobile *conduction electrons.* This is why there is a near-perfect correlation between metals' thermal conductivity and their electrical conductivity. Conduction electrons imbue metals with their extraordinary conductivity because they are *delocalized* (i.e., not tied to a specific atom) and behave rather like a sort of quantum gas due to the effects of *zero-point energy*. Furthermore, electrons are relatively light with a rest mass only $^1/_{1836}$th that of a proton. This is about the same ratio as a .22 Short bullet (29 grains or 1.88 g) compared to the rifle that shoots it. As Isaac Newton wrote with his third law of motion,

Law #3: All forces occur in pairs, and these two forces are equal in magnitude and opposite in direction.

However, a bullet accelerates faster than a rifle given an equal force. Since kinetic energy increases as the square of velocity, nearly all the kinetic energy goes into the bullet, not the rifle, even though both experience the same force from the expanding propellant gases. In the same manner, because they are much less massive, thermal energy is readily borne by mobile conduction electrons. Additionally, because they're delocalized and *very* fast, kinetic thermal energy conducts extremely quickly through metals with abundant conduction electrons.

The Diffusion of Thermal Energy: Black-body Radiation

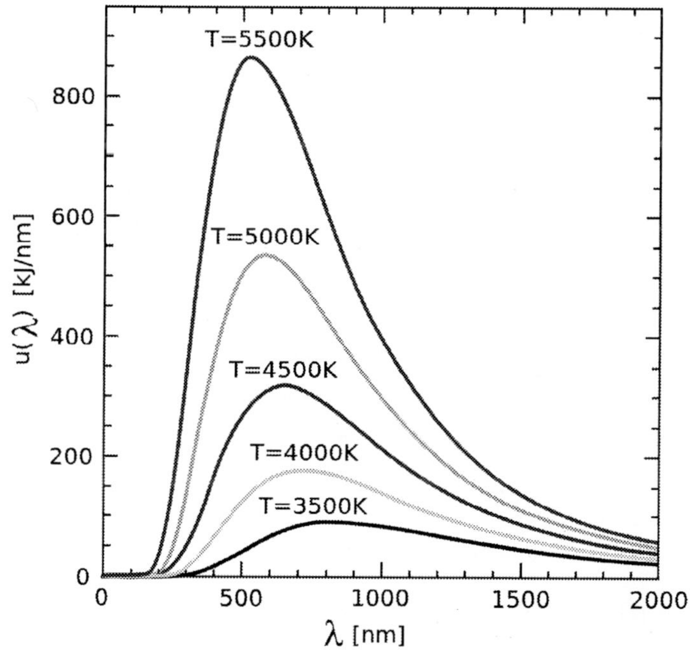

Figure: *The spectrum of black-body radiation has the form of a Planck curve. A 5500 K black-body has a peak emittance wavelength of 527 nm.*

Thermal radiation is a byproduct of the collisions arising from various vibrational motions of atoms.

These collisions cause the electrons of the atoms to emit thermal photons (known as black-body radiation).

Photons are emitted anytime an electric charge is accelerated (as happens when electron clouds of two atoms collide). Even *individual molecules* with internal temperatures greater than absolute zero also emit black-body radiation from their atoms. In any bulk quantity of a substance at equilibrium, black-body photons are emitted across a range of wavelengths in a spectrum that has a bell curve-like shape called a Planck curve.

The top of a Planck curve (the peak emittance wavelength) is located in a particular part of the electromagnetic spectrum depending on the temperature of the black-body. Substances at extreme cryogenic temperatures emit at long radio wavelengths whereas extremely hot temperatures produce short gamma rays.

Black-body radiation diffuses thermal energy throughout a substance as the photons are absorbed by neighbouring atoms, transferring momentum in the process.

Black-body photons also easily escape from a substance and can be absorbed by the ambient environment; kinetic energy is lost in the process.

As established by the Stefan–Boltzmann law, the intensity of black-body radiation increases as the fourth power of absolute temperature. Thus, a black-body at 824 K (just short of glowing dull red) emits *60 times* the radiant power as it does at 296 K (room temperature).

This is why one can so easily feel the radiant heat from hot objects at a distance. At higher temperatures, such as those found in an incandescent lamp, black-body radiation can be the principal mechanism by which thermal energy escapes a system.

Table of Thermodynamic Temperatures

The full range of the thermodynamic temperature scale, from absolute zero to absolute hot, and some notable points between them are shown in the table below.

	kelvin	*Peak emittance wavelength of black-body photons*
Absolute zero(precisely by definition)	0 K	∞
Coldest measured temperature	450 pK	6,400 kilometres
One millikelvin (precisely by definition)	0.001 K	2.897 77 metres (Radio, FM band)
Water's triple point (precisely by definition)	273.16 K	10,608.3 nm(Long wavelength I.R.)
Incandescent lamp[B]	2500 K	1160 nm(Near infrared)[C]
Sun's visible surface[C]	5778 K	501.5 nm(Green light)
Lightning bolt'schannel	28,000 K	100 nm(Far Ultraviolet light)
Sun's core	16 MK	0.18 nm (X-rays)
Thermonuclear weapon (peak temperature)	350 MK	8.3×10^{-3} nm(Gamma rays)
Sandia National Labs'Z machine[D]	2 GK	1.4×10^{-3} nm(Gamma rays)
Core of a high–massstar on its last day	3 GK	1×10^{-3} nm(Gamma rays)
Merging binary neutronstar system	350 GK	8×10^{-6} nm(Gamma rays)
Gamma-ray burstprogenitors	1 TK	3×10^{-6} nm(Gamma rays)
Relativistic HeavyIon Collider	1 TK	3×10^{-6} nm(Gamma rays)
CERN's proton vs.nucleus collisions	10 TK	3×10^{-7} nm(Gamma rays)
Universe 5.391×10^{-44} safter the Big Bang	1.417×10^{32} K	1.616×10^{-26} nm (Planck frequency)

[A] The 2500 K value is approximate.

[B] For a true blackbody (which tungsten filaments are not). Tungsten filaments' emissivity is greater at shorter wavelengths, which makes them appear whiter.

[C] Effective photosphere temperature.

[D] For a true blackbody (which the plasma was not). The Z machine's dominant emission originated from 40 MK electrons (soft x–ray emissions) within the plasma.

The Heat of Phase Changes

Figure: *Ice and water: two phases of the same substance*

The kinetic energy of particle motion is just one contributor to the total thermal energy in a substance; another is *phase transitions*, which are the potential energy of molecular bonds that can form in a substance as it cools (such as during condensing and freezing). The thermal energy required for a phase transition is called *latent heat.* This phenomenon may more easily be grasped by considering it in the reverse direction: latent heat is the energy required to *break chemical* bonds (such as during evaporation and melting). Almost everyone is familiar with the effects of phase transitions; for instance, steam at 100 °C can cause severe burns much faster than the 100 °C air from a hair dryer. This occurs because a large amount of latent heat is liberated as steam condenses into liquid water on the skin.

Even though thermal energy is liberated or absorbed during phase transitions, pure chemical elements, compounds, and eutecticalloys exhibit *no temperature change whatsoever* while they undergo them. Consider one particular type of phase transition: melting. When a solid is melting, crystal latticechemical bonds are being broken apart; the substance is transitioning from what is known as a *more ordered state* to a *less ordered state*. In above figure, the melting of ice is shown within the lower left box heading from blue to green.

At one specific thermodynamic point, the melting point (which is 0 °C across a wide pressure range in the case of water), all the atoms or molecules are, on average, at the maximum energy threshold their chemical bonds can withstand without breaking away from the lattice. Chemical bonds are all-or-nothing forces: they either hold fast, or break; there is no in-between state.

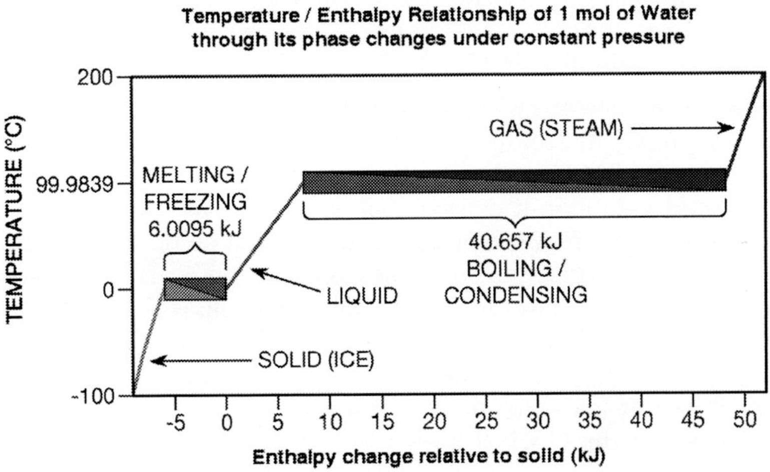

Figure: *Water's temperature does not change during phase transitions as heat flows into or out of it. The total heat capacity of a mole of water in its liquid phase (the green line) is 7.5507 kJ.*

Consequently, when a substance is at its melting point, every joule of added thermal energy only breaks the bonds of a specific quantity of its atoms or molecules, converting them into a liquid of precisely the same temperature; no kinetic energy is added to translational motion (which is what gives substances their temperature). The effect is rather like popcorn: at a certain temperature, additional thermal energy can't make the kernels any hotter until the transition (popping) is complete. If the process is reversed (as in the freezing of a liquid), thermal energy must be removed from a substance.

As stated above, the thermal energy required for a phase transition is called *latent heat.* In the specific cases of melting and freezing, it's called *enthalpy of fusion* or *heat of fusion.* If the molecular bonds in a crystal lattice are strong, the heat of fusion can be relatively great, typically in the range of 6 to 30 kJ per mole for water and most of the metallic elements. If the substance is one of the monatomic gases, (which have little tendency to form molecular bonds) the heat of fusion is more modest, ranging from 0.021 to 2.3 kJ per mole. Relatively speaking, phase transitions can be truly energetic events. To completely melt ice at 0 °C into water at 0 °C, one must add roughly 80 times the thermal energy as is required to increase the temperature of the same mass of liquid water by one degree Celsius. The metals' ratios are even greater, typically in the range of 400 to 1200 times. And the phase transition of boiling is much more energetic than freezing. For instance, the energy required to completely boil or vaporize water (what is known as *enthalpy of vaporization*) is roughly *540 times* that required for a one-degree increase.

Water's sizable enthalpy of vaporization is why one's skin can be burned so quickly as steam condenses on it (heading from red to green in *Fig. 7* above). In the opposite direction, this is why one's skin feels cool as liquid water on it evaporates (a process that occurs at a sub-ambient wet-bulb temperature that is dependent on relative humidity). Water's highly energetic enthalpy of vaporization is also an important factor underlying why *solar pool covers* (floating, insulated blankets that cover swimming pools when not in use) are so effective at reducing heating costs: they prevent evaporation. For instance, the evaporation of just 20 mm of water from a 1.29-metre-deep pool chills its water 8.4 degrees Celsius (15.1 °F).

Internal Energy

The total energy of all particle motion translational and internal, including that of conduction electrons, plus the potential energy of phase changes, plus zero-point energy comprise the internal energy of a substance.

Figure: *When many of the chemical elements, such as the noble gases and platinum-group metals, freeze to a solid — the most ordered state of matter — their crystal structures have a* closest-packed arrangement. *This yields the greatest possible packing density and the lowest energy state.*

Internal Energy at Absolute Zero

As a substance cools, different forms of internal energy and their related effects simultaneously decrease in magnitude: the latent heat of available phase transitions are liberated as a substance changes from a less ordered state to a more ordered state; the translational motions of atoms and molecules diminish (their kinetic temperature decreases); the internal motions of molecules diminish (their internal temperature decreases); conduction electrons (if the substance is an electrical conductor) travel somewhat slower; and black-body radiation's peak emittance wavelength increases (the photons' energy decreases). When the particles of a substance are as close as possible to complete rest and retain only ZPE-induced quantum mechanical motion, the substance is at the temperature of absolute zero (T=0).

Note that whereas absolute zero is the point of zero thermodynamic temperature and is also the point at which the particle constituents of matter have minimal motion, absolute zero is not necessarily the point at which a substance contains zero thermal energy; one must be very precise with what one means by internal energy. Often, all the phase changes that can occur in a substance, will have occurred by the time it reaches absolute zero. However, this is not always the case. Notably, T=0 helium remains liquid at room pressure and must be under a pressure of at least 25 bar (2.5 MPa) to crystallize. This is because helium's heat of fusion (the energy required to melt helium ice) is so low (only 21 joules per mole) that the motion-inducing effect of zero-point energy is sufficient to prevent it from freezing at lower pressures. Only if under at least 25 bar (2.5 MPa) of pressure will this latent thermal energy be liberated as helium freezes while approaching absolute zero. A further complication is that many solids change their crystal structure to more compact arrangements at extremely high pressures (up to millions of bars, or hundreds of gigapascals). These are known as solid-solid phase transitions wherein latent heat is liberated as a crystal lattice changes to a more thermodynamically favourable, compact one.

The above complexities make for rather cumbersome blanket statements regarding the internal energy in T=0 substances. Regardless of pressure though, what can be said is that at absolute zero, all solids with a lowest-energy crystal lattice such those with a closest-packed arrangement contain minimal internal energy, retaining only that due to the ever-present background of zero-point energy. One can also say that for a given substance at constant pressure, absolute zero is the point of lowest enthalpy (a measure of work potential that takes internal energy, pressure, and volume into consideration). Lastly, it is always true to say that all T=0 substances contain zero kinetic thermal energy.

Practical Applications for Thermodynamic Temperature

Figure: *Helium-4, is a superfluid when its temperature is no more than 2.17 kelvins, i.e.2.17 "Celsius degrees" above absolute zero, the starting point in the measurement of thermodynamic temperature.*

Thermodynamic temperature is useful not only for scientists, it can also be useful for lay-people in many disciplines involving gases. By expressing variables in absolute terms and applying Gay–Lussac's law of temperature/pressure proportionality, solutions to everyday problems are straightforward; for instance, calculating how a temperature change affects the pressure inside an automobile tire. If the tire has a relatively cold pressure of 200 kPa-gage, then in absolute terms (relative to a vacuum), its pressure is 300 kPa-absolute. Room temperature ("cold" in tire terms) is 296 K. If the tire pressure is 20 °C hotter (20 kelvins), the solution is calculated as $^{316\,K}/_{296\,K} = 6.8\%$ greater thermodynamic temperature *and* absolute pressure; that is, a pressure of 320 kPa-absolute, which is 220 kPa-gage.

Definition of Thermodynamic Temperature

The thermodynamic temperature is defined by the second law of thermodynamics and its consequences. The thermodynamic temperature can be shown to have special properties, and in particular can be seen to be uniquely defined (up to some constant multiplicative factor) by considering the efficiency of idealized heat engines. Thus the *ratio* T_2/T_1 of two temperatures T_1 and T_2 is the same in all absolute scales.

Strictly speaking, the temperature of a system is well-defined only if it is in thermal equilibrium. From a microscopic viewpoint, its particles (atoms, molecules, electrons, photons) are at equilibrium, so that their energies obey a Boltzmann distribution (or its quantum mechanical counterpart). There are many possible scales of temperature, derived from a variety of observations of physical phenomena.

Loosely stated, temperature differences control the flow of heat between two systems, and the universe as a whole, as with any natural system, tends to progress so as to maximize entropy. This suggests that there should be a relationship between temperature and entropy. To elucidate this, consider first the relationship between heat, work and temperature. One way to study this is to analyze a heat engine, which is a device for converting heat into mechanical work, such as the Carnot heat engine. Such a heat engine functions by using a temperature gradient between a high temperature TH and a low temperature T_C to generate work, and the work done (per cycle, say) by the heat engine is equal to the difference between the thermal energy q_H put into the system at the high temperature and the heat q_C ejected at the low temperature (in that cycle). The efficiency of the engine is the work divided by the heat put into the system or

$$\text{Efficiency} = \frac{w_{cy}}{q_H} = \frac{q_H - q_C}{q_H} = 1 - \frac{q_C}{q_H} \tag{1}$$

where w_{cy} is the work done per cycle. Thus the efficiency depends only on q_C/q_H.

Carnot's theorem states that all reversible engines operating between the same heat reservoirs are equally efficient. Thus, any reversible heat engine operating between temperatures T_1 and T_2 must have the same efficiency, that is to say, the efficiency is the function of only temperatures

$$\frac{q_C}{q_H} = f(T_H, T_C) \tag{2}$$

In addition, a reversible heat engine operating between temperatures T_1 and T_3 must have the same efficiency as one consisting of two cycles, one between T_1 and another (intermediate) temperature T_2, and the second between T_2 and T3. A quick way to see this is that should this not be the case, then energy (in the form of *Q*) will be wasted or gained, resulting in different overall efficiencies every time a cycle is split into component cycles; clearly a cycle can be composed of any number of smaller cycles.

With this understanding of Q_1, Q_2 and Q_3, we note also that mathematically,

$$f(T_1,T_3)=\frac{q_3}{q_1}=\frac{q_2q_3}{q_1q_2}=f(T_1,T_2)f(T_2,T_3).$$

But the first function is *NOT* a function of T_2, therefore the product of the final two functions *MUST* result in the removal of T_2 as a variable. The only way is therefore to define the function f as follows:

$$f(T_1,T_2)=\frac{g(T_2)}{g(T_1)}.$$

and

$$f(T_2,T_3)=\frac{g(T_3)}{g(T_2)}.$$

so that

$$f(T_1,T_3)=\frac{g(T_3)}{g(T_1)}=\frac{q_3}{q_1}.$$

i.e. The ratio of heat exchanged is a function of the respective temperatures at which they occur. We can choose any monotonic function for our $g(T)$; it is a matter of convenience and convention that we choose $g(T)=T$. Choosing then *one* fixed reference temperature (i.e. triple point of water), we establish the thermodynamic temperature scale.

It is to be noted that such a definition coincides with that of the ideal gas derivation; also it is this *definition* of the thermodynamic temperature that enables us to represent the Carnot efficiency in terms of T_H and T_C, and hence derive that the (complete) Carnot cycle is isentropic:

$$\frac{q_C}{q_H}=f(T_H,T_C)=\frac{T_C}{T_H}. \tag{3}$$

Substituting this back into our first formula for efficiency yields a relationship in terms of temperature:

$$\text{Efficiency}=1-\frac{q_C}{q_H}=1-\frac{T_C}{T_H} \tag{4}$$

Notice that for T_C=0 the efficiency is 100% and that efficiency becomes greater than 100% for T_C<0, which cases are unrealistic. Subtracting the right hand side of Equation 4 from the middle portion and rearranging gives

$$\frac{q_H}{T_H} - \frac{q_C}{T_C} = 0,$$

where the negative sign indicates heat ejected from the system. The generalization of this equation is Clausius theorem, which suggests the existence of a state functions (i.e., a function which depends only on the state of the system, not on how it reached that state) defined (up to an additive constant) by

$$dS = \frac{dq_{\text{rev}}}{T} \quad (5)$$

where the subscript indicates heat transfer in a reversible process. The function S corresponds to the entropy of the system, mentioned previously, and the change of S around any cycle is zero (as is necessary for any state function). Equation 5 can be rearranged to get an alternative definition for temperature in terms of entropy and heat (to avoid logic loop, we should first define entropy through statistical mechanics):

$$T = \frac{dq_{\text{rev}}}{dS}.$$

For a system in which the entropy S is a function $S(E)$ of its energy E, the thermodynamic temperature T is therefore given by

$$\frac{1}{T} = \frac{dS}{dE},$$

so that the reciprocal of the thermodynamic temperature is the rate of increase of entropy with energy.

History

- *Ca. 485 BC:* Parmenides in his treatise “On Nature” postulated the existente of *primum frigidum*, a hypothetical elementary substance source of all cooling or cold in the world.
- *1702–1703:* Guillaume Amontons (1663–1705) published two papers that may be used to credit him as being the first researcher to deduce the existence of a fundamental (thermodynamic) temperature scale featuring an absolute zero. He made the discovery while endeavouring to improve upon the air thermometres in use at the time. His J-tube

thermometres comprised a mercury column that was supported by a fixed mass of air entrapped within the sensing portion of the thermometre. In thermodynamic terms, his thermometres relied upon the volume / temperature relationship of gas under constant pressure. His measurements of the boiling point of water and the melting point of ice showed that regardless of the mass of air trapped inside his thermometres or the weight of mercury the air was supporting, the reduction in air volume at the ice point was always the same ratio. This observation led him to posit that a sufficient reduction in temperature would reduce the air volume to zero. In fact, his calculations projected that absolute zero was equivalent to –240 °C—only 33.15 degrees short of the true value of –273.15 °C.

- *1742:* Anders Celsius (1701–1744) created a "backwards" version of the modern Celsius temperature scale. In Celsius's original scale, zero represented the boiling point of water and 100 represented the melting point of ice. In his paper *Observations of two persistent degrees on a thermometre,* he recounted his experiments showing that ice's melting point was effectively unaffected by pressure. He also determined with remarkable precision how water's boiling point varied as a function of atmospheric pressure. He proposed that zero on his temperature scale (water's boiling point) would be calibrated at the mean barometric pressure at mean sea level.
- *1744:* Coincident with the death of Anders Celsius, the famous botanist Carolus Linnaeus (1707–1778) effectively reversed Celsius's scale upon receipt of his first thermometre featuring a scale where zero represented the melting point of ice and 100 represented water's boiling point. The custom-made *linnaeus-thermometre*, for use in his greenhouses, was made by Daniel Ekström, Sweden's leading maker of scientific instruments at the time. For the next 204 years, the scientific and thermometry communities worldwide referred to this scale as the *centigrade scale.* Temperatures on the centigrade scale were often reported simply as *degrees* or, when greater specificity was desired, *degrees centigrade.* The symbol for temperature values on this scale was °C (in several formats over the years). Because the term *centigrade* was also the French-language name for a unit of angular measurement (one-hundredth of a right angle) and had a similar connotation in other languages, the term "centesimal degree" was used

when very precise, unambiguous language was required by international standards bodies such as the Bureau international des poids et mesures (BIPM). The 9th CGPM (General Conference on Weights and Measures (*Conférence générale des poids et mesures*) and the CIPM (International Committee for Weights and Measures (*Comité international des poids et mesures*) formally adopted *degree Celsius* (symbol: °C) in 1948.

- *1777:* In his book *Pyrometrie* (Berlin: Haude & Spener, 1779) completed four months before his death, Johann Heinrich Lambert (1728–1777), sometimes incorrectly referred to as Joseph Lambert, proposed an absolute temperature scale based on the pressure/temperature relationship of a fixed volume of gas. This is distinct from the volume/temperature relationship of gas under constant pressure that Guillaume Amontons discovered 75 years earlier. Lambert stated that absolute zero was the point where a simple straight-line extrapolation reached zero gas pressure and was equal to –270 °C.
- *Circa 1787:* Notwithstanding the work of Guillaume Amontons 85 years earlier, Jacques Alexandre César Charles (1746–1823) is often credited with discovering, but not publishing, that the volume of a gas under constant pressure is proportional to its absolute temperature. The formula he created was $V_1/T_1 = V_2/T_2$.
- *1802:* Joseph Louis Gay-Lussac (1778–1850) published work (acknowledging the unpublished lab notes of Jacques Charles fifteen years earlier) describing how the volume of gas under constant pressure changes linearly with its absolute (thermodynamic) temperature. This behaviour is called Charles's Law and is one of the gas laws. His are the first known formulas to use the number *273* for the expansion coefficient of gas relative to the melting point of ice (indicating that absolute zero was equivalent to –273 °C).
- *1848:* William Thomson, (1824–1907) also known as Lord Kelvin, wrote in his paper, *On an Absolute Thermometric Scale,* of the need for a scale whereby *infinite cold* (absolute zero) was the scale's null point, and which used the degree Celsius for its unit increment. Like Gay-Lussac, Thomson calculated that absolute zero was equivalent to –273 °C on the air thermometres of the time. This absolute scale is known today as the Kelvin thermodynamic temperature scale. It's noteworthy that Thomson's value of *–273* was actually derived from 0.00366,

which was the accepted expansion coefficient of gas per degree Celsius relative to the ice point. The inverse of –0.00366 expressed to five significant digits is –273.22 °C which is remarkably close to the true value of –273.15 °C.

- *1859:* William John Macquorn Rankine (1820–1872) proposed a thermodynamic temperature scale similar to William Thomson's but which used the degree Fahrenheit for its unit increment. This absolute scale is known today as the Rankine thermodynamic temperature scale.
- *1877–1884:* Ludwig Boltzmann (1844–1906) made major contributions to thermodynamics through an understanding of the role that particle kinetics and black body radiation played. His name is now attached to several of the formulas used today in thermodynamics.
- *Circa 1930s:* Gas thermometry experiments carefully calibrated to the melting point of ice and boiling point of water showed that absolute zero was equivalent to –273.15 °C.
- *1948:* Resolution 3 of the 9th CGPM (Conférence Générale des Poids et Mesures, also known as the General Conference on Weights and Measures) fixed the triple point of water at precisely 0.01 °C. At this time, the triple point still had no formal definition for its equivalent kelvin value, which the resolution declared "will be fixed at a later date". The implication is that *if* the value of absolute zero measured in the 1930s was truly –273.15 °C, then the triple point of water (0.01 °C) was equivalent to 273.16 K. Additionally, both the CIPM (Comité international des poids et mesures, also known as the International Committee for Weights and Measures) and the CGPM formally adopted the name *Celsius* for the *degree Celsius* and the *Celsius temperature scale.*
- *1954:* Resolution 3 of the 10th CGPM gave the Kelvin scale its modern definition by choosing the triple point of water as its second defining point and assigned it a temperature of precisely 273.16 kelvin (what was actually written 273.16 *degrees Kelvin* at the time). This, in combination with Resolution 3 of the 9th CGPM, had the effect of defining absolute zero as being precisely zero kelvin and –273.15 °C.
- *1967/1968:* Resolution 3 of the 13th CGPM renamed the unit increment of thermodynamic temperature *kelvin*, symbol K, replacing *degree absolute*, symbol °K. Further, feeling it useful

to more explicitly define the magnitude of the unit increment, the 13th CGPM also decided in Resolution 4 that “The kelvin, unit of thermodynamic temperature, is the fraction 1/273.16 of the thermodynamic temperature of the triple point of water”.

- *2005:* The CIPM (Comité International des Poids et Mesures, also known as the International Committee for Weights and Measures) affirmed that for the purposes of delineating the temperature of the triple point of water, the definition of the Kelvin thermodynamic temperature scale would refer to water having an isotopic composition defined as being precisely equal to the nominal specification of Vienna Standard Mean Ocean Water.

Constant Volume Gas Thermometre

A gas thermometre measures temperature by the variation in volume or pressure of a gas. One common apparatus is a constant volume thermometre. It consists of a bulb connected by a capillary tube to a manometre. The bulb is filled with a gas such that the volume of the gas in the bulb remains constant. The volume is related to temperature by k, known as Charles’s Law. The pressure of the gas in the bulb can be obtained by measuring the level difference in the two arms of the manometre. Gas thermometres are often used to calibrate other thermometres.

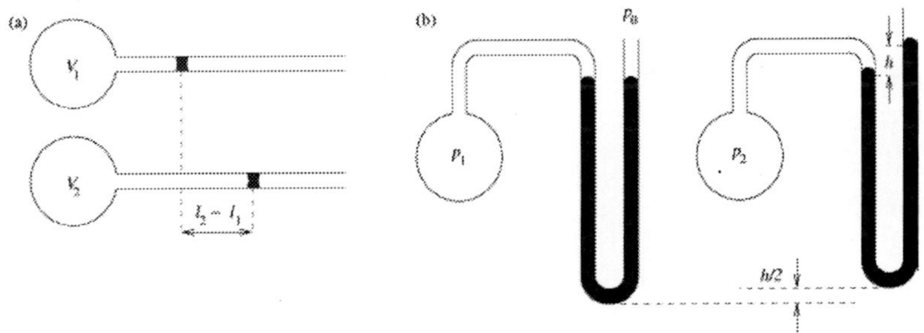

Scale of Temperature

Scale of temperature is a way to measure temperature quantitatively.

Formal Description

According to the zeroth law of thermodynamics, being in thermal equilibrium is an equivalence relation. Thus all thermal systems may be divided into a quotient set by this equivalence relation, denoted below as M. Assume the set M has the cardinality of c, then one can construct an injective function $f: M \rightarrow R$, by which every thermal system will have a number associated with it such that when and only when two thermal systems have same such value, they will be in thermal equilibrium. This is clearly the property of temperature, and the specific way of assigning numerical values as temperature is called a *scale of temperature.* In practical terms, a temperature scale is always based on usually a single physical property of a simple thermodynamic system, called a thermometre, that defines a scaling function mapping the temperature to the measurable thermometric parametre. Such temperature scales that are purely based on measurement are called *empirical temperature scales.*

The second law of thermodynamics provides a fundamental, natural definition of thermodynamic temperature starting with a null point of absolute zero. A scale for thermodynamic temperature is established similarly to the empirical temperature scales, however, needing only one additional fixing point.

Empirical Scales

Empirical scales are based on the measurement of physical parametres that express the property of interest to be measured through some formal, most commonly a simple linear, functional relationship. For the measurement of temperature, the formal definition of thermal equilibrium in terms of the thermodynamic coordinate spaces of thermodynamic systems, expressed in the zeroth law of thermodynamics, provides the framework to measure temperature.

All temperature scales, including the modern thermodynamic temperature scale used in the International System of Units, are calibrated according to thermal properties of a particular substance or device. Typically, this is established by fixing two well-defined temperature points and defining temperature increments via a linear function of the response of the thermometric device. For example, both the old Celsius scale and Fahrenheit scale were originally based on the linear expansion of a narrow mercury column within a limited range of temperature, each using different reference points and scale increments.

Different empirical scales may not be compatible with each other, except for small regions of temperature overlap. If an alcohol thermometre and a mercury thermometre have same two fixed points, namely the freezing and boiling point of water, their reading will not agree with each other except at the fixed points, as the linear 1:1 relationship of expansion between any two thermometric substances may not be guaranteed.

Empirical temperature scales are not reflective of the fundamental, microscopic laws of matter. Temperature is a universal attribute of matter, yet empirical scales map a narrow range onto a scale that is known to have a useful functional form for a particular application. Thus, their range is limited. The working material only exists in a form under certain circumstances, beyond which it no longer can serve as a scale. For example, mercury freezes below 234.32 K, so temperature lower than that cannot be measured in a scale based on mercury. Even ITS-90, which interpolates among different ranges of temperature, has only a range of 0.65 K to approximately 1358 K (–272.5 °C to 1085 °C).

Ideal Gas Scale

When pressure approaches zero, all real gas will behave like ideal gas, that is, pV of a mole of gas relying only on temperature. Therefore we can design a scale with pV as its argument. Of course any bijective function will do, but for convenience's sake linear function is the best. Therefore we define it as

$$T = \frac{1}{nR} \lim_{p \to 0} pV.$$

The ideal gas scale is in some sense a "mixed" scale. It relies on the universal properties of gas, a big advance from just a particular substance. But still it is empirical since it puts gas at a special position and thus has limited applicability—at some point no gas can exist. One distinguishing characteristic of ideal gas scale, however, is that it precisely equals thermodynamical scale when it is well defined.

International Temperature Scale of 1990

ITS-90 is designed to represent the thermodynamic temperature scale (referencing absolute zero) as closely as possible throughout its range. Many different thermometre designs are required to cover the entire range. These include helium vapour pressure thermometres, helium gas thermometres, standard platinum resistance thermometres (known as SPRTs, PRTs or Platium RTDs) and monochromatic radiation thermometres.

Although the Kelvin and Celsius scales are defined using absolute zero (0 K) and the triple point of water (273.16 K and 0.01 °C), it is impractical to use this definition at temperatures that are very different from the triple point of water. Accordingly, ITS–90 uses numerous defined points, all of which are based on various thermodynamic equilibrium states of fourteen pure chemical elements and one compound (water). Most of the defined points are based on a phase transition; specifically the melting/freezing point of a pure chemical element. However, the deepest cryogenic points are based exclusively on the vapour pressure/temperature relationship of helium and its isotopes whereas the remainder of its cold points (those less than room temperature) are based on triple points. Examples of other defining points are the triple point of hydrogen (–259.3467 °C) and the freezing point of aluminum (660.323 °C).

Thermometres calibrated per ITS–90 use complex mathematical formulas to interpolate between its defined points. ITS–90 specifies rigorous control over variables to ensure reproducibility from lab to lab. For instance, the small effect that atmospheric pressure has upon the various melting points is compensated for (an effect that typically amounts to no more than half a millikelvin across the different altitudes and barometric pressures likely to be encountered). The standard even compensates for the pressure effect due to how deeply the temperature probe is immersed into the sample. ITS–90 also draws a distinction between "freezing" and "melting" points. The distinction depends on whether heat is going *into* (melting) or *out of* (freezing) the sample when the measurement is made. Only gallium is measured while melting, all the other metals are measured while the samples are freezing.

There are often small differences between measurements calibrated per ITS–90 and thermodynamic temperature. For instance, precise measurements show that the boiling point of VSMOW water under one standard atmosphere of pressure is actually 373.1339 K (99.9839 °C) when adhering *strictly* to the two-point definition of thermodynamic temperature. When calibrated to ITS–90, where one must interpolate between the defining points of gallium and indium, the boiling point of VSMOW water is about 10 mK less, about 99.974 °C. The virtue of ITS–90 is that another lab in another part of the world will measure the very same temperature with ease due to the advantages of a comprehensive international calibration standard featuring many conveniently spaced, reproducible, defining points spanning a wide range of temperatures.

Celsius Scale

Celsius (known until 1948 as centigrade) is a temperature scale that is named after the Swedish astronomer Anders Celsius (1701–

1744), who developed a similar temperature scale two years before his death. The degree Celsius (°C) can refer to a specific temperature on the Celsius scale as well as a unit to indicate a temperature *interval* (a difference between two temperatures or an uncertainty).

From 1744 until 1954, 0 °C was defined as the freezing point of water and 100 °C was defined as the boiling point of water, both at a pressure of one standard atmosphere. Although these defining correlations are commonly taught in schools today, by international agreement the unit "degree Celsius" and the Celsius scale are currently defined by two different points: absolute zero, and the triple point of VSMOW (specially prepared water). This definition also precisely relates the Celsius scale to the Kelvin scale, which defines the SIbase unit of thermodynamic temperature (symbol: K). Absolute zero, the hypothetical but unattainable temperature at which matter exhibits zero entropy, is defined as being precisely 0 K *and* –273.15 °C. The temperature value of the triple point of water is defined as being precisely 273.16 K *and* 0.01 °C.

This definition fixes the magnitude of both the degree Celsius and the kelvin as precisely 1 part in 273.16 parts, the difference between absolute zero and the triple point of water. Thus, it sets the magnitude of one degree Celsius and that of one kelvin as exactly the same. Additionally, it establishes the difference between the two scales' null points as being precisely 273.15 degrees Celsius (–273.15 °C = 0 K and 0 °C = 273.15 K).

Thermodynamic Scale

Thermodynamic scale differs from empirical scales in that it is absolute. It is based on the fundamental laws of thermodynamics or statistical mechanics instead of some arbitrary chosen working material. Besides it covers full range of temperature and has simple relation with microscopic quantities like the average kinetic energy of particles. In experiments ITS-90 is used to approximate thermodynamic scale due to simpler realization.

Definition

Lord Kelvin devised the thermodynamic scale based on the efficiency of heat engines as shown below:

The efficiency of a engine is the work divided by the heat introduced to the system or

$$\eta = \frac{w_{cy}}{q_H} = \frac{q_H - q_C}{q_H} = 1 - \frac{q_C}{q_H} \quad (1)$$

where w_{cy} is the work done per cycle. Thus, the efficiency depends only on q_C/q_H.

Because of Carnot theorem, any reversible heat engine operating between temperatures T_1 and T_2 must have the same efficiency, meaning, the efficiency is the function of the temperatures only:

$$\frac{q_C}{q_H} = f(T_H, T_C) \qquad (2)$$

In addition, a reversible heat engine operating between temperatures T_1 and T_3 must have the same efficiency as one consisting of two cycles, one between T_1 and another (intermediate) temperature T_2, and the second between T_2andT_3. This can only be the case if

$$f(T_1, T_3) = \frac{q_3}{q_1} = \frac{q_2 q_3}{q_1 q_2} = f(T_1, T_2) f(T_2, T_3).$$

Specializing to the case that T_1 is a fixed reference temperature: the temperature of the triple point of water. Then for any T2 and T_3,

$$f(T_2, T_3) = \frac{f(T_1, T_3)}{f(T_1, T_2)} = \frac{273.16 \cdot f(T_1, T_3)}{273.16 \cdot f(T_1, T_2)}.$$

Therefore, if thermodynamic temperature is defined by

$$T = 273.16 \cdot f(T_1, T)$$

then the function *f*, viewed as a function of thermodynamic temperature, is

$$f(T_2, T_3) = \frac{T_3}{T_2},$$

and the reference temperature T_1 has the value 273.16. (Of course any reference temperature and any positive numerical value could be used—the choice here corresponds to the Kelvin scale.)

Equality to Ideal Gas Scale

It follows immediately that

$$\frac{q_C}{q_H} = f(T_H, T_C) = \frac{T_C}{T_H}. \qquad (3)$$

Substituting Equation 3 back into Equation 1 gives a relationship for the efficiency in terms of temperature:

$$\eta = 1 - \frac{q_C}{q_H} = 1 - \frac{T_C}{T_H} \qquad (4)$$

This is identical to the efficiency formula for Carnot cycle, which effectively employs the ideal gas scale. So this means the two scales equals numerically at every point.

Conversion Table Between the Different Temperature Units

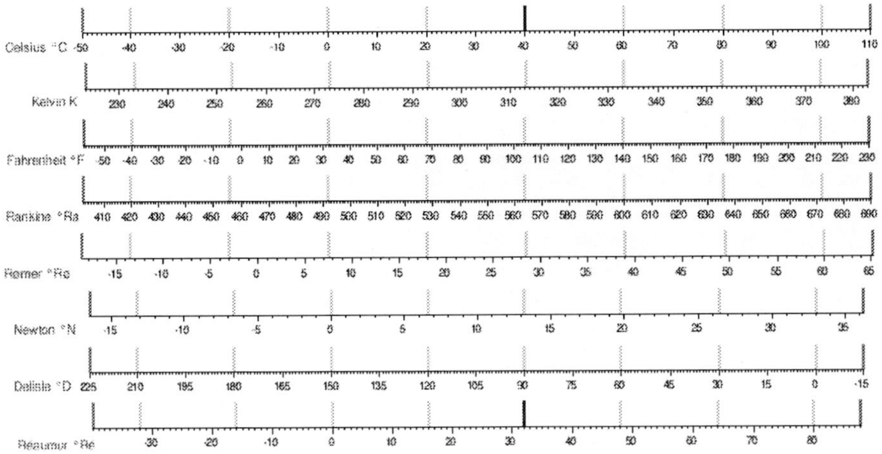

40 °C = 313.15 K = 104 °F = 563.67 °Ra = 28.5 °Rø = 13.2 °N = 90 °D = 32 °Ré

Chapter

2

Joule's Experiments

Accurate investigation of the relationship between heat developed and mechanical work spent was taken up by the British Scientist James Prescott Joule. The main aim of his investigations was to determine exactly the ratio between the work done and the quantity of heat produced.

Joule used different arrangements for doing the work, W in different ways and measured the corresponding amount of heat, H produced in each case. In all the cases he found that the expenditure of the same amount of work always produced the same amount of heat. Every time he found that 4186 joule of work was spent to produce the same amount of heat which could raise the temperature of one kg of water through 1^0C. He established the relation, W/H to be a constant quantity. The constant relation W/H was represented symbolically by the letter 'J'. J is known as Joule's mechanical equivalent of heat. 'Symbolically, we can write :

$$W/H = J \text{ or } W = JH$$

Thus he established that heat is a form of energy.

Joule's Method for Experimental Determination of J.

Details of Apparatus: The apparatus consists of a specially designed calorimetre placed in a wooden box C with felt lining to avoid heat losses to the surroundings. A number of vanes, V, V projects from the walls of the calorimetre in its interior. A spindle carrying a number of brass paddles P, P acts as a churner and it is so pivoted at the bottom that the paddles P, P are capable of turning between the fixed vanes V. The spindle can be attached to a drum D by means of screw S' whenever desired. The drum D can be rotated by the handle H or by the falling weights W, W, which are attached to two pieces of string passing over the pulleys P_1 and P_2 and wrapped round the drum. Two vertical scales are fixed to note the vertical distances through which the weights fall.

Working: The two weights W, W are allowed to fall through a height 'h'. In doing so, they turn the spindle and thereby turn the paddles P, P immersed in a known mass (m) of water contained in the calorimetre. The water is thus churned but not allowed to rotate due to fixed vanes and the potential energy of the falling weights is converted into kinetic energy of paddles. Due to friction offered by paddles KE is changed into heat and as a result of it, the temperature of the water in the calorimetre rises. The rise in temperature is measured by an accurate thermometre T inserted in the calorimetre. The process is rapidly repeated several times (every time detaching the drum, wrapping the string, tightening the screws' and allowing the weights to fall), (say n times) such that there is an accurately measurable rise in the temperature of water.

Importance of Joule's Experiments

1. Joule's experiments conclusively established that heat is a form of energy. It is not a material substance like caloric fluid.
2. Always the same amount of heat was produced by spending a given amount of mechanical work. It is immaterial what type of arrangement is used for doing mechanical work. Other alternative ways used by Joule for conversion of mechanical work into heat were (i) by mechanically stirring mercury, and (ii) by rubbing two iron rings together. Every time he found that W/H is a constant quantity, i.e., when 4.186 Joule converted to heat the temperature of 1 kg of water will rise by 1^0C.

First Law of Thermodynamics

The first law of thermodynamics is a version of the law of conservation of energy, adapted for thermodynamic systems. The law of conservation of energy states that the total energy of an isolated system is constant; energy can be transformed from one form to another, but cannot be created or destroyed. The first law of thermodynamics recognizes a particular form of energy called internal energy. It is often formulated by stating that the change in the internal energy of a closed system is equal to the amount of heat supplied to the system, minus the amount of work derived from the system. It is also often formulated by stating that when a closed system has a change of state, and its internal energy is changed only by work and not by heat transfer, then the net amount of work transferred is the same for all arrangements of work transfer that are possible for that change of state. Also, when two systems, open to each other for transfer of matter and energy, interact but are otherwise isolated, then the sum of their internal energies does not change.

Through a body that can store energy as internal energy, quantities of energy supplied or derived as work and as heat are interconvertible. A process is possible that just converts supplied work to derived heat, in which the internal energy of the storage body is left eventually unchanged, and then the ratio of net work supplied to net heat derived is a universal constant. Conversion in the other sense, from heat to work, is limited by the inverse of that same constant, but the second law of thermodynamics observes that the limit is not physically actualizable, because, in the process, some heat is eventually always also derived from the storage body so that the work eventually derived falls short of the limiting amount; this fact was known before the first law was understood. In a process that leaves its internal energy eventually unchanged, a machine that is not itself supplied with energy cannot supply energy to another body; this is traditionally expressed by saying that a perpetual motion machine of the first kind cannot exist.

History

The process of development of the first law of thermodynamics was by way of many tries and mistakes of investigation, over a period of about half a century. The first full statements of the law came in 1850 from Rudolf Clausius and by William Rankine; Rankine's statement was perhaps not quite as clear and distinct as was Clausius'. A main aspect of the struggle was to deal with the previously proposed caloric theory of heat. Germain Hess in 1840 stated a conservation law for the so-called 'heat of reaction' for chemical reactions. His law was later recognized as a consequence of the first law of thermodynamics, but Hess's statement was not explicitly concerned with the relation between energy exchanges by heat and work.

According to Truesdell (1980), Julius Robert von Mayer in 1841 made a statement that meant that "in a process at constant pressure, the heat used to produce expansion is universally interconvertible with work", but this is not a general statement of the first law.

Original Statements: the "Thermodynamic Approach"

The original nineteenth century statements of the first law of thermodynamics appeared in a conceptual framework in which transfer of energy as heat was taken as a primitive notion, not defined or constructed by the theoretical development of the framework, but rather presupposed as prior to it and already accepted. The primitive notion of heat was taken as empirically established, especially through calorimetry regarded as a subject in its own right, prior to thermodynamics. Jointly primitive with this notion of heat were the notions of empirical temperature and thermal equilibrium. This framework did not presume a concept of energy in general, but regarded it as derived or synthesized from the prior notions of heat and work. By one author, this framework has been called the "thermodynamic" approach.

The first explicit statement of the first law of thermodynamics, by Rudolf Clausius in 1850, referred to cyclic thermodynamic processes.

In all cases in which work is produced by the agency of heat, a quantity of heat is consumed which is proportional to the work done; and conversely, by the expenditure of an equal quantity of work an equal quantity of heat is produced.

Clausius also stated the law in another form, referring to the existence of a function of state of the system, the internal energy, and expressed it in terms of a differential equation for the increments of a thermodynamic process. This equation may described as follows:

In a thermodynamic process involving a closed system, the increment in the internal energy is equal to the difference between the heat accumulated by the system and the work done by it.

Because of its definition in terms of increments, the value of the internal energy of a system is not uniquely defined. It is defined only up to an arbitrary additive constant of integration, which can be adjusted to give arbitrary reference zero levels. This non-uniqueness is in keeping with the abstract mathematical nature of the internal energy. The internal energy is customarily stated relative to a conventionally chosen standard reference state of the system.

The concept of internal energy is therefore considered by Bailyn to be of "enormous interest". It cannot be immediately measured, but can only be inferred, by differencing actual immediate measurements.

Bailyn likens it to the energy states of an atom, that were revealed by Bohr's energy relation $h\nu = E_{n2} - E_{n2}$. In each case, an unmeasurable quantity (the internal energy, the atomic energy level) is revealed by considering the difference of measured quantities (increments of internal energy, quantities of emitted or absorbed radiative energy.

Conceptual Revision: the "Mechanical Approach"

Largely through the influence of Max Born, in the twentieth century, an alternative approach came to be preferred by many writers, an approach that might be called the "mechanical approach". This approach takes as its primitive notion energy transferred as work defined by mechanics. From this, it derives the notions of transfer of energy as heat, and of temperature, as theoretical developments. It regards calorimetry as a derived theory. It has an early origin in the nineteenth century, for example in the work of Helmholtz, but also in the work of many others.

Conceptually Revised Statement, According to the Mechanical Approach

The revised statement of the law takes the notions of adiabatic mechanical work, and of non-adiabatic transfer of energy, as empirically or theoretically established primitive notions. It rests on the primitive notion of *walls*, especially adiabatic walls, presupposed as physically established. Energy can pass such walls as only as adiabatic work, reversibly or irreversibly. If transfer of energy as work is not permitted between them, two systems separated by an adiabatic wall can come to their respective internal mechanical and material thermodynamic equilibrium states completely independently of one another.

The revised statement of the law postulates that a change in the internal energy of a system due to an arbitrary process of interest, that takes the system from its specified initial to its specified final state of internal thermodynamic equilibrium, can be determined through the physical existence of a reference process, for those specified states, that occurs purely through stages of adiabatic work.

The revised statement is then

> *For a closed system, in any arbitrary process of interest that takes it from an initial to a final state of internal thermodynamic equilibrium, the change of internal energy is the same as that for a reference adiabatic work process that links those two states. This is so regardless of the path of the process of interest, and regardless of*

whether it is an adiabatic or a non-adiabatic process. The reference adiabatic work process may be chosen arbitrarily from amongst the class of all such processes.

This statement is much less close to the empirical basis than are the original statements, but is often regarded as conceptually parsimonious in that it rests only on the concepts of adiabatic work and of non-adiabatic processes, not on the concepts of transfer of energy as heat and of empirical temperature that are presupposed by the original statements. Largely through the influence of Max Born, it is often regarded as theoretically preferable because of this conceptual parsimony. Born particularly observes that the revised approach avoids thinking in terms of what he calls the "imported engineering" concept of heat engines.

Basing his thinking on the mechanical approach, Born in 1921, and again in 1949, proposed to revise the definition of heat. In particular, he referred to the work of Constantin Carathéodory, who had in 1909 stated the first law without defining quantity of heat. Born's definition was specifically for closed systems, and it has been widely followed in textbooks (examples:). No definition for quantity of heat transferred between two open systems was offered by Born, who did not explicitly mention the possibility.

Description

The first law of thermodynamics was expressed in two ways by Clausius. One way referred to cyclic processes and the inputs and outputs of the system, but did not refer to increments in the internal state of the system. The other way referred to any incremental change in the internal state of the system, and did not expect the process to be cyclic. A cyclic process is one which can be repeated indefinitely often and still eventually leave the system in its original state.

In each repetition of a cyclic process, the work done by the system is proportional to the heat consumed by the system. In a cyclic process in which the system does work on its surroundings, it is necessary that some heat be taken in by the system and some be put out, and the difference is the heat consumed by the system in the process. The constant of proportionality is universal and independent of the system and was measured by James Joule in 1845 and 1847, who described it as the *mechanical equivalent of heat*.

For a closed system, in any process, the change in the internal energy is considered due to a combination of heat added to the system and work done *by* the system. Taking ΔU as a change in internal energy, one writes

$$\Delta U = Q - W \qquad \text{(sign convention of Clausius and generally in this article)},$$

where Q and W are quantities of heat supplied to the system by its surroundings and of work done by the system on its surroundings, respectively. This sign convention is implicit in Clausius' statement of the law given above, and is consistent with the use of thermodynamics to study heat engines which provide useful work, which is regarded as positive.

In modern style of teaching science, however, it is conventional to use the IUPAC convention by which the first law is formulated in terms of the work done on the system. With this alternate sign convention for work, the first law for a closed system may be written:

$$\Delta U = Q + W \qquad \text{(sign convention of IUPAC)}.$$

This convention follows physicists such as Max Planck, and considers all net energy transfers to the system as positive and all net energy transfers from the system as negative, irrespective of any use for the system as an engine or other device.

When a system expands in a fictive quasistatic process, the work done by the system on the environment is the product, $P\,dV$, of pressure, P, and volume change, dV, whereas the work done *on* the system is $-P\,dV$. Using either sign convention for work, the change in internal energy of the system is:

$$dU = \delta Q - P dV \qquad \text{(quasi-static process)},$$

where δQ denotes the infinitesimal increment of heat supplied to the system from its surroundings.

Work and heat are expressions of actual physical processes which supply or remove energy, while the internal energy U is a mathematical abstraction that keeps account of the exchanges of energy that befall the system. Thus the term heat for Q means "that amount of energy added or removed by conduction of heat or by thermal radiation", rather than referring to a form of energy within the system. Likewise, the term work energy for W means "that amount of energy gained or lost as the result of work". Internal energy is a property of the system whereas work done and heat supplied are not. A significant result of this distinction is that a given internal energy change ΔU can be achieved by, in principle, many combinations of heat and work.

Various Statements of the Law for Closed Systems

The law is of very great importance and generality and is consequently thought of from several points of view. Most careful textbook statements of the law express it for closed systems. It is stated in several ways, sometimes even by the same author.

For the thermodynamics of closed systems, the distinction between transfers of energy as work and as heat is central and is within the scope of the present article. For the thermodynamics of open systems, such a distinction is beyond the scope of the present article, but some limited comments are made on it in the section below headed 'First law of thermodynamics for open systems'.

There are two main ways of stating a law of thermodynamics, physically or mathematically. They should be logically coherent and consistent with one another.

An example of a physical statement is that of Planck (1897/1903):

It is in no way possible, either by mechanical, thermal, chemical, or other devices, to obtain perpetual motion, i.e. it is impossible to construct an engine which will work in a cycle and produce continuous work, or kinetic energy, from nothing.

This physical statement is restricted neither to closed systems nor to systems with states that are strictly defined only for thermodynamic equilibrium; it has meaning also for open systems and for systems with states that are not in thermodynamic equilibrium.

An example of a mathematical statement is that of Crawford (1963):

For a given system we let ΔE^{kin} = large-scale mechanical energy, ΔE^{pot} = large-scale potential energy, and ΔE^{tot} = total energy. The first two quantities are specifiable in terms of appropriate mechanical variables, and by definition

$$E^{\text{tot}} = E^{\text{kin}} + E^{\text{pot}} + U.$$

For any finite process, whether reversible or irreversible,

$$\Delta E^{\text{tot}} = \Delta E^{\text{kin}} + \Delta E^{\text{pot}} + \Delta U.$$

The first law in a form that involves the principle of conservation of energy more generally is

$$\Delta E^{\text{tot}} = Q + W.$$

Here Q and W are heat and work added, with no restrictions as to whether the process is reversible, quasistatic, or irreversible. [Warner, *Am. J. Phys.*, 29, 124 (1961)]

This statement by Crawford, for W, uses the sign convention of IUPAC, not that of Clausius. Though it does not explicitly say so, this statement refers to closed systems, and to internal energy U defined for bodies in states of thermodynamic equilibrium, which possess well-defined temperatures.

The history of statements of the law for closed systems has two main periods, before and after the work of Bryan (1907), of Carathéodory (1909), and the approval of Carathéodory's work given by Born (1921). The earlier traditional versions of the law for closed systems are nowadays often considered to be out of date.

Carathéodory's celebrated presentation of equilibrium thermodynamics refers to closed systems, which are allowed to contain several phases connected by internal walls of various kinds of impermeability and permeability, explicitly including walls that are permeable only to heat. Carathéodory's version of the first law of thermodynamics was stated in an axiom which refrained from defining or mentioning temperature or quantity of heat transferred. That axiom stated that the internal energy of a phase in equilibrium is a function of state, that the sum of the internal energies of the phases is the total internal energy of the system, and that the value of the total internal energy of the system is changed by the amount of work done adiabatically on it, considering work as a form of energy. That article considered this statement to be an expression of the law of conservation of energy for such systems. This version is nowadays widely accepted as authoritative, but is stated in slightly varied ways by different authors. The Carathéodory statement of the law in axiomatic form does not mention heat or temperature, but the equilibrium states to which it refers are explicitly defined by variable sets that necessarily include "non-deformation variables", such as pressures, which, within reasonable restrictions, can be rightly interpreted as empirical temperatures, and the walls connecting the phases of the system are explicitly defined as possibly impermeable to heat or permeable only to heat. According to Münster (1970), "A somewhat unsatisfactory aspect of Carathéodory's theory is that a consequence of the Second Law must be considered at this point [in the statement of the first law], i.e. that it is not always possible to reach any state 2 from any other state 1 by means of an adiabatic process." Münster instances that no adiabatic process can reduce the internal energy of a system at constant volume. Carathéodory's paper asserts that its statement of the first law corresponds exactly to Joule's experimental arrangement, regarded as an instance of adiabatic work. It does not point out that Joule's experimental arrangement performed essentially irreversible work, through friction of paddles in a liquid, or passage of electric current through a resistance inside the system, driven by motion of a coil and inductive heating, or by an external current source, which can access the system only by the passage of electrons, and so is not strictly adiabatic, because electrons are a form of matter, which cannot penetrate adiabatic walls. The paper goes on to base its main argument on the possibility of quasi-static adiabatic work, which is essentially reversible. The paper asserts that it will avoid reference to Carnot cycles, and then proceeds to base its argument on cycles of forward and backward quasi-static adiabatic stages, with isothermal stages of zero magnitude.

Some respected modern statements of the first law for closed systems assert the existence of internal energy as a function of state defined in terms of adiabatic work and accept the idea that heat is not defined in its own right, that is to say calorimetrically or as due to temperature difference; they define heat as a residual difference between change of internal energy and work done on the system, when that work does not account for the whole of the change of internal energy and the system is not adiabatically isolated.

Sometimes the concept of internal energy is not made explicit in the statement.

Sometimes the existence of the internal energy is made explicit but work is not explicitly mentioned in the statement of the first postulate of thermodynamics. Heat supplied is then defined as the residual change in internal energy after work has been taken into account, in a non-adiabatic process.

A respected modern author states the first law of thermodynamics as "Heat is a form of energy", which explicitly mentions neither internal energy nor adiabatic work. Heat is defined as energy transferred by thermal contact with a reservoir, which has a temperature, and is generally so large that addition and removal of heat do not alter its temperature. A current student text on chemistry defines heat thus: "*heat* is the exchange of thermal energy between a system and its surroundings caused by a temperature difference." The author then explains how heat is defined or measured by calorimetry, in terms of heat capacity, specific heat capacity, molar heat capacity, and temperature.

A respected text disregards the Carathéodory's exclusion of mention of heat from the statement of the first law for closed systems, and admits heat calorimetrically defined along with work and internal energy. Another respected text defines heat exchange as determined by temperature difference, but also mentions that the Born (1921) version is "completely rigorous". These versions follow the traditional approach that is now considered out of date, exemplified by that of Planck (1897/1903).

Evidence for the First Law of Thermodynamics for Closed Systems

The first law of thermodynamics for closed systems was originally induced from empirically observed evidence, however, it is now taken to be the definition of heat via the law of conservation of energy and the definition of work in terms of changes in the external parametres of a system. The original discovery of the law was gradual over a period of perhaps half a century or more, and some early studies were in terms of cyclic processes. The following is an account in terms of changes of state through compound processes that are not necessarily cyclic. This account first considers processes for which the first law is easily verified because of their simplicity, namely adiabatic processes (in which no heat is transferred) and adynamic processes (in which no work is transferred).

Adiabatic Processes

Given a closed system in an initial state, if work is done on the system in an adiabatic (i.e. no heat transfer) way, and given the final state after a process, the amount of work required to be transferred to the system is the same, irrespective of how this work is performed. The work done on the system is defined and measured by changes in mechanical or quasi-mechanical variables external to the system. Physically, adiabatic transfer of energy as work requires the existence of adiabatic enclosures.

For instance, in Joule's experiment, the initial system is a tank of water with a paddle wheel inside. If we isolate thermally the tank and move the paddle wheel with a pulley and a weight we can relate the increase in temperature with the height descended by the mass. Now the system is returned to its initial state, isolated again, and the same amount of work is done on the tank using different devices (an electric motor, a chemical battery, a spring,...). In every case, the amount of work can be measured independently. The return to the initial state is not conducted by doing adiabatic work on the system. The evidence shows that the final state of the water (in particular, its temperature and volume) is the same in every case. It is irrelevant if the work is electrical, mechanical, chemical,... or if done suddenly or slowly, as long as it is performed in an adiabatic way, that is to say, without heat transfer into or out of the system.

Evidence of this kind shows that to increase the temperature of the water in the tank, the qualitative kind of adiabatically performed work does not matter. No qualitative kind of adiabatic work has ever been observed to decrease the temperature of the water in the tank.

A change from one state to another, for example an increase of both temperature and volume, may be conducted in several stages, for example by externally supplied electrical work on a resistor in the body, and adiabatic expansion allowing the body to do work on the surroundings. It needs to be shown that the time order of the stages, and their relative magnitudes, does not affect the amount of adiabatic work that needs to be done for the change of state. According to one respected scholar: "Unfortunately, it does not seem that experiments of this kind have ever been carried out carefully.... We must therefore admit that the statement which we have enunciated here, and which is equivalent to the first law of thermodynamics, is not well founded on direct experimental evidence."

This kind of evidence, of independence of sequence of stages, combined with the above-mentioned evidence, of independence of qualitative kind of work, would show the existence of a very important state variable that corresponds with adiabatic work, but not that such a state variable represented a conserved quantity. For the latter, another step of evidence is needed, which may be related to the concept of reversibility, as mentioned below.

That very important state variable was first recognized and denoted U by Clausius in 1850, but he did not then name it, and he defined it in terms not only of work but also of heat transfer in the same process. It was also independently recognized in 1850 by Rankine, who also denoted it U; and in 1851 by Kelvin who then called it "mechanical energy", and later "intrinsic energy". In 1865, after some hestitation, Clausius began calling his state function U"energy". In 1882 it was named as the *internal energy* by Helmholtz. If only adiabatic processes were of interest, and heat could be ignored, the concept of internal energy would hardly arise or be needed. The relevant physics would be largely covered by the concept of potential energy, as was intended in the 1847 paper of Helmholtz on the principle of conservation of energy, though that did not deal with forces that cannot be described by a potential, and thus did not fully justify the principle; moreover that paper was very critical of the early work of Joule which had by then been performed. A great merit of the internal energy concept is that it frees thermodynamics from a restriction to cyclic processes, and allows a treatment in terms of thermodynamic states. In an adiabatic process, adiabatic work takes the system either from a reference state O with internal energy $U(O)$ to an arbitrary one A with internal energy $U(A)$, or from the state A to the state O:

$$U(A) = U(O) - W_{O\to A}^{\text{adiabatic}} \text{ or } U(O) = U(A) - W_{A\to O}^{\text{adiabatic}}.$$

Except under the special, and strictly speaking, fictional, condition of reversibility, only one of the processes adiabatic,$O \to A$ or adiabatic,$A \to O$ is empirically feasible by a simple application of externally supplied work. The reason for this is given as the second law of thermodynamics and is not considered in the present article.

The fact of such irreversibility may be dealt with in two main ways, according to different points of view. Since the work of Bryan (1907),

- To deal with it nowadays, the most accepted way, followed by Carathéodory, is to rely on the previously established concept of quasi-static processes, as follows. Actual physical processes of transfer of energy as work are always at least to some degree irreversible. The irreversibility is often due to mechanisms known as dissipative, that transform bulk kinetic energy into internal energy. Examples are friction and viscosity. If the process is performed more slowly, the frictional or viscous dissipation is less. In the limit of infinitely slow performance, the dissipation tends to zero and then the limiting process, though fictional rather than actual, is notionally reversible, and is called quasi-static. Throughout the course of the fictional limiting quasi-static process, the internal intensive variables of the system are equal to the external intensive variables, those which describe the reactive forces exerted by the surroundings. This can be taken to justify the formula

$$W_{A\to O}^{\text{adiabatic,quasi-static}} = -W_{O\to A}^{\text{adiabatic,quasi-static}}. \tag{1}$$

- Another way to deal with it is to allow that experiments with processes of heat transfer to or from the system may be used to justify the formula (1) just above. Moreover, it deals to some extent with the problem of lack of direct experimental evidence that the time order of stages of a process does not matter in the determination of internal energy. This way does not provide theoretical purity in terms of adiabatic work processes, but is empirically feasible, and is in accord with experiments actually done, such as the Joule experiments mentioned just above, and with older traditions.

The formula (1) above allows that to go by processes of quasi-static adiabatic work from the state A to the state B we can take a path that goes through the reference state O, since the quasi-static adiabatic work is independent of the path

$$-W_{A\to B}^{\text{adiabatic,quasi-static}} = -W_{A\to O}^{\text{adiabatic,quasi-static}} - W_{O\to B}^{\text{adiabatic,quasi-static}} = W_{O\to A}^{\text{adiabatic,quasi-static}} - W_{O\to B}^{\text{adiabatic,quasi-static}} = -U(A) + U(B) = \Delta U$$

This kind of empirical evidence, coupled with theory of this kind, largely justifies the following statement:

For all adiabatic processes between two specified states of a closed system of any nature, the net work done is the same regardless the details of the process, and determines a state function called internal energy, U."

Adynamic Processes

A complementary observable aspect of the first law is about heat transfer. Adynamic transfer of energy as heat can be measured empirically by changes in the surroundings of the system of interest by calorimetry. This again requires the existence of adiabatic enclosure of the entire process, system and surroundings, though the separating wall between the surroundings and the system is thermally conductive or radiatively permeable, not adiabatic. A calorimetre can rely on measurement of sensible heat, which requires the existence of thermometres and measurement of temperature change in bodies of known sensible heat capacity under specified conditions; or it can rely on the measurement of latent heat, through measurement of masses of material which change phase, at temperatures fixed by the occurrence of phase changes under specified conditions in bodies of known latent heat of phase change. The calorimetre can be calibrated by adiabatically doing externally determined work on it. The most accurate method is by passing an electric current from outside through a resistance inside the calorimetre. The calibration allows comparison of calorimetric measurement of quantity of heat transferred with quantity of energy transferred as work. According to one textbook, "The most common device for measuring ΔU is an adiabatic bomb calorimetre." According to another textbook, "Calorimetry is widely used in present day laboratories." According to one opinion, "Most thermodynamic data come from calorimetry..." According to another opinion, "The most common method of measuring "heat" is with a calorimetre."

When the system evolves with transfer of energy as heat, without energy being transferred as work, in an adynamic process, the heat transferred to the system is equal to the increase in its internal energy:

$$Q_{A\to B}^{\text{adynamic}} = \Delta U.$$

General Case for Reversible Processes

Heat transfer is practically reversible when it is driven by practically negligibly small temperature gradients. Work transfer is practically reversible when it occurs so slowly that there are no frictional effects within the system; frictional effects outside the system should also be zero if the process is to be globally reversible. For a particular reversible process in general, the work done reversibly on the system, $W_{A\to B}^{\text{path } P_0,\text{reversible}}$, and the heat transferred reversibly to the system, $Q_{A\to B}^{\text{path } P_0,\text{reversible}}$ are not required to occur respectively adiabatically or adynamically, but they must belong to the same particular process defined by its particular reversible path, P_0, through the space of thermodynamic states. Then the work and heat transfers can occur and be calculated simultaneously.

Putting the two complementary aspects together, the first law for a particular reversible process can be written

$$-W_{A\to B}^{\text{path } P_0,\text{reversible}} + Q_{A\to B}^{\text{path } P_0,\text{reversible}} = \Delta U.$$

This combined statement is the expression the first law of thermodynamics for reversible processes for closed systems.

In particular, if no work is done on a thermally isolated closed system we have

$$\Delta U = 0\ .$$

This is one aspect of the law of conservation of energy and can be stated:

The internal energy of an isolated system remains constant.

General Case for Irreversible Processes

If, in a process of change of state of a closed system, the energy transfer is not under a practically zero temperature gradient and practically frictionless, then the process is irreversible. Then the heat and work transfers may be difficult to calculate, and irreversible thermodynamics is called for. Nevertheless, the first law still holds and provides a check on the measurements and calculations of the work done irreversibly on the system, $W_{A\to B}^{\text{path } P_1,\text{irreversible}}$, and the heat transferred irreversibly to the system, $Q_{A\to B}^{\text{path } P_1,\text{irreversible}}$, which belong to the same particular process defined by its particular irreversible path, P_1, through the space of thermodynamic states.

$$-W_{A\to B}^{\text{path } P_1,\text{irreversible}} + Q_{A\to B}^{\text{path } P_1,\text{irreversible}} = \Delta U.$$

This means that the internal energy U is a function of state and that the internal energy change ΔU between two states is a function only of the two states.

Overview of the Weight of Evidence for the Law

The first law of thermdynamics is very general and makes so many predictions that they can hardly all be directly tested by experiment. Nevertheless, very very many of its predictions have been found empirically accurate. And very importantly, no accurately and properly conducted experiment has ever detected a violation of the law. Consequently, within its scope of applicability, the law is so reliably established, that, nowadays, rather than experiment being considered as testing the accuracy of the law, it is far more practical and realistic to think of the law as testing the accuracy of experiment. An experimental result that seems to violate the law may be assumed to be inaccurate or wrongly conceived, for example due to failure to consider an important physical factor.

State Functional Formulation for Infinitesimal Processes

When the heat and work transfers in the equations above are infinitesimal in magnitude, they are often denoted by δ, rather than exact differentials denoted by "d", as a reminder that heat and work do not describe the *state* of any system. The integral of an inexact differential depends upon the particular path taken through the space of thermodynamic parametres while the integral of an exact differential depends only upon the initial and final states. If the initial and final states are the same, then the integral of an inexact differential may or may not be zero, but the integral of an exact differential will always be zero. The path taken by a thermodynamic system through a chemical or physical change is known as a thermodynamic process.

For a homogeneous system, with a well-defined temperature and pressure, the expression for dU can be written in terms of exact differentials, if the work that the system does is equal to its pressure times the infinitesimal increase in its volume. Here one assumes that the changes are quasistatic, so slow that there is at each instant negligible departure from thermodynamic equilibrium within the system. In other words, $\delta W = -PdV$ where P is pressure and V is volume. As such a quasistatic process in a homogeneous system is reversible, the total amount of heat added to a closed system can be expressed as $\delta Q = TdS$ where T is the temperature and S the entropy of the system. Therefore, for closed, homogeneous systems:

$$dU = TdS - PdV.$$

The above equation is known as the fundamental thermodynamic relation, for which the independent variables are taken as S and V, with respect to which T and P are partial derivatives of U. While this has been derived for quasistatic changes, it is valid in general, as U can be considered as a thermodynamic state function of the independent variables S and V.

E.g., suppose that the system is initially in a state of thermal equilibrium defined by S and V, and then the system is suddenly perturbed so that thermal equilibrium breaks down and no temperature and pressure can be defined. Then the system settles down again to a state of thermal equilibrium, defined by an entropy and a volume which differ infinitesimally from the initial values. The infinitesimal difference in internal energy between the initial and final state will then satisfy the above equation. The work done and heat added to the system will then not satisfy the above expressions, they will instead satisfy the inequalities: $\delta Q < TdS'$ *and* $\delta W < PdV'$.

In the case of a closed system in which the particles of the system are of different types and, because chemical reactions may occur, their respective numbers are not necessarily constant, the expression for dU becomes:

$$dU = \delta Q - \delta W + \sum_i \mu_i dN_i$$

where dN_i is the (small) increase in amount of type-i particles in the reaction, and μ_i is known as the chemical potential of the type-i particles in the system. If dN_i is expressed in mol then μ_i is expressed in J/mol. The statement of the first law, using exact differentials is now:

$$dU = TdS - PdV + \sum_i \mu_i dN_i.$$

If the system has more external mechanical variables than just the volume that can change, the fundamental thermodynamic relation generalizes to:

$$dU = TdS - \sum_i X_i dx_i + \sum_j \mu_j dN_j.$$

Here the X_i are the generalized forces corresponding to the external variables x_i. The parametres X_i are independent of the size of the system and are called intensive parametres and the x_i are proportional to the size and called extensive parametres.

For an open system, there can be transfers of particles as well as energy into or out of the system during a process. For this case, the first law of thermodynamics still holds, in the form that the internal energy is a function of state and the change of internal energy in a process is a function only of its initial and final states, as noted in the section below headed First law of thermodynamics for open systems.

A useful idea from mechanics is that the energy gained by a particle is equal to the force applied to the particle multiplied by the displacement of the particle while that force is applied. Now consider the first law without the heating term: $dU = -PdV$. The pressure P can be viewed as a force (and in fact has units of force per unit area) while dV is the displacement (with units of distance times area). We may say, with respect to this work term, that a pressure difference forces a transfer of volume, and that the product of the two (work) is the amount of energy transferred out of the system as a result of the process. If one were to make this term negative then this would be the work done on the system.

It is useful to view the TdS term in the same light: here the temperature is known as a "generalized" force (rather than an actual mechanical force) and the entropy is a generalized displacement.

Similarly, a difference in chemical potential between groups of particles in the system drives a chemical reaction that changes the numbers of particles, and the corresponding product is the amount of chemical potential energy transformed in process. For example, consider a system consisting of two phases: liquid water and water vapour. There is a generalized "force" of evaporation which drives water molecules out of the liquid. There is a generalized "force" of condensation which drives vapour molecules out of the vapour. Only when these two "forces" (or chemical potentials) are equal will there be equilibrium, and the net rate of transfer will be zero.

The two thermodynamic parametres which form a generalized force-displacement pair are termed "conjugate variables". The two most familiar pairs are, of course, pressure-volume, and temperature-entropy.

Spatially Inhomogeneous Systems

Classical thermodynamics is initially focused on closed homogeneous systems (e.g. Planck 1897/1903), which might be regarded as 'zero-dimensional' in the sense that they have no spatial variation. But it is desired to study also systems with distinct internal motion and spatial inhomogeneity. For such systems, the principle of conservation of energy is expressed in terms not only of internal energy as defined for homogeneous systems, but also in terms of kinetic energy and potential energies of parts of the inhomogeneous system with respect to each other and with respect to long-range external forces.

How the total energy of a system is allocated between these three more specific kinds of energy varies according to the purposes of different writers; this is because these components of energy are to some extent mathematical artefacts rather than actually measured physical quantities. For any closed homogeneous component of an inhomogeneous closed system, if E denotes the total energy of that component system, one may write

$$E = E^{\text{kin}} + E^{\text{pot}} + U$$

where E^{kin} and E^{pot} denote respectively the total kinetic energy and the total potential energy of the component closed homogeneous system, and U denotes its internal energy.

Potential energy can be exchanged with the surroundings of the system when the surroundings impose a force field, such as gravitational or electromagnetic, on the system.

A compound system consisting of two interacting closed homogeneous component subsystems has a potential energy of interaction E_{12}^{pot} between the subsystems. Thus, in an obvious notation, one may write

$$E = E_1^{\text{kin}} + E_1^{\text{pot}} + U_1 + E_2^{\text{kin}} + E_2^{\text{pot}} + U_2 + E_{12}^{\text{pot}}$$

The quantity E_{12}^{pot} in general lacks an assignment to either subsystem in a way that is not arbitrary, and this stands in the way of a general non-arbitrary definition of transfer of energy as work. On occasions, authors make their various respective arbitrary assignments.

The distinction between internal and kinetic energy is hard to make in the presence of turbulent motion within the system, as friction gradually dissipates macroscopic kinetic energy of localised bulk flow into molecular random motion of molecules that is classified as internal energy. The rate of dissipation by friction of kinetic energy of localised bulk flow into internal energy, whether in turbulent or in streamlined flow, is an important quantity in non-equilibrium thermodynamics. This is a serious difficulty for attempts to define entropy for time-varying spatially inhomogeneous systems.

First Law of Thermodynamics for Open Systems

For the first law of thermodynamics, there is no trivial passage of physical conception from the closed system view to an open system view. For closed systems, the concepts of an adiabatic enclosure and of an adiabatic wall are fundamental. Matter and internal energy cannot permeate or penetrate such a wall. For an open system, there is a wall that allows penetration by matter. In general, matter in motion will carry with it some internal energy, and some potential energy changes will accompany the motion. An open system is not adiabatically enclosed.

There are some cases in which a process for an open system can, for particular purposes, be considered as if it were for a closed system. In an open system, by definition hypothetically or potentially, matter can pass between the system and its surroundings. But when, in a particular case, the process of interest involves only hypothetical or potential but no actual passage of matter, the process can be considered as if it were for a closed system.

Internal Energy for an Open System

Since the revised and more rigorous definition of the internal energy of a closed system rests upon the possibility of processes by which adiabatic work takes the system from one state to another, this leaves a problem for the definition of internal energy for an open system, for which adiabatic work is not in general possible. According to Max Born, the transfer of matter and energy across an open connection "cannot be reduced to mechanics". In contrast to the case of closed systems, for open systems, in the presence of diffusion, there is no unconstrained and unconditional physical distinction between convective transfer of internal energy by bulk flow of matter, the transfer of internal energy without transfer of matter (usually called heat conduction and work transfer), and change of various potential energies. The older traditional way and the conceptually revised (Carathéodory) way agree that there is no physically unique definition of heat and work transfer processes between open systems.

In particular, between two otherwise isolated open systems an adiabatic wall is by definition impossible. This problem is solved by recourse to the principle of conservation of energy. This principle allows a composite isolated system to be derived from two other component non-interacting isolated systems, in such a way that the total energy of the composite isolated system is equal to the sum of the total energies of the two component isolated systems. Two previously isolated systems can be subjected to the thermodynamic operation of placement between them of a wall permeable to matter and energy, followed by a time for establishment of a new thermodynamic state of internal equilibrium in the new single unpartitioned system. The internal energies of the initial two systems and of the final new system, considered respectively as closed systems as above, can be measured. Then the law of conservation of energy requires that

$$\Delta U_s + \Delta U_o = 0,$$

where ΔU_s and ΔU_o denote the changes in internal energy of the system and of its surroundings respectively. This is a statement of the first law of thermodynamics for a transfer between two otherwise isolated open systems, that fits well with the conceptually revised and rigorous statement of the law stated above. For the thermodynamic operation of adding two systems with internal energies U_1 and U_2, to produce a new system with internal energy U, one may write $U = U_1 + U_2$; the reference states for U, U_1 and U_2 should be specified accordingly, maintaining also that the internal energy of a system be proportional to its mass, so that the internal energies are extensive variables. There is a sense in which this expresses a fundamental postulate that goes beyond the simplest ideas of classical closed system thermodynamics; the extensivity of some variables is not obvious, and needs explicit expression; indeed one author goes so far as to say that it could be recognized as a fourth law of thermodynamics, though this is not repeated by other authors.

Also of course

$$\Delta N_s + \Delta N_o = 0,$$

where ΔN_s and ΔN_o denote the changes in mole number of a component substance of the system and of its surroundings respectively. This is a statement of the law of conservation of mass.

Process of Transfer of Matter Between an Open System and Its Surroundings

A system connected to its surroundings only through contact by a single permeable wall, but otherwise isolated, is an open system. If it is initially in a state of contact equilibrium with a surrounding subsystem, a thermodynamic process of transfer of matter can be made to occur between them if the surrounding subsystem is subjected to some thermodynamic operation, for example, removal of a partition between it and some further surrounding subsystem. The removal of the partition in the surroundings initiates a process of exchange between the system and its contiguous surrounding subsystem.

An example is evaporation. One may consider an open system consisting of a collection of liquid, enclosed except where it is allowed to evaporate into or to receive condensate from its vapour above it, which may be considered as its contiguous surrounding subsystem, and subject to control of its volume and temperature. A thermodynamic process might be initiated by a thermodynamic operation in the surroundings, that mechanically increases in the controlled volume of the vapour. Some mechanical work will be done within the surroundings by the vapour, but also some of the parent liquid will evaporate and enter the vapour collection which is the contiguous surrounding subsystem. Some internal energy will accompany the vapour that leaves the system, but it will not make sense to try to uniquely identify part of that internal energy as heat and part of it as work. Consequently, the energy transfer that accompanies the transfer of matter between the system and its surrounding subsystem cannot be uniquely split into heat and work transfers to or from the open system. The component of total energy transfer that accompanies the transfer of vapour into the surrounding subsystem is customarily called 'latent heat of evaporation', but this use of the word heat is a quirk of customary historical language, not in strict compliance with the thermodynamic definition of transfer of energy as heat. In this example, kinetic energy of bulk flow and potential energy with respect to long-range external forces such as gravity are both considered to be zero. The first law of thermodynamics refers to the change of internal energy of the open system, between its initial and final states of internal equilibrium.

Open System with Multiple Contacts

An open system can be in contact equilibrium with several other systems at once.

This includes cases in which there is contact equilibrium between the system, and several subsystems in its surroundings, including separate connections with subsystems through walls that are permeable to the transfer of matter and internal energy as heat and allowing friction of passage of the transferred matter, but immovable, and separate connections through adiabatic walls with others, and separate connections through diathermic walls impermeable to matter with yet others. Because there are separate connections that are permeable to energy but impermeable to matter, between the system and its surroundings, energy transfers between them can occur with definite heat and work characters. The total increase of internal energy in the process is then determined as the sum of the internal energy transferred from the surroundings with the transfer of matter through the walls that are permeable to it, and of the internal energy transferred to the system as heat through the diathermic walls, and of the energy transferred to the system as work through the adiabatic walls, including the energy transferred to the system by long-range forces. These simultaneously transferred quantities of energy are defined by events in the surroundings of the system. Because the internal energy transferred with matter is not in general uniquely resolvable into heat and work components, the total energy transfer cannot in general be uniquely resolved into heat and work components. Under these conditions, the following formula can describe the process in terms of externally defined thermodynamic variables, as a statement of the first law of thermodynamics:

$$\Delta U_0 = Q - W - \sum_{i=1}^{m} \Delta U_i \tag{1}$$

(suitably defined surrounding subsystems, general process, quasi-static or irreversible),

where ΔU_0 denotes the change of internal energy of the system, and ΔU_i denotes the change of internal energy of the *i*th of the *m* surrounding subsystems that are in open contact with the system, due to transfer between the system and that *i*th surrounding subsystem, and Q denotes the internal energy transferred as heat from the heat reservoir of the surroundings to the system, and W denotes the energy transferred from the system to the surrounding subsystems that are in adiabatic connection with it. The case of a wall that is permeable to matter and can move so as to allow transfer of energy as work is not considered here.

Combination of First and Second Laws

If the system is described by the energetic fundamental equation, $U_0 = U_0(S, V, N_j)$, *and if the process can be described in the quasi-static formalism, in terms of the internal state variables of the system, then the process can also be described by a combination of the first and second laws of thermodynamics, by the formula*

$$dU_0 = T\,dS - P\,dV + \sum_{j=1}^{n} \mu_j\, dN_j \tag{2}$$

where there are n chemical constituents of the system and permeably connected surrounding subsystems, and where T, S, P, V, N_j, and μ_j, are defined as above.

For a general natural process, there is no simple termwise correspondence between equations (1) and (2), because they describe the process in different conceptual frames.

Nevertheless, for the special fictive case of quasi-static transfers, there is a simple correspondence. For this, it is supposed that the system has multiple areas of contact with its surroundings. There are pistons that allow adiabatic work, purely diathermal walls, and open connections with surrounding subsystems of completely controllable chemical potential (or equivalent controls for charged species). Then, for a suitable fictive quasi-static transfer, one can write

$$\delta Q = T\,\mathrm{d}S \quad \text{and} \quad \delta W = P\,\mathrm{d}V \quad \text{(suitably defined surrounding subsystems, quasi-static transfers of energy).}$$

For fictive quasi-static transfers for which the chemical potentials in the connected surrounding subsystems are suitably controlled, these can be put into equation (2) to yield

$$\mathrm{d}U_0 = \delta Q - \delta W + \sum_{j=1}^{n} \mu_j\,\mathrm{d}N_j \quad \text{(suitably defined surrounding subsystems, quasi-static transfers).} \qquad (3)$$

The reference does not actually write equation (3), but what it does write is fully compatible with it.

There are several other accounts of this, in apparent mutual conflict.

Non-equilibrium Transfers

The transfer of energy between an open system and a single contiguous subsystem of its surroundings is considered also in non-equilibrium thermodynamics. The problem of definition arises also in this case. It may be allowed that the wall between the system and the subsystem is not only permeable to matter and to internal energy, but also may be movable so as to allow work to be done when the two systems have different pressures. In this case, the transfer of energy as heat is not defined.

Methods for study of non-equilibrium processes mostly deal with spatially continuous flow systems. In this case, the open connection between system and surroundings is usually taken to fully surround the system, so that there are no separate connections impermeable to matter but permeable to heat. Except for the special case mentioned above when there is no actual transfer of matter, which can be treated as if for a closed system, in strictly defined thermodynamic terms, it follows that transfer of energy as heat is not defined. In this sense, there is no such thing as 'heat flow' for a continuous-flow open system. Properly, for closed systems, one speaks of transfer of internal energy as heat, but in general, for open systems, one can speak safely only of transfer of internal energy. A factor here is that there are often cross-effects between distinct transfers, for example that transfer of one substance may cause transfer of another even when the latter has zero chemical potential gradient.

Usually transfer between a system and its surroundings applies to transfer of a state variable, and obeys a balance law, that the amount lost by the donor system is equal to the amount gained by the receptor system. Heat is not a state variable. For his 1947 definition of "heat transfer" for discrete open systems, the author Prigogine carefully explains at some length that his definition of it does not obey a balance law. He describes this as paradoxical.

The situation is clarified by Gyarmati, who shows that his definition of "heat transfer", for continuous-flow systems, really refers not specifically to heat, but rather to transfer of internal energy, as follows. He considers a conceptual small cell in a situation of continuous-flow as a system defined in the so-called Lagrangian way, moving with the local centre of mass. The flow of matter across the boundary is zero when considered as a flow of total mass. Nevertheless if the material constitution is of several chemically distinct components that can diffuse with respect to one another, the system is considered to be open, the diffusive flows of the components being defined with respect to the centre of mass of the system, and balancing one another as to mass transfer. Still there can be a distinction between bulk flow of internal energy and diffusive flow of internal energy in this case, because the internal energy density does not have to be constant per unit mass of material, and allowing for non-conservation of internal energy because of local conversion of kinetic energy of bulk flow to internal energy by viscosity.

Gyarmati shows that his definition of "the heat flow vector" is strictly speaking a definition of flow of internal energy, not specifically of heat, and so it turns out that his use here of the word heat is contrary to the strict thermodynamic definition of heat, though it is more or less compatible with historical custom, that often enough did not clearly distinguish between heat and internal energy; he writes "that this relation must be considered to be the exact definition of the concept of heat flow, fairly loosely used in experimental physics and heat technics." Apparently in a different frame of thinking from that of the above-mentioned paradoxical usage in the earlier sections of the historic 1947 work by Prigogine, about discrete systems, this usage of Gyarmati is consistent with the later sections of the same 1947 work by Prigogine, about continuous-flow systems, which use the term "heat flux" in just this way. This usage is also followed by Glansdorff and Prigogine in their 1971 text about continuous-flow systems. They write: "Again the flow of internal energy may be split into a convection flow ρuv and a conduction flow. This conduction flow is by definition the heat flow W. Therefore: j[U] = ρuv + W where u denotes the [internal] energy per unit mass. [These authors actually use the symbols E and e to denote internal energy but their notation has been changed here to accord with the notation of the present article. These authors actually use the symbol U to refer to total energy, including kinetic energy of bulk flow.]" This usage is followed also by other writers on non-equilibrium thermodynamics such as Lebon, Jou, and Casas-Vásquez, and de Groot and Mazur. This usage is described by Bailyn as stating the non-convective flow of internal energy, and is listed as his definition number 1, according to the first law of thermodynamics. This usage is also followed by workers in the kinetic theory of gases. This is not the *ad hoc* definition of "reduced heat flux" of Haase.

In the case of a flowing system of only one chemical constituent, in the Lagrangian representation, there is no distinction between bulk flow and diffusion of matter. Moreover, the flow of matter is zero into or out of the cell that moves with the local centre of mass. In effect, in this description, one is dealing with a system effectively closed to the transfer of matter. But still one can validly talk of a distinction between bulk flow and diffusive flow of internal energy, the latter driven by a temperature gradient within the flowing material, and being defined with respect to the local centre of mass of the bulk flow. In this case of a virtually closed system, because of the zero matter transfer, as noted above, one can safely distinguish between transfer of energy as work, and transfer of internal energy as heat.

The Steady Flow Energy Equation (SFEE)

A very large class of devices of interest to engineers such as turbines, compressors, nozzles, boiler and condensers operat under long-term steady-state condition i.e. after the initial start up period they operate in a way that there are no variation of properties with time i.e. All changes in the measurable quantities inside the control volume are non-existent or cyclic. Then Et+dt = Et and so

$$\frac{\text{dEc.v.}}{\text{d}_t} \textit{term} \Rightarrow o.\text{Also}\,\Sigma\,\text{m}_i = \Sigma\,\overset{\vee}{m}_e$$

or there would be an accumulation or fall in masses in the control volume which would nullify the original condition for steady flow. So to obtain steady flow conditions we need to satisfy three conditions:

1. The streams of material crossing the control surface must not change their state or flow rate with time.
2. Each point within the control volume must not change its state with time or only cyclic state variation occur.
3. The heat and work transfer rates must not change with time or the mean rates in the case of cyclic behaviour must not change.

Thus we get the Steady Flow Energy Equation (SFEE)

$$\text{Q} - \text{W}\times = \sum_{\text{out}} m\left(h + \frac{\text{V}^2}{2} + g\underline{z}\right) - \sum_{\text{in}} \text{m}\left(h + \frac{\text{V}^2}{2} + g\underline{z}\right)$$

Or for single streams:

$$\overset{\vee}{\text{Q}} - \overset{\vee}{\text{W}}_\times = \overset{\vee}{\text{m}}\left\{\left[\left(h_2 + \frac{\text{V}_2{}^2}{2} + gz_2\right) - \left(h_1 + \frac{\text{V}^2}{2} + gz_1\right)\right]\right\}$$

Non-Uniform Streams

In developing the energy equations we assumed that any given entry or exit stream has a unique velocity. But in all practical cases the velocity at entry is non-uniform and a mean velocity that satisfies the simple continuity equation is used:

$$\underline{\text{V}} = \frac{\overset{\vee}{\text{m}}}{\rho\text{A}} \leftarrow \frac{\overset{\vee}{\text{mv}}}{\text{A}}$$

Second Law of Thermodynamics

The second law of thermodynamics states that the entropy of an isolated system never decreases, because isolated systems spontaneously evolve towards thermodynamic equilibrium—the state of maximum entropy. Equivalently, perpetual motion machines of the second kind are impossible.

The second law is an empirically validated postulate of thermodynamics, but it can be understood and explained using the underlying quantum statistical mechanics, together with the assumption of low-entropy initial conditions in the distant past (possibly at the beginning of the universe). In the language of statistical mechanics, entropy is a measure of the number of microscopic configurations corresponding to a macroscopic state. Because equilibrium corresponds to a vastly greater number of microscopic configurations than any non-equilibrium state, it has the maximum entropy, and the second law follows because random chance alone practically guarantees that the system will evolve towards equilibrium.

It is an expression of the fact that over time, differences in temperature, pressure, and chemical potential decrease in an isolated non-gravitational physical system, leading eventually to a state of thermodynamic equilibrium.

The second law may be expressed in many specific ways, but the first formulation is credited to the French scientist Sadi Carnot in 1824.

The second law has been shown to be equivalent to the internal energy U being a weakly convex function, when written as a function of extensive properties (mass, volume, entropy,...).

Description

The first law of thermodynamics provides the basic definition of thermodynamic energy, also called internal energy, associated with all thermodynamic systems, but unknown in classical mechanics, and states the rule of conservation of energy in nature.

The concept of energy in the first law does not, however, account for the observation that natural processes have a preferred direction of progress. For example, heat always flows spontaneously from regions of higher temperature to regions of lower temperature, and never the reverse, unless external work is performed on the system. The first law is completely symmetrical with respect to the initial and final states of an evolving system. The key concept for the explanation of this phenomenon through the second law of thermodynamics is the definition of a new physical property, the entropy.

In a reversible process, an infinitesimal increment in the entropy (dS) of a system results from an infinitesimal transfer of heat (δQ)

to a closed system divided by the common temperature (T) of the system and the surroundings which supply the heat.

$$dS = \frac{\delta Q}{T}$$

The entropy of an isolated system in its own internal thermodynamic equilibrium does not change with time. An isolated system may consist initially of several subsystems, separated from one another by partitions, but still each in its own internal thermodynamic equilibrium. If the partitions are removed, the former subsystems will in general interact and produce a new common final system in its own internal thermodynamic equilibrium. The sum of the entropies of the initial subsystems is in general less than the entropy of the final common system. If all of the initial subsystems have all the same values of their intensive variables, then the sum of the initial entropies will be equal to the final common entropy, and the final common system will have the same values of its intensive variables.

For a body in thermal equilibrium with another, there are indefinitely many empirical temperature scales, in general respectively depending on the properties of a particular reference thermometric body. Thermal equilibrium between two bodies entails that they have equal temperatures. The zeroth law of thermodynamics in its usual short statement allows recognition that two bodies have the same temperature, especially that a test body has the same temperature as a reference thermometric body. The second law allows a distinguished temperature scale, which defines an absolute, thermodynamic temperature, independent of the properties of any particular thermometric body.

The second law of thermodynamics may be expressed in many specific ways, the most prominent classical statements being the statement by Rudolph Clausius (1854), the statement by Lord Kelvin (1851), and the statement in axiomatic thermodynamics by Constantin Carathéodory (1909). These statements cast the law in general physical terms citing the impossibility of certain processes. The Clausius and the Kelvin statements have been shown to be equivalent.

Clausius Statement

The German scientist Rudolf Clausius laid the foundation for the second law of thermodynamics in 1850 by examining the relation between heat transfer and work. His formulation of the second law, which was published in German in 1854, is known as the *Clausius statement*:

Heat can never pass from a colder to a warmer body without some other change, connected therewith, occurring at the same time.

Heat cannot spontaneously flow from cold regions to hot regions without external work being performed on the system, which is evident from ordinary experience of refrigeration, for example. In a refrigerator, heat flows from cold to hot, but only when forced by an external agent, the refrigeration system.

Kelvin Statement

Lord Kelvin Expressed the Second Law as: It is impossible, by means of inanimate material agency, to derive mechanical effect from any portion of matter by cooling it below the temperature of the coldest of the surrounding objects.

Planck's Principle

In 1926 Max Planck wrote an important paper on the basics of thermodynamics. He indicated the principle. The internal energy of a closed system is increased by an adiabatic process that is isochoric throughout its course.

This formulation does not mention heat and does not mention temperature, nor even entropy, and does not necessarily implicitly rely on those concepts, but it implies the content of the second law. A closely related statement is that "Frictional pressure never does positive work." Using a now obsolete form of words, Planck himself wrote: "The production of heat by friction is irreversible."

Principle of Carathéodory

Constantin Carathéodory formulated thermodynamics on a purely mathematical axiomatic foundation. His statement of the second law is known as the Principle of Carathéodory, which may be formulated as follows:

In every neighbourhood of any state S of an adiabatically isolated system there are states inaccessible from S.

With this formulation he described the concept of adiabatic accessibility for the first time and provided the foundation for a new subfield of classical thermodynamics, often called geometrical thermodynamics. It follows from Carathéodory's principle that quantity of energy quasi-statically transferred as heat is a holonomic process function, in other words, $\delta Q = TdS$.

Though it is almost customary in textbooks to say that Carathéodory's principle expresses the second law and to treat it as equivalent to the Clausius or to the Kelvin-Planck statements, such is not the case. To get all the content of the second law, Carathéodory's principle needs to be supplemented by Planck's principle, that isochoric work always increases the internal energy of a closed system that was initially in its own internal thermodynamic equilibrium.

Equivalence of the Clausius and the Kelvin Statements

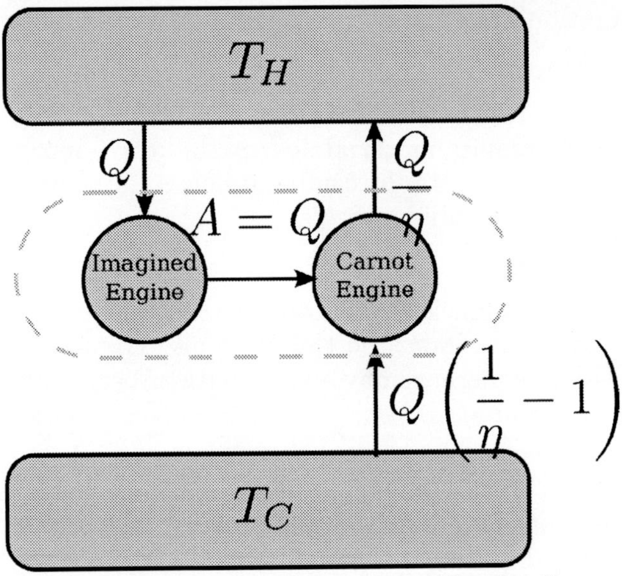

Figure: *Derive Kelvin Statement from Clausius Statement*

Suppose there is an engine violating the Kelvin statement: i.e., one that drains heat and converts it completely into work in a cyclic fashion without any other result. Now pair it with a reversed Carnot engine as shown by the graph. The net and sole effect of this newly created engine consisting of the two engines mentioned is transferring heat $\Delta Q = Q\left(\frac{1}{\eta} - 1\right)$ from the cooler reservoir to the hotter one, which violates the Clausius statement. Thus a violation of the Kelvin statement implies a violation of the Clausius statement, i.e. the Clausius statement implies the Kelvin statement. We can prove in a similar manner that the Kelvin statement implies the Clausius statement, and hence the two are equivalent.

Gravitational Systems

In non-gravitational systems, objects always have positive heat capacity, meaning that the temperature rises with energy. Therefore, when energy flows from a high-temperature object to a low-temperature object, the source temperature is decreased while the sink temperature is increased; hence temperature differences tend to diminish over time.

However, this is not always the case for systems in which the gravitational force is important. The most striking examples are black holes, which - according to theory - have negative heat capacity. The larger the black hole, the more energy it contains, but the lower its temperature. Thus, the supermassive black hole in the centre of the milky way is supposed to have a temperature of 10^{-14} K, much lower than the cosmic microwave background temperature of 2.7K, but as it absorbs photons of the cosmic microwave background its mass is increasing so that its low temperature further decreases with time.

For this reason, gravitational systems tend towards non-even distribution of mass and energy.

Corollaries

Perpetual Motion of the Second Kind: Before the establishment of the Second Law, many people who were interested in inventing a perpetual motion machine had tried to circumvent the restrictions of First Law of Thermodynamics by extracting the massive internal energy of the environment as the power of the machine. Such a machine is called a "perpetual motion machine of the second kind". The second law declared the impossibility of such machines.

Carnot Theorem

Carnot's theorem (1824) is a principle that limits the maximum efficiency for any possible engine. The efficiency solely depends on the temperature difference between the hot and cold thermal reservoirs. Carnot's theorem states:

- All irreversible heat engines between two heat reservoirs are less efficient than a Carnot engine operating between the same reservoirs.
- All reversible heat engines between two heat reservoirs are equally efficient with a Carnot engine operating between the same reservoirs.

In his ideal model, the heat of caloric converted into work could be reinstated by reversing the motion of the cycle, a concept subsequently known as thermodynamic reversibility. Carnot however further postulated that some caloric is lost, not being converted to mechanical work. Hence no real heat engine could realise the Carnot cycle's reversibility and was condemned to be less efficient.

Though formulated in terms of caloric, rather than entropy, this was an early insight into the second law.

Clausius Inequality

The Clausius Theorem (1854) states that in a cyclic process

$$\oint \frac{\delta Q}{T} \leq 0.$$

The equality holds in the reversible case and the '<' is in the irreversible case. The reversible case is used to introduce the state function entropy. This is because in cyclic processes the variation of a state function is zero from state functionality.

Thermodynamic Temperature

For an arbitrary heat engine, the efficiency is:

$$\eta = \frac{A}{q_H} = \frac{q_H - q_C}{q_H} = 1 - \frac{q_C}{q_H} \tag{1}$$

where A is the work done per cycle. Thus the efficiency depends only on q_C/q_H.

Carnot's theorem states that all reversible engines operating between the same heat reservoirs are equally efficient. Thus, any reversible heat engine operating between temperatures T_1 and T_2 must have the same efficiency, that is to say, the efficiency is the function of temperatures only: $\frac{q_C}{q_H} = f(T_H, T_C)$ (2)

In addition, a reversible heat engine operating between temperatures T_1 and T_3 must have the same efficiency as one consisting of two cycles, one between T_1 and another (intermediate) temperature T_2, and the second between T_2 and T3. This can only be the case if

$$f(T_1, T_3) = \frac{q_3}{q_1} = \frac{q_2 q_3}{q_1 q_2} = f(T_1, T_2) f(T_2, T_3).$$

Now consider the case where T_1 is a fixed reference temperature: the temperature of the triple point of water. Then for any T_2 and T_3,

$$f(T_2, T_3) = \frac{f(T_1, T_3)}{f(T_1, T_2)} = \frac{273.16 \cdot f(T_1, T_3)}{273.16 \cdot f(T_1, T_2)}.$$

Therefore if thermodynamic temperature is defined by

$$T = 273.16 \cdot f(T_1, T)$$

then the function *f*, viewed as a function of thermodynamic temperature, is simply

$$f(T_2,T_3)=\frac{T_3}{T_2},$$

and the reference temperature T_1 will have the value 273.16. (Of course any reference temperature and any positive numerical value could be used—the choice here corresponds to the Kelvin scale.)

Entropy

According to the Clausius equality, for a reversible process

$$\oint\frac{\delta Q}{T}=0$$

That means the line integral $\int_L\frac{\delta Q}{T}$ is path independent.

So we can define a state function S called entropy, which satisfies

$$dS=\frac{\delta Q}{T}$$

With this we can only obtain the difference of entropy by integrating the above formula. To obtain the absolute value, we need the Third Law of Thermodynamics, which states that S=0 at absolute zero for perfect crystals.

For any irreversible process, since entropy is a state function, we can always connect the initial and terminal status with an imaginary reversible process and integrating on that path to calculate the difference in entropy.

Now reverse the reversible process and combine it with the said irreversible process. Applying Clausius inequality on this loop,

$$-\Delta S+\int\frac{\delta Q}{T}=\oint\frac{\delta Q}{T}<0$$

Thus,

$$\Delta S\geq\int\frac{\delta Q}{T}$$

where the equality holds if the transformation is reversible.

Notice that if the process is an adiabatic process, then $\delta Q=0$, so $\Delta S\geq 0$.

Exergy, Available Useful Work

An important and revealing idealized special case is to consider applying the Second Law to the scenario of an isolated system (called the total system or universe), made up of two parts: a sub-system of interest, and the sub-system's surroundings. These surroundings are imagined to be so large that they can be considered as an *unlimited* heat reservoir at temperature T_R and pressure P_R — so that no matter how much heat is transferred to (or from) the sub-system, the temperature of the surroundings will remain T_R; and no matter how much the volume of the sub-system expands (or contracts), the pressure of the surroundings will remain P_R.

Whatever changes to dS and dS_R occur in the entropies of the sub-system and the surroundings individually, according to the Second Law the entropy S_{tot} of the isolated total system must not decrease:

$$dS_{\text{tot}} = dS + dS_R \geq 0$$

According to the First Law of Thermodynamics, the change dU in the internal energy of the sub-system is the sum of the heat δq added to the sub-system, *less* any work δw done *by* the sub-system, *plus* any net chemical energy entering the sub-system $d\,\Sigma\mu_{iR}N_i$, so that:

$$dU = \delta q - \delta w + d\left(\sum \mu_{iR} N_i\right)$$

where μ_{iR} are the chemical potentials of chemical species in the external surroundings.

Now the heat leaving the reservoir and entering the sub-system is

$$\delta q = T_R(-dS_R) \leq T_R dS$$

where we have first used the definition of entropy in classical thermodynamics (alternatively, in statistical thermodynamics, the relation between entropy change, temperature and absorbed heat can be derived); and then the Second Law inequality from above.

It therefore follows that any net work δw done by the sub-system must obey

$$\delta w \leq -dU + T_R dS + \sum \mu_{iR} dN_i$$

It is useful to separate the work δw done by the subsystem into the *useful* work δw_u that can be done *by* the sub-system, over and beyond the work $p_R\,dV$ done merely by the sub-system expanding against the surrounding external pressure, giving the following relation for the useful work (exergy) that can be done:

$$\delta w_u \leq -d\left(U - T_R S + p_R V - \sum \mu_{iR} N_i\right)$$

It is convenient to define the right-hand-side as the exact derivative of a thermodynamic potential, called the *availability* or *exergyE* of the subsystem,

$$E = U - T_R S + p_R V - \sum \mu_{iR} N_i$$

The Second Law therefore implies that for any process which can be considered as divided simply into a subsystem, and an unlimited temperature and pressure reservoir with which it is in contact,

$$dE + \delta w_u \le 0$$

i.e. the change in the subsystem's exergy plus the useful work done *by* the subsystem (or, the change in the subsystem's exergy less any work, additional to that done by the pressure reservoir, done *on* the system) must be less than or equal to zero.

In sum, if a proper *infinite-reservoir-like* reference state is chosen as the system surroundings in the real world, then the Second Law predicts a decrease in E for an irreversible process and no change for a reversible process.

$dS_{tot} \ge 0$ Is equivalent to $dE + \delta w_u \le 0$

This expression together with the associated reference state permits a design engineer working at the macroscopic scale (above the thermodynamic limit) to utilize the Second Law without directly measuring or considering entropy change in a total isolated system. Those changes have already been considered by the assumption that the system under consideration can reach equilibrium with the reference state without altering the reference state. An efficiency for a process or collection of processes that compares it to the reversible ideal may also be found.

This approach to the Second Law is widely utilized in engineering practice, environmental accounting, systems ecology, and other disciplines.

History

The first theory of the conversion of heat into mechanical work is due to Nicolas Léonard Sadi Carnot in 1824. He was the first to realize correctly that the efficiency of this conversion depends on the difference of temperature between an engine and its environment.

Recognizing the significance of James Prescott Joule's work on the conservation of energy, Rudolf Clausius was the first to formulate the second law during 1850, in this form: heat does not flow *spontaneously* from cold to hot bodies. While common knowledge now, this was contrary to the caloric theory of heat popular at the time, which considered heat as a fluid. From there he was able to infer the principle of Sadi Carnot and the definition of entropy (1865).

Established during the 19th century, the Kelvin-Planck statement of the Second Law says, "It is impossible for any device that operates on a cycle to receive heat from a single reservoir and produce a net amount of work." This was shown to be equivalent to the statement of Clausius. The ergodic hypothesis is also important for the Boltzmann approach. It says that, over long periods of time, the time spent in some region of the phase space of microstates with the same energy is proportional to the volume of this region, i.e. that all accessible microstates are equally probable over a long period of time. Equivalently, it says that time average and average over the statistical ensemble are the same.

It has been shown that not only classical systems but also quantum mechanical ones tend to maximize their entropy over time. Thus the second law follows, given initial conditions with low entropy. More precisely, it has been shown that the local von Neumann entropy is at its maximum value with a very high probability. The result is valid for a large class of isolated quantum systems (e.g. a gas in a container). While the full system is pure and therefore does not have any entropy, the entanglement between gas and container gives rise to an increase of the local entropy of the gas. This result is one of the most important achievements of quantum thermodynamics.

Today, much effort in the field is attempting to understand why the initial conditions early in the universe were those of low entropy, as this is seen as the origin of the second law.

Informal Descriptions

The second law can be stated in various succinct ways, including:

- It is impossible to produce work in the surroundings using a cyclic process connected to a single heat reservoir (Kelvin, 1851).
- It is impossible to carry out a cyclic process using an engine connected to two heat reservoirs that will have as its only effect the transfer of a quantity of heat from the low-temperature reservoir to the high-temperature reservoir (Clausius, 1854).
- If thermodynamic work is to be done at a finite rate, free energy must be expended. (Stoner, 2000)

Mathematical Descriptions

In 1856, the German physicist Rudolf Clausius stated what he called the "second fundamental theorem in the mechanical theory of heat" in the following form:

$$\int \frac{\delta Q}{T} = -N$$

where Q is heat, T is temperature and N is the "equivalence-value" of all uncompensated transformations involved in a cyclical process. Later, in 1865, Clausius would come to define "equivalence-value" as entropy. On the heels of this definition, that same year, the most famous version of the second law was read in a presentation at the Philosophical Society of Zurich on April 24, in which, in the end of his presentation, Clausius concludes:

The Entropy of the Universe Tends to a Maximum.

This statement is the best-known phrasing of the second law. Moreover, owing to the general broadness of the terminology used here, e.g. universe, as well as lack of specific conditions, e.g. open, closed, or isolated, to which this statement applies, many people take this simple statement to mean that the second law of thermodynamics applies virtually to every subject imaginable. This, of course, is not true; this statement is only a simplified version of a more complex description.

In terms of time variation, the mathematical statement of the second law for an isolated system undergoing an arbitrary transformation is:

$$\frac{dS}{dt} \geq 0$$

where

S is the entropy of the system and

t is time.

The equality sign holds in the case that only reversible processes take place inside the system. If irreversible processes take place (which is the case in real systems in operation) the >-sign holds. An alternative way of formulating of the second law for isolated systems is:

$$\frac{dS}{dt} = \dot{S}_i \text{ with } \dot{S}_i \geq 0$$

with $\dot{S}_i$ the sum of the rate of entropy production by all processes inside the system. The advantage of this formulation is that it shows the effect of the entropy production. The rate of entropy production is a very important concept since it determines (limits) the efficiency of thermal machines. Multiplied with ambient temperature T_a it gives the so-called dissipated energy $P_{diss} = T_a\dot{S}_i$.

The expression of the second law for closed systems (so, allowing heat exchange and moving boundaries, but not exchange of matter) is:

$$\frac{dS}{dt} = \frac{\dot{Q}}{T} + \dot{S}_i \text{ with } \dot{S}_i \geq 0$$

Here

$\dot{Q}$ is the heat flow into the system

T is the temperature at the point where the heat enters the system.

If heat is supplied to the system at several places we have to take the algebraic sum of the corresponding terms.

For open systems (also allowing exchange of matter):

$$\frac{dS}{dt} = \frac{\dot{Q}}{T} + \dot{S} + \dot{S}_i \text{ with } \dot{S}_i \geq 0$$

Here $\dot{S}$ is the flow of entropy into the system associated with the flow of matter entering the system. It should not be confused with the time derivative of the entropy. If matter is supplied at several places we have to take the algebraic sum of these contributions.

Statistical mechanics gives an explanation for the second law by postulating that a material is composed of atoms and molecules which are in constant motion. A particular set of positions and velocities for each particle in the system is called a microstate of the system and because of the constant motion, the system is constantly changing its microstate. Statistical mechanics postulates that, in equilibrium, each microstate that the system might be in is equally likely to occur, and when this assumption is made, it leads directly to the conclusion that the second law must hold in a statistical sense. That is, the second law will hold on average, with a statistical variation on the order of 1√N where N is the number of particles in the system. For everyday (macroscopic) situations, the probability that the second law will be violated is practically zero. However, for systems with a small number of particles, thermodynamic parametres, including the entropy, may show significant statistical deviations from that predicted by the second law. Classical thermodynamic theory does not deal with these statistical variations.

Derivation from Statistical Mechanics

Due to Loschmidt's paradox, derivations the Second Law have to make an assumption regarding the past, namely that the system is uncorrelated at some time in the past; this allows for simple probabilistic treatment. This assumption is usually thought as a boundary condition, and thus the second Law is ultimately a consequence of the initial conditions somewhere in the past, probably at the beginning of the universe (the Big Bang), though other scenarios have also been suggested.

Given these assumptions, in statistical mechanics, the Second Law is not a postulate, rather it is a consequence of the fundamental postulate, also known as the equal prior probability postulate, so long as one is clear that simple probability arguments are applied only to the future, while for the past there are auxiliary sources of information which tell us that it was low entropy. The first part of the second law, which states that the entropy of a thermally isolated system can only increase is a trivial consequence of the equal prior probability postulate, if we restrict the notion of the entropy to systems in thermal equilibrium. The entropy of an isolated system in thermal equilibrium containing an amount of energy of E is:

$$S = k\log\left[\Omega(E)\right]$$

where $\Omega(E)$ is the number of quantum states in a small interval between E and $E+\delta E$. Here δE is a macroscopically small energy interval that is kept fixed. Strictly speaking this means that the entropy depends on the choice of δE. However, in the thermodynamic limit (i.e. in the limit of infinitely large system size), the specific entropy (entropy per unit volume or per unit mass) does not depend on δE.

Suppose we have an isolated system whose macroscopic state is specified by a number of variables. These macroscopic variables can, e.g., refer to the total volume, the positions of pistons in the system, etc. Then Ω will depend on the values of these variables. If a variable is not fixed, (e.g. we do not clamp a piston in a certain position), then because all the accessible states are equally likely in equilibrium, the free variable in equilibrium will be such that Ω is maximized as that is the most probable situation in equilibrium.

If the variable was initially fixed to some value then upon release and when the new equilibrium has been reached, the fact the variable will adjust itself so that Ω is maximized, implies that the entropy will have increased or it will have stayed the same (if the value at which the variable was fixed happened to be the equilibrium value).

The entropy of a system that is not in equilibrium can be defined as:

$$S = -k_B \sum_j P_j \log(P_j)$$

Here the P_j is the probabilities for the system to be found in the states libelled by the subscript j. In thermal equilibrium, the probabilities for states inside the energy interval δE are all equal to $1/\Omega$, and in that case the general definition coincides with the previous definition of S that applies to the case of thermal equilibrium.

Suppose we start from an equilibrium situation and we suddenly remove a constraint on a variable. Then right after we do this, there are a number Ω of accessible microstates, but equilibrium has not yet been reached, so the actual probabilities of the system being in some accessible state are not yet equal to the prior probability of $1/\Omega$. We have already seen that in the final equilibrium state, the entropy will have increased or have stayed the same relative to the previous equilibrium state. Boltzmann's H-theorem, however, proves that the entropy will increase continuously as a function of time during the intermediate out of equilibrium state.

Derivation of the Entropy Change for Reversible Processes

The second part of the Second Law states that the entropy change of a system undergoing a reversible process is given by:

$$dS = \frac{\delta Q}{T}$$

where the temperature is defined as:

$$\frac{1}{kT} \equiv \beta \equiv \frac{d\log[\Omega(E)]}{dE}$$

See here for the justification for this definition. Suppose that the system has some external parametre, x, that can be changed. In general, the energy eigenstates of the system will depend on x. According to the adiabatic theorem of quantum mechanics, in the limit of an infinitely slow change of the system's Hamiltonian, the system will stay in the same energy eigenstate and thus change its energy according to the change in energy of the energy eigenstate it is in.

The generalized force, X, corresponding to the external variable x is defined such that Xdx is the work performed by the system if x is increased by an amount dx. E.g., if x is the volume, then X is the pressure. The generalized force for a system known to be in energy eigenstate E_r is given by:

$$X = -\frac{dE_r}{dx}$$

Since the system can be in any energy eigenstate within an interval of δE, we define the generalized force for the system as the expectation value of the above expression:

$$X = -\left\langle \frac{dE_r}{dx} \right\rangle$$

To evaluate the average, we partition the $\Omega(E)$ energy eigenstates by counting how many of them have a value for $\frac{dE_r}{dx}$ within a range between Y and $Y + \delta Y$. Calling this number $\Omega_Y(E)$, we have:

$$\Omega(E) = \sum_Y \Omega_Y(E)$$

The average defining the generalized force can now be written:

$$X = -\frac{1}{\Omega(E)} \sum_Y Y\Omega_Y(E)$$

We can relate this to the derivative of the entropy w.r.t. x at constant energy E as follows. Suppose we change x to x + dx. Then $\Omega(E)$ will change because the energy eigenstates depend on x, causing energy eigenstates to move into or out of the range between E and $E + \delta E$. Let's focus again on the energy eigenstates for which $\frac{dE_r}{dx}$ lies within the range between Y and $Y + \delta Y$. Since these energy eigenstates increase in energy by Y dx, all such energy eigenstates that are in the interval ranging from E – Y dx to E move from below E to above E. There are

$$N_Y(E) = \frac{\Omega_Y(E)}{\delta E} Ydx$$

such energy eigenstates. If $Ydx \leq \delta E$, all these energy eigenstates will move into the range between E and $E + \delta E$ and contribute to an increase in Ω. The number of energy eigenstates that move from below $E + \delta E$ to above $E + \delta E$ is, of course, given by $N_Y(E + \delta E)$. The difference

$$N_Y(E) - N_Y(E + \delta E)$$

is thus the net contribution to the increase in Ω. Note that if Y dx is larger than δE there will be the energy eigenstates that move from below E to above $E + \delta E$. They are counted in both $N_Y(E)$ and $N_Y(E + \delta E)$, therefore the above expression is also valid in that case.

Expressing the above expression as a derivative w.r.t. E and summing over Y yields the expression:

$$\left(\frac{\partial \Omega}{\partial x}\right)_E = -\sum_Y Y \left(\frac{\partial \Omega_Y}{\partial E}\right)_x = \left(\frac{\partial (\Omega X)}{\partial E}\right)_x$$

The logarithmic derivative of Ω w.r.t. x is thus given by:

$$\left(\frac{\partial \log(\Omega)}{\partial x}\right)_E = \beta X + \left(\frac{\partial X}{\partial E}\right)_x$$

The first term is intensive, i.e. it does not scale with system size. In contrast, the last term scales as the inverse system size and will thus vanishes in the thermodynamic limit. We have thus found that:

$$\left(\frac{\partial S}{\partial x}\right)_E = \frac{X}{T}$$

Combining this with

$$\left(\frac{\partial S}{\partial E}\right)_x = \frac{1}{T}$$

Gives:

$$dS = \left(\frac{\partial S}{\partial E}\right)_x dE + \left(\frac{\partial S}{\partial x}\right)_E dx = \frac{dE}{T} + \frac{X}{T}dx = \frac{\delta Q}{T}$$

Derivation for Systems Described by the Canonical Ensemble

If a system is in thermal contact with a heat bath at some temperature T then, in equilibrium, the probability distribution over the energy eigenvalues are given by the canonical ensemble:

$$P_j = \frac{\exp\left(-\frac{E_j}{kT}\right)}{Z}$$

Here Z is a factor that normalizes the sum of all the probabilities to 1, this function is known as the partition function. We now consider an infinitesimal reversible change in the temperature and in the external parametres on which the energy levels depend. It follows from the general formula for the entropy:

$$S = -k\sum_j P_j \log(P_j)$$

that

$$dS = -k\sum_j \log(P_j)\, dP_j$$

Inserting the formula for P_j for the canonical ensemble in here gives:

$$dS = \frac{1}{T}\sum_j E_j dP_j = \frac{1}{T}\sum_j d\left(E_j P_j\right) - \frac{1}{T}\sum_j P_j dE_j = \frac{dE + \delta W}{T} = \frac{\delta Q}{T}$$

General Derivation from Unitarity of Quantum Mechanics

The time development operator in quantum theory is unitary, because the Hamiltonian is hermitian. Consequently, the transition probability matrix is doubly stochastic, which implies the Second Law of Thermodynamics. This derivation is quite general, based on the Shannon entropy, and does not require any assumptions beyond unitarity, which is universally accepted. It is a *consequence* of the irreversibility or singular nature of the general transition matrix.

Non-equilibrium States

It is only by convention, for the purposes of thermodynamic analysis, that any arbitrary occasion of space-time is said to be in thermodynamic equilibrium. In general, an occasion of space-time found in nature is not in thermodynamic equilibrium, read in the most stringent terms. In looser terms, nothing in the entire universe is or has ever been truly in exact thermodynamic equilibrium. If it is assumed, for the purposes of physical analysis, that one is dealing with a system in thermodynamic equilibrium, then statistically it is possible for that system to achieve moments of non-equilibrium. In some statistically unlikely events, hot particles "steal" the energy of cold particles, enough that the cold side gets colder and the hot side gets hotter, for a very brief time.

The physics involved in such events is beyond the scope of classical equilibrium thermodynamics, and is the topic of the fluctuation theorem (not to be confused with the fluctuation-dissipation theorem). This was first proved by Bochov and Kuzovlev, and later by Evans and Searles. It gives a numerical estimate of the probability that a system away from equilibrium will have a certain change in entropy over a certain amount of time. The theorem is proved with the exact time reversible dynamical equations of motion but assumes the Axiom of Causality, which is equivalent to assuming uncorrelated initial conditions (namely, uncorrelated past). Such events have been observed at a small enough scale where the likelihood of such a thing happening is significant. Quantitative predictions of this theorem have been confirmed in laboratory experiments by use of optical tweezers apparatus.

Arrow of Time

The second law of thermodynamics is a physical law that is not symmetric to reversal of the time direction.

The second law has been proposed to supply an explanation of the difference between moving forward and backwards in time, such as why the cause precedes the effect (the causal arrow of time).

Controversies

Maxwell's Demon: James Clerk Maxwell imagined one container divided into two parts, *A* and *B*. Both parts are filled with the same gas at equal temperatures and placed next to each other. Observing the molecules on both sides, an imaginary demon guards a trapdoor between the two parts. When a faster-than-average molecule from *A* flies towards the trapdoor, the demon opens it, and the molecule will fly from *A* to *B*. The average speed of the molecules in *B* will have increased while in *A* they will have slowed down on average. Since average molecular speed corresponds to temperature, the temperature decreases in *A* and increases in *B*, contrary to the second law of thermodynamics. One of the most famous responses to this question was suggested in 1929 by Leσ Szilárd and later by Léon Brillouin. Szilárd pointed out that a real-life Maxwell's demon would need to have some means of measuring molecular speed, and that the act of acquiring information would require an expenditure of energy. But later exceptions were found.

Loschmidt's Paradox

Loschmidt's paradox, also known as the reversibility paradox, is the objection that it should not be possible to deduce an irreversible process from time-symmetric dynamics. This puts the time reversal symmetry of nearly all known low-level fundamental physical processes at odds with any attempt to infer from them the second law of thermodynamics which describes the behaviour of macroscopic systems. Both of these are well-accepted principles in physics, with sound observational and theoretical support, yet they seem to be in conflict; hence the paradox.

Due to this paradox, derivations of the Second Law have to make an assumption regarding the past, namely that the system is uncorrelated at some time in the past, or - equivalently - that the entropy in the past was lower than in the future. This assumption is usually thought as a boundary condition, and thus the second Law is ultimately derived from the initial conditions of the Big Bang.

Gibbs Paradox

In statistical mechanics, a simple derivation of the entropy of an ideal gas based on the Boltzmann distribution yields an expression for the entropy which is not extensive (is not proportional to the amount of gas in question). This leads to an apparent paradox known as the Gibbs paradox, allowing, for instance, the entropy of closed systems to decrease, violating the second law of thermodynamics.

The paradox is averted by recognizing that the identity of the particles does not influence the entropy. In the conventional explanation, this is associated with an indistinguishability of the particles associated with quantum mechanics. However, a growing number of papers now take the perspective that it is merely the definition of entropy that is changed to ignore particle permutation (and thereby avert the paradox). The resulting equation for the entropy (of a classical ideal gas) is extensive, and is known as the Sackur-Tetrode equation.

Poincaré Recurrence Theorem

The Poincaré recurrence theorem states that certain systems will, after a sufficiently long time, return to a state very close to the initial state. The Poincaré recurrence time is the length of time elapsed until the recurrence, which is of the order of $\sim \exp(S/k)$. The result applies to physical systems in which energy is conserved. The Recurrence theorem apparently contradicts the Second law of thermodynamics, which says that large dynamical systems evolve irreversibly towards the state with higher entropy, so that if one starts with a low-entropy state, the system will never return to it. There are many possible ways to resolve this paradox, but none of them is universally accepted. The most reasonable argument is that for typical thermodynamical systems the recurrence time is so large (many many times longer than the lifetime of the universe) that, for all practical purposes, one cannot observe the recurrence.

Future of the Universe

It has been suggested in the past that since the entropy in the universe is continuously rising, the amount of free energy diminishes and the universe will arrive at a heat death, in which no work can be done and life cannot exist. However, an expanding universe is not in a thermodynamical equilibrium, and the simple considerations leading to the heat death scenario are no longer valid.

Taking the current view of the universe into account, it has been shown that the universe will probably exhibit a future in which all known energy sources (such as stars) will decay. Nevertheless, it may be the case that work in smaller and smaller energy scales will still be possible, so that "interesting things can continue to happen at ... increasingly low levels of energy".

PMM of Second Kind

Perpetual motion describes motion that continues indefinitely without any external source of energy. This is impossible in practice because of friction and other sources of energy loss. Furthermore, the term is often used to in a stronger sense to describe a perpetual motion machine of the first kind, a "hypothetical machine which, once activated, would continue to function and produce work" indefinitely with no input of energy. There is a scientific consensus that perpetual motion is impossible, as it would violate the first or second law of thermodynamics.

Nikola Tesla noted that we are immersed in a variety of energetic fields, so that cases of apparent perpetual motion can exist in nature, but such motions either are not truly perpetual or cannot be used to do work without changing the nature of the motion (as occurs in energy harvesting). For example, the motion or rotation of celestial bodies such as planets may appear perpetual, but are actually subjected to many forces such as solar winds, interstellar medium resistance, gravitation thermal radiation and electro-magnetic radiation. The flow of electric current in a superconducting loop may be perpetual and could be used as an energy storage medium, but following the principle of energy conservation the *source* of energy output would in fact originate from the *energy input* with which it was previously charged.

Machines which extract energy from seemingly perpetual sources—such as ocean currents—are capable of moving "perpetually" (for as long as that energy source itself endures), but they are not considered to be perpetual motion machines because they are consuming energy from an external source and are not isolated systems. Similarly, machines which comply with both laws of thermodynamics but access energy from obscure sources are sometimes referred to as perpetual motion machines, although they also do not meet the criteria for the name.

Despite the fact that successful perpetual motion devices are physically impossible in terms of the current understanding of the laws of physics, the pursuit of perpetual motion remains popular.

Basic Principles

There is a scientific consensus that perpetual motion in an isolated system violates either the first law of thermodynamics, the second law of thermodynamics, or both. The first law of thermodynamics is essentially a statement of conservation of energy. The second law can be phrased in several different ways, the most intuitive of which is that heat flows spontaneously from hotter to colder places; the most well known statement is that entropy tends to increase, or at the least stay the same; another statement is that no heat engine (an engine which produces work while moving heat from a high temperature to a low temperature) can be more efficient than a Carnot heat engine.

In other words:

1. In any isolated system, one cannot create new energy (first law of thermodynamics)
2. The output power of heat engines is always smaller than the input heating power. The rest of the energy is removed as heat at ambient temperature. The efficiency (this is the produced power divided by the input heating power) has a maximum, given by the Carnot efficiency. It is always lower than one
3. The efficiency of real heat engines is even lower than the Carnot efficiency due to irreversible processes.

The statements 2 and 3 only apply to heat engines. Other types of engines, which convert e.g. mechanical into electromagnetic energy, can, in principle, operate with 100% efficiency.

Machines which comply with both laws of thermodynamics by accessing energy from unconventional sources are sometimes referred to as perpetual motion machines, although they do not meet the standard criteria for the name. By way of example, clocks and other low-power machines, such as Cox's timepiece, have been designed to run on the differences in barometric pressure or temperature between night and day. These machines have a source of energy, albeit one which is not readily apparent so that they only seem to violate the laws of thermodynamics.

Machines which extract energy from seemingly perpetual sources - such as ocean currents - are indeed capable of moving "perpetually" until that energy source runs down. They are not considered to be perpetual motion machines because they are consuming energy from an external source and are not isolated systems.

Classification

One classification of perpetual motion machines refers to the particular law of thermodynamics the machines purport to violate:

- A perpetual motion machine of the first kind produces work without the input of energy. It thus violates the first law of thermodynamics: the law of conservation of energy.
- A perpetual motion machine of the second kind is a machine which spontaneously converts thermal energy into mechanical work. When the thermal energy is equivalent to the work done, this does not violate the law of conservation of energy. However it does violate the more subtle second law of thermodynamics. The signature of a perpetual motion machine of the second kind is that there is only one heat reservoir involved, which is being spontaneously cooled without involving a transfer of heat to a cooler reservoir. This conversion of heat into useful work, without any side effect, is impossible, according to the second law of thermodynamics.
- A more obscure category is a perpetual motion machine of the third kind, usually (but not always) defined as one that completely eliminates friction and other dissipative forces, to maintain motion forever (due to its mass inertia). *Third* in this case refers solely to the position in the above classification scheme, not the third law of thermodynamics. Although it is impossible to make such a machine, as dissipation can never be 100% eliminated in a mechanical system, it is nevertheless possible to get very close to this ideal. Such a machine would not serve as a source of energy but would have utility as a perpetual energy storage device.

Use of the term "Impossible" and Perpetual Motion

"Epistemic impossibility" describes things which absolutely cannot occur within our *current* formulation of the physical laws. This interpretation of the word "impossible" is what is intended in discussions of the impossibility of perpetual motion in a closed system.

The conservation laws are particularly robust from a mathematical perspective. Noether's theorem, which was proven mathematically in 1915, states that any conservation law can be derived from a corresponding continuous symmetry of the action of a physical system. For example, if the true laws of physics remain invariant over time then the conservation of energy follows. On the other hand, if the conservation laws are invalid, then the foundations of physics would need to change.

Scientific investigations as to whether the laws of physics are invariant over time use telescopes to examine the universe in the distant past to discover, to the limits of our measurements, whether ancient stars were identical to stars today. Combining different measurements such as spectroscopy, direct measurement of the speed of light in the past and similar measurements demonstrates that physics has remained substantially the same, if not identical, for all of observable history spanning billions of years.

The principles of thermodynamics are so well established, both theoretically and experimentally, that proposals for perpetual motion machines are universally met with disbelief on the part of physicists. Any proposed perpetual motion design offers a potentially instructive challenge to physicists: one is almost completely certain that it cannot work, so one must explain *how* it fails to work. The difficulty (and the value) of such an exercise depends on the subtlety of the proposal; the best ones tend to arise from physicists' own thought experiments and often shed light upon certain aspects of physics. So, for example, the thought experiment of a Brownian ratchet as a perpetual motion machine was first discussed by Gabriel Lippmann in 1900 but it was not until 1912 that Marian Smoluchowski gave an adequate explanation for why it cannot work. However, during that twelve-year period scientists did not believe that the machine was possible. They were merely unaware of the exact mechanism by which it would inevitably fail.

The law that entropy always increases, holds, I think, the supreme position among the laws of Nature. If someone points out to you that your pet theory of the universe is in disagreement with Maxwell's equations — then so much the worse for Maxwell's equations. If it is found to be contradicted by observation — well, these experimentalists do bungle things sometimes. But if your theory is found to be against the second law of thermodynamics I can give you no hope; there is nothing for it but to collapse in deepest humiliation.

—Sir Arthur Stanley Eddington, The Nature of the Physical World (1927)

In the mid 19th-century Henry Dircks investigated the history of perpetual motion experiments, writing a vitriolic attack on those who continued to attempt what he believed to be impossible:

> *"There is something lamentable, degrading, and almost insane in pursuing the visionary schemes of past ages with dogged determination, in paths of learning which have been investigated by superior minds, and with which such adventurous persons are totally unacquainted. The history of Perpetual Motion is a*

history of the fool-hardiness of either half-learned, or totally ignorant persons."

—Henry Dircks, *Perpetuum Mobile: Or, A History of the Search for Self-motive* (1861)

Techniques

One day man will connect his apparatus to the very wheelwork of the universe [...] and the very forces that motivate the planets in their orbits and cause them to rotate will rotate his own machinery.

—*Nikola Tesla*

Some common ideas reoccur repeatedly in perpetual motion machine designs. Many ideas that continue to appear today were stated as early as 1670 by John Wilkins, Bishop of Chester and an official of the Royal Society. He outlined three potential sources of power for a perpetual motion machine, "Chymical Extractions", "Magnetical Virtues" and "the Natural Affection of Gravity".

The seemingly mysterious ability of magnets to influence motion at a distance without any apparent energy source has long appealed to inventors. One of the earliest examples of a magnetic motor was proposed by Wilkins and has been widely copied since: it consists of a ramp with a magnet at the top, which pulled a metal ball up the ramp. Near the magnet was a small hole that was supposed to allow the ball to drop under the ramp and return to the bottom, where a flap allowed it to return to the top again.

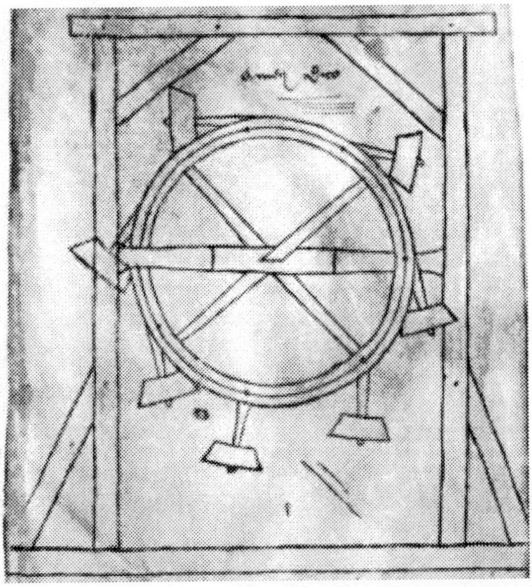

Figure: *Perpetuum Mobile of Villard de Honnecourt (about 1230).*

The device simply could not work: any magnet strong enough to pull the ball up the ramp would necessarily be too powerful to allow it to drop through the hole. Faced with this problem, more modern versions typically use a series of ramps and magnets, positioned so the ball is to be handed off from one magnet to another as it moves. The problem remains the same.

Gravity also acts at a distance, without an apparent energy source. But to get energy out of a gravitational field (for instance, by dropping a heavy object, producing kinetic energy as it falls) one has to put energy in (for instance, by lifting the object up), and some energy is always dissipated in the process. A typical application of gravity in a perpetual motion machine is Bhaskara's wheel in the 12th century, whose key idea is itself a recurring theme, often called the overbalanced wheel: Moving weights are attached to a wheel in such a way that they fall to a position further from the wheel's centre for one half of the wheel's rotation, and closer to the centre for the other half. Since weights further from the centre apply a greater torque, the result is (or would be, if such a device worked) that the wheel rotates forever. The moving weights may be hammers on pivoted arms, or rolling balls, or mercury in tubes; the principle is the same.

Figure: *Perpetual motion wheels from a drawing of Leonardo da Vinci.*

Yet another theoretical machine involves a frictionless environment for motion. This involves the use of diamagnetic or electromagnet levitation to float an object. This is done in a vacuum to eliminate air friction and friction from an axle. The levitated object is then free to rotate around its centre of gravity without interference.

However, this machine has no practical purpose because the rotated object cannot do any work as work requires the levitated object to cause motion in other objects, bringing friction into the problem. Furthermore, a *perfect* vacuum is an unattainable goal since both the container and the object itself would slowly vaporize, thereby degrading the vacuum.

To extract work from heat, thus producing a perpetual motion machine of the second kind, the most common approach (dating back at least to Maxwell's demon) is *unidirectionality*. Only molecules moving fast enough and in the right direction are allowed through the demon's trap door. In a Brownian ratchet, forces tending to turn the ratchet one way are able to do so while forces in the other direction are not. A diode in a heat bath allows through currents in one direction and not the other. These schemes typically fail in two ways: either maintaining the unidirectionality costs energy (It would require Maxwell's demon to perform more thermodynamic work to gauge the speed of the molecules than the amount of energy gained by the difference of temperature caused) or the unidirectionality is an illusion and occasional big violations make up for the frequent small non-violations (the Brownian ratchet will be subject to internal Brownian forces and therefore will sometimes turn the wrong way).

Buoyancy is another frequently-misunderstood phenomenon. Some proposed perpetual-motion machines miss the fact that to push a volume of air down in a fluid takes the same work as to raise a corresponding volume of fluid up against gravity. These types of machines may involve two chambers with pistons, and a mechanism to squeeze the air out of the top chamber into the bottom one, which then becomes buoyant and floats to the top. The squeezing mechanism in these designs would not be able to do enough work to move the air down, or would leave no excess work available to be extracted.

Patents

Proposals for such inoperable machines have become so common that the United States Patent and Trademark Office (USPTO) has made an official policy of refusing to grant patents for perpetual motion machines without a working model. The USPTO Manual of Patent Examining Practice states:

With the exception of cases involving perpetual motion, a model is not ordinarily required by the Office to demonstrate the operability of a device. If operability of a device is questioned, the applicant must establish it to the satisfaction of the examiner, but he or she may choose his or her own way of so doing.

And, Further, That: A rejection [of a patent application] on the ground of lack of utility includes the more specific grounds of inoperativeness, involving perpetual motion. A rejection under 35 U.S.C. 101 for lack of utility should not be based on grounds that the invention is frivolous, fraudulent or against public policy.

The filing of a patent application is a clerical task, and the USPTO will not refuse filings for perpetual motion machines; the application will be filed and then most probably rejected by the patent examiner, after he has done a formal examination. Even if a patent is granted, it does not mean that the invention actually works, it just means that the examiner believes that it works, or was unable to figure out why it would not work.

The USPTO maintains a collection of Perpetual Motion Gimmicks. The United Kingdom Patent Office has a specific practice on perpetual motion; Section 4.05 of the UKPO Manual of Patent Practice states:

Processes or articles alleged to operate in a manner which is clearly contrary to well-established physical laws, such as perpetual motion machines, are regarded as not having industrial application.

Examples of decisions by the UK Patent Office to refuse patent applications for perpetual motion machines include:

Decision BL O/044/06, John Frederick Willmott's application no. 0502841

Decision BL O/150/06, Ezra Shimshi's application no. 0417271

The European Patent Classification (ECLA) has classes including patent applications on perpetual motion systems: ECLA classes "F03B17/04: Alleged perpetua mobilia ..." and "F03B17/00B: [... machines or engines] (with closed loop circulation or similar : ... Installations wherein the liquid circulates in a closed loop; Alleged perpetua mobilia of this or similar kind ...".

Apparent Perpetual Motion Machines

While "perpetual motion" can only exist in isolated systems, and true isolated systems don't exist, there aren't any real "perpetual motion" devices. However there are concepts and technical drafts that propose "perpetual motion", but on closer analysis it's revealed that they actually "consume" some sort of natural resource or latent energy, such as the phase changes of water or other fluids or small natural temperature gradients, or simply can't sustain indefinite operation. In general, extracting large amounts of work using these devices is difficult to impossible.

Resource Consuming

Some examples of such devices include:

- The drinking bird toy functions using small ambient temperature gradients and evaporation.
- A capillarity based water pump functions using small ambient temperature gradients and vapour pressure differences.
- A Crookes radiometre consists of a partial vacuum glass container with a lightweight propeller moved by (light-induced) temperature gradients.
- Any device picking up minimal amounts of energy from the natural electromagnetic radiation around it, such as a solar powered motor.
- Any device powered by changes in air pressure, such as some clocks. The motion leeches energy from moving air which in turn gained its energy from being acted on.
- The Atmos clock uses changes in the vapour pressure of ethyl chloride with temperature to wind the clock spring.
- A device powered by radioactive decay from an isotope with a relatively long half-life; such a device could plausibly operate for hundreds or thousands of years.
- The Oxford Electric Bell and Karpen Pile driven by dry pile batteries.
- Cox's timepiece and the Beverly Clock powered by changes in atmospheric pressure.

Low Friction

- In flywheel energy storage, "modern flywheels can have a zero-load rundown time measurable in years."
- Once spun up, objects in the vacuum of space—stars, black holes, planets, moons, spin-stabilized satellites, etc.—continue spinning almost indefinitely with no further energy input. Tide on Earth is dissipating the gravitational energy of the Moon/Earth system at an average rate of about 3.75 terawatts.
- In certain quantum-mechanical systems (such as superfluidity and superconductivity), dissipation-free "motion" is possible.

Thought Experiments

In some cases a thought (or "gedanken") experiment appears to suggest that perpetual motion may be possible through accepted and understood physical processes. However, in all cases, a flaw has been found when all of the relevant physics is considered. Examples include:

- Maxwell's Demon: This was originally proposed to show that the Second Law of Thermodynamics applied in the statistical sense only, by postulating a "demon" that could select energetic molecules and extract their energy. Subsequent analysis (and experiment) have shown there is no way to physically implement such a system that does not result in an overall increase in entropy.
- Brownian Ratchet: In this thought experiment, one imagines a paddle wheel connected to a ratchet. Brownian motion would cause surrounding gas molecules to strike the paddles, but the ratchet would only allow it to turn in one direction. A more thorough analysis showed that when a physical ratchet was considered at this molecular scale, Brownian motion would also affect the ratchet and cause it to randomly fail resulting in no net gain. Thus, the device would not violate the Laws of thermodynamics.

Carnot's Principle

Carnot's theorem, developed in 1824, also called *Carnot's rule* is a principle that specifies limits on the maximum efficiency any heat engine can obtain, which thus solely depends on the difference between the hot and cold temperature reservoirs.

Carnot's theorem states:

- All heat engines between two heat reservoirs are less efficient than a Carnot engine operating between the same reservoirs.
- Every Carnot engine between a pair of heat reservoirs is equally efficient, regardless of the working substance employed or the operation details.

The formula for this maximum efficiency is

$$\eta_{\max} = \eta_{\text{Carnot}} = 1 - \frac{T_C}{T_H}$$

where TC is the absolute temperature of the cold reservoir, T_H is the absolute temperature of the hot reservoir, and the efficiency η is the ratio of the work done by the engine to the heat drawn out of the hot reservoir.

Based on modern thermodynamics, Carnot's theorem is a result of the second law of thermodynamics. Historically, however, it was based on contemporary caloric theory and preceded the establishment of the second law.

Proof

The theorem may be proved in the following way for the cases of the irreversible and the reversible heat engines.

Irreversible Engine

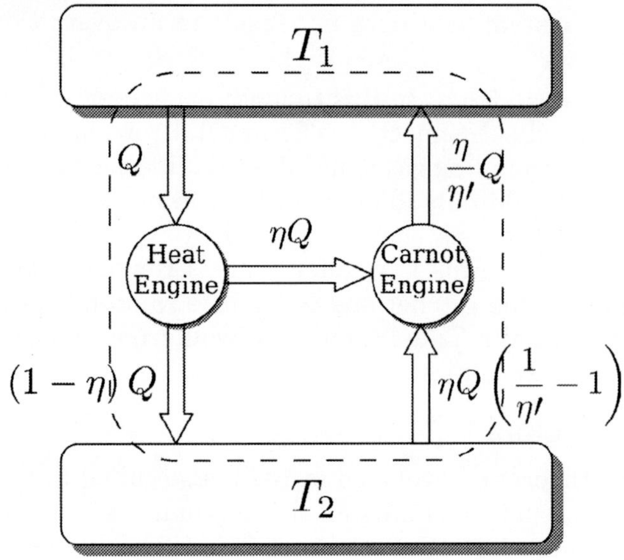

Figure: *Proving Carnot's Theorem*

The derivation assumes an irreversible heat engine that has an efficiency η, operating between reservoirs *T1 and* T_2. It is combined with a reversed Carnot engine that has efficiency η′ as shown by the diagram to the right.

If $\eta = \eta'$, then the combined engine does nothing in a cycle, which contradicts the irreversibility.

If $\eta > \eta'$, then the net effect of the combined engine is draining heat

$$\Delta Q = \eta Q\left(\frac{1}{\eta'} - 1\right) - (1-\eta)Q = Q\left(\frac{\eta}{\eta'} - 1\right) > 0$$

from the colder reservoir T_2 and releases the same amount to the hotter reservoir T_1 without affecting any other changes. This is clearly in violation of the Clausius statement of the second law.

The conclusion is that the only possibility remaining is $\eta < \eta'$.

Reversible Engine

In the case of a reversible heat engine, the proof may be conducted similarly, in that $\eta \leqslant \eta'$. The said engine is reversed and combined with a regular Carnot engine. In a similar manner the conclusion is obtained: $\eta' \leqslant \eta$. Thus,

$$\eta = \eta'.$$

Definition of Thermodynamic Temperature

The efficiency of the engine is the work divided by the heat introduced to the system or

$$\eta = \frac{w_{cy}}{q_H} = \frac{q_H - q_C}{q_H} = 1 - \frac{q_C}{q_H} \qquad (1)$$

where w_{cy} is the work done per cycle. Thus, the efficiency depends only on q_C/q_H. Because all reversible engines operating between the same heat reservoirs are equally efficient, any reversible heat engine operating between temperatures T_1 and T_2 must have the same efficiency, meaning, the efficiency is the function of the temperatures only:

$$\frac{q_C}{q_H} = f(T_H, T_C) \qquad (2)$$

In addition, a reversible heat engine operating between temperatures T_1 and T_3 must have the same efficiency as one consisting of two cycles, one between T_1 and another (intermediate) temperature T_2, and the second between T_2andT_3. This can only be the case if

$$f(T_1, T_3) = \frac{q_3}{q_1} = \frac{q_2 q_3}{q_1 q_2} = f(T_1, T_2) f(T_2, T_3).$$

Specializing to the case that T_1 is a fixed reference temperature: the temperature of the triple point of water. Then for any T2 and T_3,

$$f(T_2, T_3) = \frac{f(T_1, T_3)}{f(T_1, T_2)} = \frac{273.16 \cdot f(T_1, T_3)}{273.16 \cdot f(T_1, T_2)}.$$

Therefore, if thermodynamic temperature is defined by

$$T = 273.16 \cdot f(T_1, T)$$

then the function *f*, viewed as a function of thermodynamic temperature, is

$$f(T_2, T_3) = \frac{T_3}{T_2},$$

and the reference temperature T_1 has the value 273.16. (Of course any reference temperature and any positive numerical value could be used—the choice here corresponds to the Kelvin scale.)

It follows immediately that

$$\frac{q_C}{q_H} = f(T_H, T_C) = \frac{T_C}{T_H} \tag{3}$$

Substituting Equation 3 back into Equation 1 gives a relationship for the efficiency in terms of temperature:

$$\eta = 1 - \frac{q_C}{q_H} = 1 - \frac{T_C}{T_H} \tag{4}$$

Applicability to Fuel Cells and Batteries

Since fuel cells and batteries can generate useful power when all components of the system are at the same temperature ($T = T_H = T_C$), they are clearly not limited by Carnot's theorem, which states that no power can be generated when $T_H = T_C$. This is because Carnot's theorem applies to engines converting thermal energy to work, whereas fuel cells and batteries instead convert chemical energy to work. Nevertheless, the second law of thermodynamics still provides restrictions on fuel cell and battery energy conversion.

Carnot Cycle

The Carnot cycle is a theoretical thermodynamic cycle proposed by Nicolas Léonard Sadi Carnot in 1823 and expanded by in the 1830s and 40s. It can be shown that it is the most efficient cycle for converting a given amount of thermal energy into work, or conversely, creating a temperature difference (e.g. refrigeration) by doing a given amount of work. Every single thermodynamic system exists in a particular state. When a system is taken through a series of different states and finally returned to its initial state, a thermodynamic cycle is said to have occurred. In the process of going through this cycle, the system may perform work on its surroundings, thereby acting as a heat engine. A system undergoing a Carnot cycle is called a Carnot heat engine, although such a 'perfect' engine is only a theoretical limit and cannot be built in practice.

Stages of the Carnot Cycle

The Carnot cycle when acting as a heat engine consists of the following steps:

1. Reversible isothermal expansion of the gas at the "hot" temperature, T_1 (isothermal heat addition or absorption). During this step the gas is allowed to expand and it does work on the surroundings. The temperature of the gas does not change during the process, and thus the expansion is isothermal.

The gas expansion is propelled by absorption of heat energy Q_1 and of entropy $\Delta S = Q_1 / T_1$ from the high temperature reservoir.

2. Isentropic (reversible adiabatic) expansion of the gas (isentropic work output). For this step the piston and cylinder are assumed to be thermally insulated, thus they neither gain nor lose heat. The gas continues to expand, doing work on the surroundings, and losing an equivalent amount of internal energy. The gas expansion causes it to cool to the "cold" temperature, T_2. The entropy remains unchanged.
3. Reversible isothermal compression of the gas at the "cold" temperature, T_2. (isothermal heat rejection). Now the surroundings do work on the gas, causing an amount of heat energy Q_2 and of entropy $\Delta S = Q_2 / T_2$ to flow out of the gas to the low temperature reservoir. (This is the same amount of entropy absorbed in step 1, as can be seen from the Clausius inequality.)
4. Isentropic compression of the gas (isentropic work input). Once again the piston and cylinder are assumed to be thermally insulated. During this step, the surroundings do work on the gas, increasing its internal energy and compressing it, causing the temperature to rise to T_1. The entropy remains unchanged. At this point the gas is in the same state as at the start of step 1.

The Pressure-volume Graph

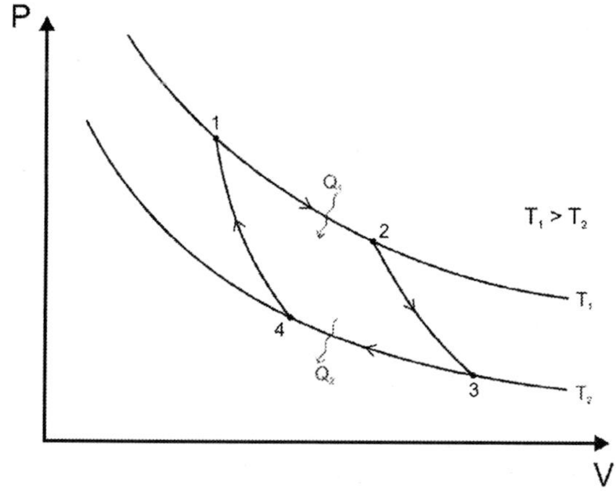

Figure: *A Carnot cycle illustrated on a pressure-volume graph to illustrate the work done.*

When the Carnot cycle is plotted on a pressure-volume graph, the isothermal stages follow the isotherm lines for the working fluid, adiabatic stages move between isotherms and the area bounded by the complete cycle path represents the total work that can be done during one cycle.

Properties and Significance

The temperature-entropy diagram:

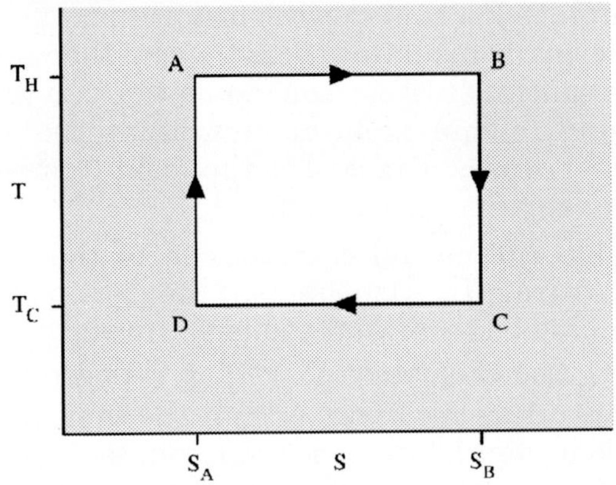

Figure: *A Carnot cycle acting as a heat engine, illustrated on a temperature-entropy diagram. The cycle takes place between a hot reservoir at temperature T_H and a cold reservoir at temperature T_C. The vertical axis is temperature, the horizontal axis is entropy.*

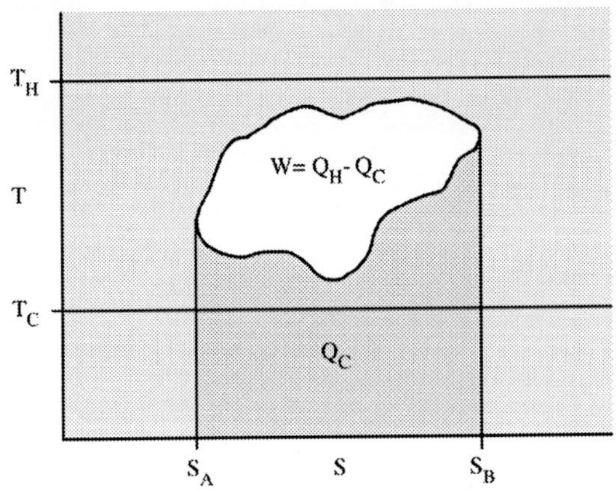

Figure: *A generalized thermodynamic cycle taking place between a hot reservoir at temperature TH and a cold reservoir at temperature TC. By the second law of thermodynamics, the cycle cannot extend outside the temperature band from TC to TH. The area in red QC is the amount of energy exchanged between the system and the cold reservoir. The area in white W is the amount of work energy exchanged by the system with its surroundings. The amount of heat exchanged with the hot reservoir is the sum of the two. If the system is behaving as an engine, the process moves clockwise around the loop, and moves counter-clockwise if it is behaving as a refrigerator. The efficiency of the cycle is the ratio of the white area (work) divided by the sum of the white and red areas (heat absorbed from the hot reservoir).*

The behaviour of a Carnot engine or refrigerator is best understood by using a temperature-entropy diagram (TS diagram), in which the thermodynamic state is specified by a point on a graph with entropy (S) as the horizontal axis and temperature (T) as the vertical axis. For a simple system with a fixed number of particles, any point on the graph will represent a particular state of the system. A thermodynamic process will consist of a curve connecting an initial state (A) and a final state (B). The area under the curve will be:

$$Q = \int_A^B T\,dS \tag{1}$$

which is the amount of thermal energy transferred in the process. If the process moves to greater entropy, the area under the curve will be the amount of heat absorbed by the system in that process. If the process moves towards lesser entropy, it will be the amount of heat removed. For any cyclic process, there will be an upper portion of the cycle and a lower portion. For a clockwise cycle, the area under the upper portion will be the thermal energy absorbed during the cycle, while the area under the lower portion will be the thermal energy removed during the cycle. The area inside the cycle will then be the difference between the two, but since the internal energy of the system must have returned to its initial value, this difference must be the amount of work done by the system over the cycle. Referring to figure, mathematically, for a reversible process we may write the amount of work done over a cyclic process as:

$$W = \oint P dV = \oint (dQ - dU) = \oint (TdS - dU) \tag{2}$$

Since dU is an exact differential, its integral over any closed loop is zero and it follows that the area inside the loop on a T-S diagram is equal to the total work performed if the loop is traversed in a clockwise direction, and is equal to the total work done on the system as the loop is traversed in a counterclockwise direction.

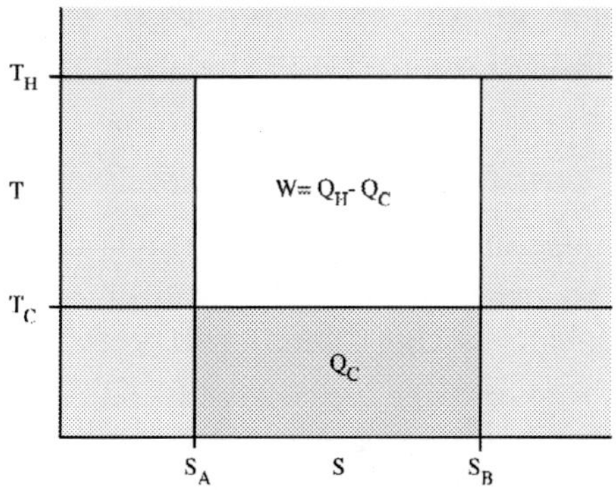

***Figure:** A Carnot cycle taking place between a hot reservoir at temperature* T*H and a cold reservoir at temperature* TC.

The Carnot Cycle

Evaluation of the above integral is particularly simple for the Carnot cycle. The amount of energy transferred as work is

$$W = \oint PdV = (T_H - T_C)(S_B - S_A)$$

The total amount of thermal energy transferred between the hot reservoir and the system will be

$$Q_H = T_H(S_B - S_A)$$

and the total amount of thermal energy transferred between the system and the cold reservoir will be

$$Q_C = T_C(S_B - S_A)$$

The efficiency η is defined to be:

$$\eta = \frac{W}{Q_H} = 1 - \frac{T_C}{T_H} \quad (3)$$

where

W is the work done by the system (energy exiting the system as work),

Q_H is the heat put into the system (heat energy entering the system),

T_C is the absolute temperature of the cold reservoir, and

T_H is the absolute temperature of the hot reservoir.

S_B is the maximum system entropy

S_A is the minimum system entropy

This efficiency makes sense for a heat engine, since it is the fraction of the heat energy extracted from the hot reservoir and converted to mechanical work. A Rankine cycle is usually the practical approximation.

The Reversed Carnot Cycle

The Carnot heat-engine cycle described is a totally reversible cycle. Therefore, all the processes that comprised it can be reversed, in which case it becomes the Carnot refrigeration cycle. This time, the cycle remains exactly the same, except that the directions of any heat and work interactions are reversed: Heat is absorbed from the low-temperature reservoir, heat in is rejected to a high-temperature reservoir, and a work input is required to accomplish all this. The P-

V diagram of the reversed Carnot cycle is the same as for the Carnot cycle, except that the directions of the processes are reversed.

Carnot's Theorem

It can be seen from the above diagram, that for any cycle operating between temperatures T_H and T_C, none can exceed the efficiency of a Carnot cycle.

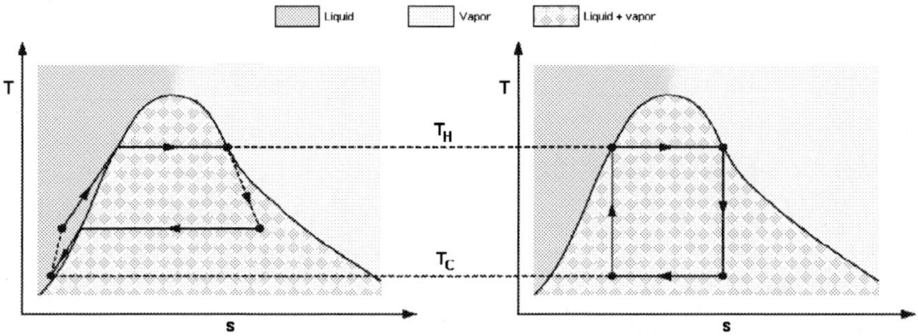

***Figure:** A real engine (left) compared to the Carnot cycle (right). The entropy of a real material changes with temperature. This change is indicated by the curve on a T-S diagram. For this figure, the curve indicates a vapour-liquid equilibrium. Irreversible systems and losses of heat (for example, due to friction) prevent the ideal from taking place at every step.*

Carnot's theorem is a formal statement of this fact: *No engine operating between two heat reservoirs can be more efficient than a Carnot engine operating between those same reservoirs.* Thus, Equation 3 gives the maximum efficiency possible for any engine using the corresponding temperatures. A corollary to Carnot's theorem states that: *All reversible engines operating between the same heat reservoirs are equally efficient.* Rearranging the right side of the equation gives what may be a more easily understood form of the equation. Namely that the theoretical maximum efficiency of a heat engine equals the difference in temperature between the hot and cold reservoir divided by the absolute temperature of the hot reservoir. To find the absolute temperature in kelvin, add 273.15 degrees to the Celsius temperature. Looking at this formula an interesting fact becomes apparent. Lowering the temperature of the cold reservoir will have more effect on the ceiling efficiency of a heat engine than raising the temperature of the hot reservoir by the same amount. In the real world, this may be difficult to achieve since the cold reservoir is often an existing ambient temperature.

In other words, maximum efficiency is achieved if and only if no new entropy is created in the cycle. Otherwise, since entropy is a state function, the required dumping of heat into the environment to dispose of excess entropy leads to a reduction in efficiency. So Equation 3 gives the efficiency of any reversibleheat engine. In mesoscopic heat engines, work per cycle of operation fluctuates due to thermal noise. For the case when work and heat fluctuations are counted, there is exact equality that relates average of exponents of work performed by any heat engine and the heat transfer from the hotter heat bath. This relation transforms the Carnot's inequality into exact equality that is applied to an arbitrary heat engine coupled to two heat reservoirs and operating at arbitrary rate.

Efficiency of Real Heat Engines

Carnot realized that in reality it is not possible to build a thermodynamically reversible engine, so real heat engines are less efficient than indicated by Equation 3. In addition, real engines that operate along this cycle are rare. Nevertheless, Equation 3 is extremely useful for determining the maximum efficiency that could ever be expected for a given set of thermal reservoirs. Although Carnot's cycle is an idealisation, the expression of Carnot efficiency is still useful. Consider the average temperatures,

$$\langle T_H \rangle = \frac{1}{\Delta S} \int_{Q_{in}} T dS$$

$$\langle T_C \rangle = \frac{1}{\Delta S} \int_{Q_{out}} T dS$$

at which heat is input and output, respectively. Replace T_H and T_C in Equation (3) by)$\langle T_H \rangle$ and)$\langle T_C \rangle$ respectively.

For the Carnot cycle, or its equivalent,)$\langle T_H \rangle$ is the highest temperature available and)$\langle T_C \rangle$ the lowest. For other less efficient cycles,)$\langle T_H \rangle$ will be lower than T_H, and)$\langle T_C \rangle$ will be higher than T_C. This can help illustrate, for example, why a reheater or a regenerator can improve the thermal efficiency of steam power plants—and why the thermal efficiency of combined-cycle power plants (which incorporate gas turbines operating at even higher temperatures) exceeds that of conventional steam plants.

Clausius Inequality

The Clausius theorem (1855) states that for a system undergoing a cyclic process (i.e. a process which ultimately returns a system to its original state):

$$\oint \frac{\delta Q}{T} \leq 0,$$

where δQ is the amount of heat absorbed by the system. The equality holds in the reversible case and the inequality holds in the irreversible case. The reversible case is used to introduce the entropy state function. This is because in cyclic process the variation of a state function is zero.

History

The Clausius Theorem is a mathematical explanation of the Second Law of Thermodynamics. Also referred to as the "Inequality of Clausius", the theorem was developed by Rudolph Clausius who intended to explain the relationship between the heat flow in a system and the entropy of the system and its surroundings. Clausius developed this in his efforts to explain entropy and define it quantitatively. In more direct terms, the theorem gives us a way to determine if a cyclical process is reversible or irreversible. The Clausius Theorem provides a quantitative formula for understanding the second law.

Clausius was one of the first to work on the idea of entropy and is even responsible for giving it that name. What is now known as the Clausius Theorem was first published in 1862 in Clausius' sixth memoir, "On the Application of the Theorem of the Equivalence of Transformations to Interior Work". Clausius sought to show a proportional relationship between entropy and the energy flow by heating (δQ) into a system. In a system, this heat energy can be transformed into work, and work can be transformed into heat through a cyclical process. Clausius writes that "The algebraic sum of all the transformations occurring in a cyclical process can only be less than zero, or, as an extreme case, equal to nothing." In other words, the equation

$$\oint \frac{\delta Q}{T} = 0$$

where δQ is energy flow into the system due to heating and T is absolute temperature of the body when that energy is absorbed is found to be true for any process that is cyclical and reversible. Clausius then took this a step further and determined that the following equation must be found true for any cyclical process that is possible, reversible or not. This equation is the "Clausius Inequality".

$$\oint \frac{\delta Q}{T} \leq 0$$

Now that this is known, there must be a relation developed between the Clausius Inequality and entropy. The amount of entropy S added to the system during the cycle is defined as

$$\Delta S = \oint \frac{\delta Q}{T}$$

It has been determined, as stated in the second law of thermodynamics, that the entropy is a state function: It depends only upon the state that the system is in, and not what path the system took to get there. This is in contrast to the amount of energy added as heat (δQ) and as work (δW), which may vary depending on the path. In a cyclic process, therefore, the entropy of the system at the beginning of the cycle must equal the entropy at the end of the cycle. In the irreversible case, entropy will be created in the system, and more entropy must be extracted than was added $(\Delta S < 0)$ in order to return the system to its original state. In the reversible case, no entropy is created and the amount of entropy added is equal to the amount extracted $(\Delta S = 0)$.

If the amount of energy added by heating can be measured during the process, and the temperature can be measured during the process, the Clausius inequality can be used to determine whether the process is reversible or irreversible by carrying out the integration in the Clausius inequality.

Proof

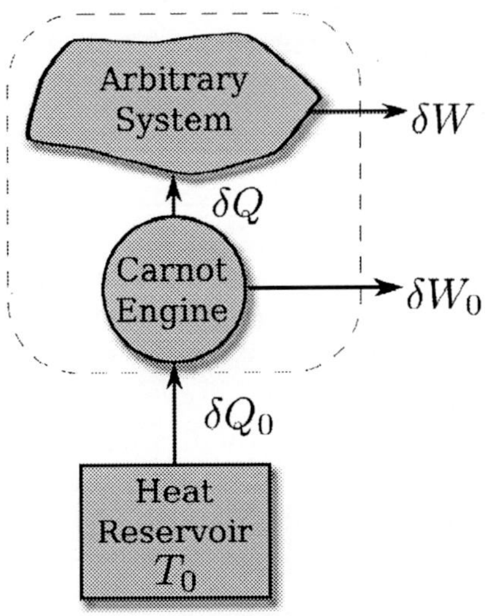

Figure: *Proving Clausius Inequality*

Suppose a system absorbs heat δQ at temperature T. Since the value of $\frac{\delta Q}{T}$ does not depend on the details of how the heat is transferred, we can assume it is from a Carnot engine, which in turn absorbs heat δQ_0 from a heat reservoir with constant temperature T_0.

According to the nature of Carnot cycle,

$$\frac{\delta Q}{T} = \frac{\delta Q_0}{T_0}$$

$$\Rightarrow \delta Q_0 = T_0 \frac{\delta Q}{T}$$

Therefore in one cycle, the total heat absorbed from the reservoir is

$$Q_0 = T_0 \oint \frac{\delta Q}{T}$$

Since after a cycle, the system and the Carnot engine as a whole return to its initial status, the difference of the internal energy is zero. Thus according to First Law of Thermodynamics,

$$Q_0 = \Delta U + W + W_0 = W + W_0 = W_{\text{total}}$$

According to the Kelvin-Planck statement of Second Law of thermodynamics, we cannot drain heat from one reservoir and convert them entirely into work without making any other changes, so

$$W_{\text{total}} \leq 0$$

Combine all the above and we get Clausius inequality

$$\oint \frac{\delta Q}{T} \leq 0$$

If the system is reversible, then reverse its path and do the experiment again we can get

$$-\oint \frac{\delta Q}{T} \leq 0$$

Thus

$$\oint \frac{\delta Q}{T} = 0$$

Entropy

Entropy is a measure of the number of specific ways in which a system may be arranged, often taken to be a measure of disorder. The entropy of an isolated system never decreases, because isolated systems spontaneously evolve towards thermodynamic equilibrium, which is the state of maximum entropy.

Entropy was originally defined for a thermodynamically reversible process as

$$\Delta S = \int \frac{dQ_{rev}}{T}$$

where the entropy (S) is found from the uniform thermodynamic temperature (T) of a closed system divided into an incremental reversible transfer of heat into that system (dQ). The above definition is sometimes called the macroscopic definition of entropy because it can be used without regard to any microscopic picture of the contents of a system. In thermodynamics, entropy has been found to be more generally useful and it has several other formulations. Entropy was discovered when it was noticed to be a quantity that behaves as a function of state. Entropy is an extensive property, but it is often given as an intensive property of specific entropy as entropy per unit mass or entropy per mole.

In the modern microscopic interpretation of entropy in statistical mechanics, entropy is the amount of additional information needed to specify the exact physical state of a system, given its thermodynamic specification. The role of thermodynamic entropy in various thermodynamic processes can thus be understood by understanding how and why that information changes as the system evolves from its initial condition. It is often said that entropy is an expression of the disorder, or randomness of a system, or of our lack of information about it (which on some views of probability, amounts to the same thing as randomness). The second law is now often seen as an expression of the fundamental postulate of statistical mechanics via the modern definition of entropy.

History

The analysis which led to the concept of entropy began with the work of French mathematician Lazare Carnot who in his 1803 paper *Fundamental Principles of Equilibrium and Movement* proposed that in any machine the accelerations and shocks of the moving parts represent losses of *moment of activity*. In other words, in any natural process there exists an inherent tendency towards the dissipation of useful energy. Building on this work, in 1824 Lazare's son Sadi Carnot published *Reflections on the Motive Power of Fire* which posited that in all heat-engines whenever "caloric", or what is now known as heat, falls through a temperature difference, work or motive power can be produced from the actions of the "fall of caloric" between a hot and cold body. He made the analogy with that of how water falls in a water wheel. This was an early insight into the second law of thermodynamics. Carnot based his views of heat partially on the early 18th century "Newtonian hypothesis" that both heat and light were types of indestructible forms of matter, which are attracted and repelled by other matter, and partially on the contemporary views of Count Rumford who showed (1789) that heat could be created by friction as when cannon bores are machined. Carnot reasoned that if the body of the working substance, such as a body of steam, is returned to its original state at the end of a complete engine cycle, that "no change occurs in the condition of the working body".

The first law of thermodynamics, formalized based on the heat-friction experiments of James Joule in 1843, deals with the concept of energy, which is conserved in all processes; the first law, however, is unable to quantify the effects of friction and dissipation.

In the 1850s and 1860s, German physicist Rudolf Clausius objected to the supposition that no change occurs in the working body, and gave this "change" a mathematical interpretation by questioning the nature of the inherent loss of usable heat when work is done, e.g. heat produced by friction. Clausius described entropy as the *transformation-content*, i.e. dissipative energy use, of a thermodynamic system or working body of chemical speciesduring a change of state. This was in contrast to earlier views, based on the theories of Isaac Newton, that heat was an indestructible particle that had mass.

Later, scientists such as Ludwig Boltzmann, Josiah Willard Gibbs, and James Clerk Maxwell gave entropy a statistical basis. In 1877 Boltzmann visualized a probabilistic way to measure the entropy of an ensemble of ideal gas particles, in which he defined entropy to be proportional to the logarithm of the number of microstates such a gas could occupy. Henceforth, the essential problem in statistical thermodynamics, i.e. according to Erwin Schrödinger, has been to determine the distribution of a given amount of energy E over N identical systems. Carathéodory linked entropy with a mathematical definition of irreversibility, in terms of trajectories and integrability.

Definitions and Descriptions

Any method involving the notion of entropy, the very existence of which depends on the second law of thermodynamics, will doubtless seem to many far-fetched, and may repel beginners as obscure and difficult of comprehension.

There are two related definitions of entropy: the thermodynamic definition and the statistical mechanics definition. Historically, the classical thermodynamics definition developed first, and it has more recently been extended in the area of non-equilibrium thermodynamics. Entropy was defined from a classical thermodynamics viewpoint, in which the details of the system's constituents are not directly considered, with their behaviour only showing up in macroscopically averaged properties, e.g. heat capacity. Later, thermodynamic entropy was more generally defined from a statistical thermodynamics viewpoint, in which the detailed constituents—modelled at first classically, e.g. Newtonian particles constituting a gas, and later quantum-mechanically (photons, phonons, spins, etc.)—were explicitly considered.

Function of State

There are many thermodynamic properties that are functions of state. This means that at a particular thermodynamic state (which should not be confused with the microsopic state of a system), these properties have a certain value. Often, if two properties have a particular value, then the state is determined and the other properties values are set. For instance, an ideal gas, at a particular temperature and pressure, has a particular volume according to the ideal gas equation. As another instance, a pure substance of single phase at a particular uniform temperature and pressure (and thus a particular state) is at not only a particular volume but also at a particular entropy. That entropy is a function of state is one reason it is useful. In the Carnot cycle, the working fluid returns to the same state at a particular stage of the cycle, hence the line integral of any state function, such as entropy, over the cycle is zero.

Reversible Process

Entropy is defined for a reversible process and for a system that, at all times, can be treated as being at a uniform state and thus at a uniform temperature. Reversibility is an ideal that some real processes approximate and that is often presented in study exercises. For a reversible process, entropy behaves as a conserved quantity and no change occurs in total entropy. More specifically, total entropy is conserved in a reversible process and not conserved in an irreversible process. One has to be careful about system boundaries. For example, in the Carnot cycle, while the heat flow from the hot reservoir to the cold reservoir represents an increase in entropy, the work output, if reversibly and perfectly stored in some energy storage mechanism, represents a decrease in entropy that could be used to operate the heat engine in reverse and return to the previous state, thus the *total* entropy change is still zero at all times if the entire process is reversible. Any process that does not meet the requirements of a reversible process must be treated as an irreversible process, which is usually a complex task. An irreversible process increases entropy.

Heat *transfer* situations require two or more non-isolated systems in thermal contact. In irreversible heat transfer, heat energy is irreversibly transferred from the higher temperature system to the lower temperature system, and the combined entropy of the systems increases. Each system, by definition, must have its own absolute temperature applicable within all areas in each respective system in order to calculate the entropy transfer. Thus, when a system at higher temperature T_H transfers heat dQ to a system of lower temperature T_C, the former loses entropy dQ/T_H and the latter gains entropy dQ/T_C. The combined entropy change is dQ/T_C-dQ/T_H which is positive, reflecting an increase in the combined entropy. When calculating entropy, the same requirement of having an absolute temperature for each system in thermal contact exchanging heat also applies to the entropy change of an isolated system having no thermal contact.

The concept of entropy arose from Rudolf Clausius's study of the Carnot cycle. In a Carnot cycle, heat (Q_1) is absorbed from a 'hot' reservoir, isothermally at the higher temperature T_1, and given up isothermally to a 'cold' reservoir, Q_2, at a lower temperature, T_2. According to Carnot's principle, work can only be done when there is a temperature difference, and the work should be some function of the difference in temperature and the heat absorbed. Carnot did not distinguish between Q_1 and Q_2, since he was working under the incorrect hypothesis that caloric theory was valid, and hence heat was conserved (the incorrect assumption that and Q_1 were Q_2 equal) when, in fact, $Q_1 - Q_2 > 0$. Through the efforts of Clausius and Kelvin, it is now known that the maximum work that can be done is the product of the Carnot efficiency and the heat absorbed at the hot reservoir: $W = \left(1 - \frac{T_2}{T_1}\right) Q_1$. In order to derive the Carnot efficiency, $1 - \frac{T_2}{T_1}$, Kelvin had to evaluate the ratio of the work done to the heat absorbed in the isothermal expansion with the help of the Carnot-Clapeyron equation which contained an unknown function, known as the Carnot function. The fact that the Carnot function could be the temperature, measured from zero, was suggested by Joule in a letter to Kelvin, and this allowed Kelvin to establish his absolute temperature scale. It is also known that the work is the difference in the heat absorbed at the hot reservoir and rejected at the cold one: $W = Q_1 - Q_2$. Since the latter is valid over the entire cycle, this gave Clausius the hint that at each stage of the cycle, work and heat would not be equal, but rather their difference would be a state function that would vanish upon completion of the cycle. The state function was called the internal energy and it became the first law of thermodynamics.

Now equating the two expressions gives

$$\frac{Q_1}{T_1} - \frac{Q_2}{T_2} = 0.$$

If we allow Q_2 to incorporate the algebraic sign, this becomes a sum and implies that there is a function of state which is conserved over a complete cycle. Clausius called this state function *entropy*. One can see that entropy was discovered through mathematics rather than through laboratory results. It is a mathematical construct and has no easy physical analogy. This makes the concept somewhat obscure or abstract, akin to how the concept of energy arose.

Then Clausius asked what would happen if there would be less work done than that predicted by Carnot's principle. The right-hand side of the first equation would be the upper bound of the work, which would now be converted into an inequality $W < \left(1 - \frac{T_2}{T_1}\right) Q_1$. When the second equation is used to express the work as a difference in heats, we get $Q_1 - Q_2 < \left(1 - \frac{T_2}{T_1}\right) Q_1$ or $Q_2 > \frac{T_2}{T_1} Q_1$ So more heat is given off to the cold reservoir than in the Carnot cycle. If we denote the entropies by $S_i = Q_i / T_i$ for the two states, then the above inequality can be written as a decrease in the entropy $S_1 - S_2 < 0$. The wasted heat implies that irreversible processes must have prevented the cycle from carrying out maximum work.

Classical Thermodynamics

Conjugate Variablesof Thermodynamics	
Pressure	Volume
(Stress)	(Strain)
Temperature	Entropy
Chemical potential	Particle number

The thermodynamic definition was developed in the early 1850s by Rudolf Clausius and essentially describes how to measure the entropy of an isolated system in thermodynamic equilibrium. Clausius created the term entropy in 1865 as an extensive thermodynamic variable was shown to be useful in characterizing the Carnot cycle. Heat transfer along the isotherm steps of the Carnot cycle was found to be proportional to the temperature of a system (known as its absolute temperature). This relationship was expressed in increments of entropy equal to the ratio of incremental heat transfer divided by temperature, which was found to vary in the thermodynamic cycle but eventually return to the same value at the end of every cycle. Thus it was found to be a function of state, specifically a thermodynamic state of the system. Clausius based the term *entropy* on the Greek ενôρïπβα [entropía], *a turning towards*, from εν- [en-] (in) and τροπή [tropç] (turn, conversion).

While Clausius based his definition on a reversible process, there are also irreversible processes that change entropy. Following the second law of thermodynamics, entropy of an isolated system always increases. The difference between an isolated system and closed system is that heat may *not* flow to and from an isolated system, but heat flow to and from a closed system is possible. Nevertheless, for both closed and isolated systems, and indeed, also in open systems, irreversible thermodynamics processes may occur.

According to the Clausius equality, for a reversible cyclic process: $\oint \frac{\delta Q}{T} = 0$. This means the line integral $\int_L \frac{\delta Q}{T}$ is path independent.

So we can define a state function S called entropy, which satisfies:

$$dS = \frac{\delta Q}{T}$$

With this we can only obtain the difference of entropy by integrating the above formula. To obtain the absolute value, we need the Third Law of Thermodynamics, which states that S=0 at absolute zero for perfect crystals.

From a macroscopic perspective, in classical thermodynamics the entropy is interpreted as a state function of a thermodynamic system: that is, a property depending only on the current state of the system, independent of how that state came to be achieved. In any process where the system gives up energy ΔE, and its entropy falls by ΔS, a quantity at least $T_R\ \Delta S$ of that energy must be given up to the system's surroundings as unusable heat (T_R is the temperature of the system's external surroundings). Otherwise the process will not go forward. In classical thermodynamics, the entropy of a system is defined only if it is in thermodynamic equilibrium.

Statistical Mechanics

The statistical definition was developed by Ludwig Boltzmann in the 1870s by analyzing the statistical behaviour of the microscopic components of the system. Boltzmann showed that this definition of entropy was equivalent to the thermodynamic entropy to within a constant number which has since been known as Boltzmann's constant. In summary, the thermodynamic definition of entropy provides the experimental definition of entropy, while the statistical definition of entropy extends the concept, providing an explanation and a deeper understanding of its nature.

The interpretation of entropy in statistical mechanics is the measure of uncertainty, or *mixedupness* in the phrase of Gibbs, which remains about a system after its observable macroscopic properties, such as temperature, pressure and volume, have been taken into account. For a given set of macroscopic variables, the entropy measures the degree to which the probability of the system is spread out over different possible microstates. In contrast to the macrostate, which characterizes plainly observable average quantities, a microstate specifies all molecular details about the system including the position and velocity of every molecule. The more such states available to the system with appreciable probability, the greater the entropy. In statistical mechanics, entropy is a measure of the number of ways in which a system may be arranged, often taken to be a measure of "disorder" (the higher the entropy, the higher the disorder). This definition describes the entropy as being proportional to the natural logarithm of the number of possible microscopic configurations of the individual atoms and molecules of the system (microstates) which could give rise to the observed macroscopic state (macrostate) of the system. The constant of proportionality is the Boltzmann constant.

Specifically, entropy is a logarithmic measure of the number of states with significant probability of being occupied:

$$S = -k_B \sum_i p_i \ln p_i,$$

where k_B is the Boltzmann constant, equal to 1.38065×10^{-23} J K^{-1}. The summation is over all the possible microstates of the system, and p_i is the probability that the system is in the *i*th microstate. This definition assumes that the basis set of states has been picked so that there is no information on their relative phases. In a different basis set, the more general expression is

$$S = -k_B Tr(\hat{\rho} \ln(\hat{\rho}))$$

where $\hat{\rho}$ is the density matrix. This density matrix formulation is not needed in cases of thermal equilibrium so long as the basis states are chosen to be energy eigenstates. For most practical purposes, this can be taken as the fundamental definition of entropy since all other formulas for S can be mathematically derived from it, but not vice versa.

In what has been called *the fundamental assumption of statistical thermodynamics* or *the fundamental postulate in statistical mechanics*, the occupation of any microstate is assumed to be equally probable (i.e. $P_i=1/\Omega$ where Ω is the number of microstates); this assumption is usually justified for an isolated system in equilibrium. Then the previous equation reduces to:

$$S = k_B \ln \Omega,$$

In thermodynamics, such a system is one in which the volume, number of molecules, and internal energy are fixed (the microcanonical ensemble). The most general interpretation of entropy is as a measure of our uncertainty about a system. The equilibrium state of a system maximizes the entropy because we have lost all information about the initial conditions except for the conserved variables; maximizing the entropy maximizes our ignorance about the details of the system. This uncertainty is not of the everyday subjective kind, but rather the uncertainty inherent to the experimental method and interpretative model.

The interpretative model has a central role in determining entropy. The qualifier "for a given set of macroscopic variables" above has deep implications: if two observers use different sets of macroscopic variables, they will observe different entropies. For example, if observer A uses the variables U, V and W, and observer B uses U, V, W, X, then, by changing X, observer B can cause an effect that looks like a violation of the second law of thermodynamics to observer A. In other words: the set of macroscopic variables one chooses must include everything that may change in the experiment, otherwise one might see decreasing entropy!

Entropy can be defined for any Markov processes with reversible dynamics and the detailed balance property.

In Boltzmann's 1896 *Lectures on Gas Theory*, he showed that this expression gives a measure of entropy for systems of atoms and molecules in the gas phase, thus providing a measure for the entropy of classical thermodynamics.

Entropy of a System

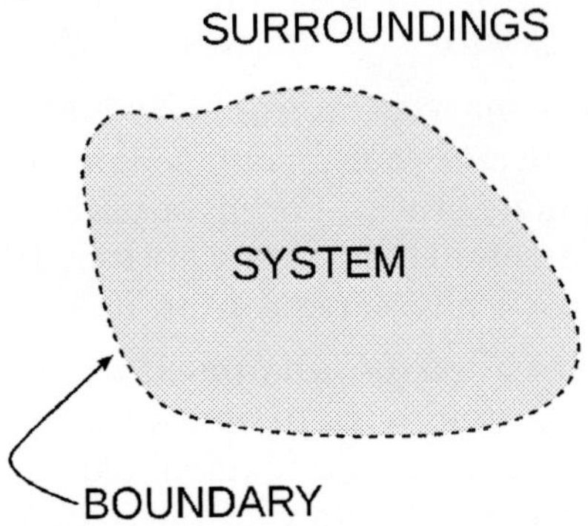

Figure: *A thermodynamic system*

Entropy is the above-mentioned unexpected and, to some, obscure integral that arises directly from the Carnot cycle. It is reversible heat divided by temperature. It is, remarkably, a function of state and it is fundamental and very useful.

In a thermodynamic system, pressure, density, and temperature tend to become uniform over time because this equilibrium state has higher probability (more possible combinations of microstates) than any other. As an example, for a glass of ice water in air at room temperature, the difference in temperature between a warm room (the surroundings) and cold glass of ice and water (the system and not part of the room), begins to be equalized as portions of the thermal energy from the warm surroundings spread to the cooler system of ice and water. Over time the temperature of the glass and its contents and the temperature of the room become equal. The entropy of the room has decreased as some of its energy has been dispersed to the ice and water. However, as calculated in the example, the entropy of the system of ice and water has increased more than the entropy of the surrounding room has decreased. In an isolated system such as the room and ice water taken together, the dispersal of energy from warmer to cooler always results in a net increase in entropy. Thus, when the "universe" of the room and ice water system has reached a temperature equilibrium, the entropy change from the initial state is at a maximum. The entropy of the thermodynamic system is a measure of how far the equalization has progressed.

Thermodynamic entropy is a non-conserved state function that is of great importance in the sciences of physics and chemistry. Historically, the concept of entropy evolved in order to explain why some processes (permitted by conservation laws) occur spontaneously while their time reversals (also permitted by conservation laws) do not; systems tend to progress in the direction of increasing entropy. For isolated systems, entropy never decreases. This fact has several important consequences in science: first, it prohibits "perpetual motion" machines; and second, it implies the arrow of entropy has the same direction as the arrow of time. Increases in entropy correspond to irreversible changes in a system, because some energy is expended as waste heat, limiting the amount of work a system can do.

Unlike many other functions of state, entropy cannot be directly observed but must be calculated. Entropy can be calculated for a substance as the standard molar entropy from absolute zero (also known as absolute entropy) or as a difference in entropy from some other reference state which is defined as zero entropy. Entropy has the dimension of energy divided by temperature, which has a unit of joules per kelvin (J/K) in the International System of Units. While these are the same units as heat capacity, the two concepts are distinct. Entropy is not a conserved quantity: for example, in an isolated system with non-uniform temperature, heat might irreversibly flow and the temperature become more uniform such that entropy increases. The second law of thermodynamics, states that a closed system has entropy which may increase or otherwise remain constant. Chemical reactions cause changes in entropy and entropy plays an important role in determining in which direction a chemical reaction spontaneously proceeds.

One dictionary definition of entropy is that it is "a measure of thermal energy per unit temperature that is not available for useful work". For instance, a substance at uniform temperature is at maximum entropy and cannot drive a heat engine. A substance at non-uniform temperature is at a lower entropy (than if the heat distribution is allowed to even out) and some of the thermal energy can drive a heat engine.

A special case of entropy increase, the entropy of mixing, occurs when two or more different substances are mixed. If the substances are at the same temperature and pressure, there will be no net exchange of heat or work – the entropy change will be entirely due to the mixing of the different substances. At a statistical mechanical level, this results due to the change in available volume per particle with mixing.

Applications

The Fundamental Thermodynamic Relation: The entropy of a system depends on its internal energy and the external parametres, such as the volume. In the thermodynamic limit this fact leads to an equation relating the change in the internal energy to changes in the entropy and the external parametres. This relation is known as the fundamental thermodynamic relation. If the volume is the only external parametre, this relation is: $dU = TdS - PdV$

Since the internal energy is fixed when one specifies the entropy and the volume, this relation is valid even if the change from one state of thermal equilibrium to another with infinitesimally larger entropy and volume happens in a non-quasistatic way (so during this change the system may be very far out of thermal equilibrium and then the entropy, pressure and temperature may not exist). The fundamental thermodynamic relation implies many thermodynamic identities that are valid in general, independent of the microscopic details of the system. Important examples are the Maxwell relations and the relations between heat capacities.

Entropy in Chemical Thermodynamics

Thermodynamic entropy is central in chemical thermodynamics, enabling changes to be quantified and the outcome of reactions predicted. The second law of thermodynamics states that entropy in an isolated system – the combination of a subsystem under study and its surroundings – increases during all spontaneous chemical and physical processes. The Clausius equation of $\delta q_{rev}/T = \Delta S$ introduces the measurement of entropy change, ΔS. Entropy change describes the direction and quantifies the magnitude of simple changes such as heat transfer between systems – always from hotter to cooler spontaneously.

The thermodynamic entropy therefore has the dimension of energy divided by temperature, and the unit joule per kelvin (J/K) in the International System of Units (SI).

Thermodynamic entropy is an extensive property, meaning that it scales with the size or extent of a system. In many processes it is useful to specify the entropy as an intensive property independent of the size, as a specific entropy characteristic of the type of system studied. Specific entropy may be expressed relative to a unit of mass, typically the kilogram (unit: $Jkg^{-1}K^{-1}$). Alternatively, in chemistry, it is also referred to one mole of substance, in which case it is called the *molar entropy* with a unit of $Jmol^{-1}K^{-1}$.

Thus, when one mole of substance at about 0K is warmed by its surroundings to 298K, the sum of the incremental values of q_{rev}/T constitute each element's or compound's standard molar entropy, an indicator of the amount of energy stored by a substance at 298K.

Entropy change also measures the mixing of substances as a summation of their relative quantities in the final mixture.

Entropy is equally essential in predicting the extent and direction of complex chemical reactions. For such applications, ΔS must be incorporated in an expression that includes both the system and its surroundings, $\Delta S_{universe} = \Delta S_{surroundings} + \Delta S_{system}$. This expression becomes, via some steps, the Gibbs free energy equation for reactants and products in the system: ΔG [the Gibbs free energy change of the system] = ΔH [the enthalpy change] $-T\,\Delta S$ [the entropy change].

Entropy Balance Equation for Open Systems

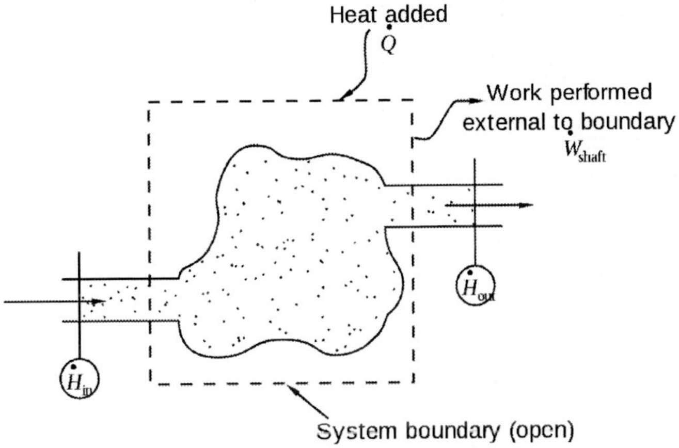

Figure: *During steady-state continuous operation, an entropy balance applied to an open system accounts for system entropy changes related to heat flow and mass flow across the system boundary.*

In chemical engineering, the principles of thermodynamics are commonly applied to "open systems", i.e. those in which heat, work, and mass flow across the system boundary. In a system in which there are flows of both heat ($\dot{Q}$) and work, i.e. $\dot{W}_S$ (shaft work) and $P(dV/dt)$ (pressure-volume work), across the system boundaries, the heat flow, but not the work flow, causes a change in the entropy of the system. This rate of entropy change is $\dot{Q}/T$, where T is the absolute thermodynamic temperature of the system at the point of the heat flow. If, in addition, there are mass flows across the system boundaries, the total entropy of the system will also change due to this convected flow.

To derive a generalized entropy balanced equation, we start with the general balance equation for the change in any extensive quantity Θ in a thermodynamic system, a quantity that may be either conserved, such as energy, or non-conserved, such as entropy. The basic generic balance expression states that $d\Theta/dt$, i.e. the rate of change of Θ in the system, equals the rate at which Θ enters the system at the boundaries, minus the rate at which Θ leaves the system across the system boundaries, plus the rate at which Θ is generated within the system. Using this generic balance equation, with respect to the rate of change with time of the extensive quantity entropy *S*, the entropy balance equation for an open thermodynamic system is:

$$\frac{dS}{dt} = \sum_{k=1}^{K} \dot{M}_k \hat{S}_k + \frac{\dot{Q}}{T} + \dot{S}_{gen}$$

where

$\sum_{k=1}^{K} \dot{M}_k \hat{S}_k$ = the net rate of entropy flow due to the flows of mass into and out of the system (where $\hat{s}$ = entropy per unit mass).

$\frac{\dot{Q}}{T}$ = the rate of entropy flow due to the flow of heat across the system boundary.

$\dot{S}_{gen}$ = the rate of entropy production within the system.

Note, also, that if there are multiple heat flows, the term $\dot{Q}/T$ is to be replaced by $\sum \dot{Q}_j / T_j$, where $\dot{Q}_j$ is the heat flow and T_j is the temperature at the *jth* heat flow port into the system.

Entropy and Other Forms of Energy Beyond Work

The fundamental equation of thermodynamics for a system containing *n* constituent species, with the *i*-th species having N_i particles is, with additional terms:

$$dU = TdS - PdV + \sum_{i}^{n} \mu_i dN_i + \Phi dQ + vdp$$

U is internal energy, T is temperature, P is pressure, V is volume, μ_i and N_i are the chemical potential and number of molecules of the chemical, Φ and Q are electric potential and charge, v and p are velocity and momentum.

Solving for the change in entropy we get:

$$dS = \frac{1}{T}dU + \frac{P}{T}dV - \sum_{i}^{n} \frac{\mu_i}{T} dN_i - \frac{\Phi}{T} dQ - \frac{v}{T} dp$$

There is a minuscule change in internal energy for any change in entropy (ds will change by 1/T*dU). But in theory, the entropy of a system can be changed without changing its energy. That is done by keeping all variables constant, including temperature (isothermally) and entropy (adiabatically). That is easy to see, but typically the energy of the system will change. e.g. You can attempt to keep volume constant but you will always do work on the system, and work changes the energy. Other potentials such as the gravitational potential can also be taken into account.

Entropy Change Formulas for Simple Processes

For certain simple transformations in systems of constant composition, the entropy changes are given by simple formulas.

Isothermal Expansion or Compression of an Ideal Gas

For the expansion (or compression) of an ideal gas from an initial volume V_0 and pressure P_0 to a final volume V and pressure P at any constant temperature, the change in entropy is given by:

$$\Delta S = nR\ln\frac{V}{V_0} = -nR\ln\frac{P}{P_0}.$$

Here n is the number of moles of gas and R is the ideal gas constant. These equations also apply for expansion into a finite vacuum or a throttling process, where the temperature, internal energy and enthalpy for an ideal gas remain constant.

Heating and Cooling

For heating (or cooling) of any system (gas, liquid or solid) at constant pressure from an initial temperature T_0 to a final temperature T, the entropy change is

$$\Delta S = nC_P\ln\frac{T}{T_0}.$$

provided that the constant-pressure molar heat capacity (or specific heat) C_P is constant and that no phase transition occurs. in this temperature interval.

Similarly at constant volume, the entropy change is

$$\Delta S = nC_v\ln\frac{T}{T_0},$$

where the constant-volume heat capacity C_v is constant and there is no phase change.

At low temperatures near absolute zero, heat capacities of solids quickly drop off to near zero, so the assumption of constant heat capacity does not apply.

Since entropy is a state function, the entropy change of any process in which temperature and volume both vary is the same as for a path divided into two steps - heating at constant volume and expansion at constant temperature. For an ideal gas, the total entropy change is

$$\Delta S = nC_v \ln\frac{T}{T_0} + nR\ln\frac{V}{V_0}.$$

Similarly if the temperature and pressure of an ideal gas both vary,

$$\Delta S = nC_P \ln\frac{T}{T_0} - nR\ln\frac{P}{P_0}.$$

Phase Transitions

Reversible phase transitions occur at constant temperature and pressure. The reversible heat is the enthalpy change for the transition, and the entropy change is the enthalpy change divided by the thermodynamic temperature. For fusion (melting) of a solid to a liquid at the melting point T_m, the entropy of fusion is

$$\Delta S_{fus} = \frac{\Delta H_{fus}}{T_m}.$$

Similarly for vaporization of a liquid to a gas at the boiling point T_b, the entropy of vaporization is

$$\Delta S_{vap} = \frac{\Delta H_{vap}}{T_b}.$$

Approaches to Understanding Entropy

As a fundamental aspect of thermodynamics and physics, several different approaches to entropy beyond that of Clausius and Boltzmann are valid.

Standard Textbook Definitions

The following is a list of additional definitions of entropy from a collection of textbooks:

- a measure of energy dispersal at a specific temperature.
- a measure of disorder in the universe or of the availability of the energy in a system to do work.

- a measure of a system's thermal energy per unit temperature that is unavailable for doing useful work.

In Boltzmann's definition, entropy is a measure of the number of possible microscopic states (or microstates) of a system in thermodynamic equilibrium. Consistent with the Boltzmann definition, the second law of thermodynamics needs to be re-worded as such that entropy decreases over time, though the underlying principle remains the same.

Order and Disorder

Entropy has often been loosely associated with the amount of order, disorder, and/or chaos in a thermodynamic system. The traditional qualitative description of entropy is that it refers to changes in the status quo of the system and is a measure of "molecular disorder" and the amount of wasted energy in a dynamical energy transformation from one state or form to another. In this direction, several recent authors have derived exact entropy formulas to account for and measure disorder and order in atomic and molecular assemblies. One of the simpler entropy order/disorder formulas is that derived in 1984 by thermodynamic physicist Peter Landsberg, based on a combination of thermodynamics and information theory arguments. He argues that when constraints operate on a system, such that it is prevented from entering one or more of its possible or permitted states, as contrasted with its forbidden states, the measure of the total amount of "disorder" in the system is given by:

$$\text{Disorder} = \frac{C_D}{C_I}.$$

Similarly, the total amount of "order" in the system is given by:

$$\text{Order} = 1 - \frac{C_O}{C_I}.$$

In which C_D is the "disorder" capacity of the system, which is the entropy of the parts contained in the permitted ensemble, C_I is the "information" capacity of the system, an expression similar to Shannon's channel capacity, and C_O is the "order" capacity of the system.

Energy Dispersal

The concept of entropy can be described qualitatively as a measure of energy dispersal at a specific temperature. Similar terms have been in use from early in the history of classical thermodynamics, and with the development of statistical thermodynamics and quantum theory, entropy changes have been described in terms of the mixing or "spreading" of the total energy of each constituent of a system over its particular quantized energy levels.

Ambiguities in the terms *disorder* and *chaos*, which usually have meanings directly opposed to equilibrium, contribute to widespread confusion and hamper comprehension of entropy for most students. As the second law of thermodynamics shows, in an isolated system internal portions at different temperatures will tend to adjust to a single uniform temperature and thus produce equilibrium. A recently developed educational approach avoids ambiguous terms and describes such spreading out of energy as dispersal, which leads to loss of the differentials required for work even though the total energy remains constant in accordance with the first law of thermodynamics (compare discussion in next section). Physical chemist Peter Atkins, for example, who previously wrote of dispersal leading to a disordered state, now writes that "spontaneous changes are always accompanied by a dispersal of energy".

Relating Entropy to Energy Usefulness

Following on from the above, it is possible (in a thermal context) to regard entropy as an indicator or measure of the *effectiveness* or *usefulness* of a particular quantity of energy. This is because energy supplied at a high temperature (i.e. with low entropy) tends to be more useful than the same amount of energy available at room temperature. Mixing a hot parcel of a fluid with a cold one produces a parcel of intermediate temperature, in which the overall increase in entropy represents a "loss" which can never be replaced.

Thus, the fact that the entropy of the universe is steadily increasing, means that its total energy is becoming less useful: eventually, this will lead to the "heat death of the Universe".

Entropy and Adiabatic Accessibility

A definition of entropy based entirely on the relation of adiabatic accessibility between equilibrium states was given by E.H.Lieb and J. Yngvason in 1999. This approach has several predecessors, including the pioneering work of Constantin Carathéodory from 1909 and the monograph by R. Giles from 1964. In the setting of Lieb and Yngvason one starts by picking, for a unit amount of the substance under consideration, two reference states X_0 and X_1 such that the latter is adiabatically accessible from the former but not vice versa. Defining the entropies of the reference states to be 0 and 1 respectively the entropy of a state X is defined as the largest number λ such that X

is adiabatically accessible from a composite state consisting of an amount λ in the state X_1 and a complementary amount, $(1-\lambda)$, in the state X_0. A simple but important result within this setting is that entropy is uniquely determined, apart from a choice of unit and an additive constant for each chemical element, by the following properties: It is monotonic with respect to the relation of adiabatic accessibility, additive on composite systems, and extensive under scaling.

Entropy in Quantum Mechanics

In quantum statistical mechanics, the concept of entropy was developed by John von Neumann and is generally referred to as "von Neumann entropy", $S = -k_B \mathrm{Tr}(\rho \log \rho)$

where ρ is the density matrix and Tr is the trace operator.

This upholds the correspondence principle, because in the classical limit, when the phases between the basis states used for the classical probabilities are purely random, this expression is equivalent to the familiar classical definition of entropy, $S = -k_B \sum_i p_i \log p_i$, i.e. in such a basis the density matrix is diagonal.

Von Neumann established a rigorous mathematical framework for quantum mechanics with his work *Mathematische Grundlagen der Quantenmechanik*. He provided in this work a theory of measurement, where the usual notion of wave function collapse is described as an irreversible process (the so-called von Neumann or projective measurement). Using this concept, in conjunction with the density matrix he extended the classical concept of entropy into the quantum domain.

Information Theory

I thought of calling it 'information', but the word was overly used, so I decided to call it 'uncertainty'. [...] Von Neumann told me, 'You should call it entropy, for two reasons. In the first place your uncertainty function has been used in statistical mechanics under that name, so it already has a name. In the second place, and more important, nobody knows what entropy really is, so in a debate you will always have the advantage.'

Conversation between Claude Shannon and John von Neumann regarding what name to give to the attenuation in phone-line signals

When viewed in terms of information theory, the entropy state function is simply the amount of information (in the Shannon sense)

that would be needed to specify the full microstate of the system. This is left unspecified by the macroscopic description.

In information theory, *entropy* is the measure of the amount of information that is missing before reception and is sometimes referred to as *Shannon entropy*. Shannon entropy is a broad and general concept which finds applications in information theory as well as thermodynamics. It was originally devised by Claude Shannon in 1948 to study the amount of information in a transmitted message. The definition of the information entropy is, however, quite general, and is expressed in terms of a discrete set of probabilities p_i:

$$H(X) = -\sum_{i=1}^{n} p(x_i) \log p(x_i).$$

In the case of transmitted messages, these probabilities were the probabilities that a particular message was actually transmitted, and the entropy of the message system was a measure of the average amount of information in a message. For the case of equal probabilities (i.e. each message is equally probable), the Shannon entropy (in bits) is just the number of yes/no questions needed to determine the content of the message.

The question of the link between information entropy and thermodynamic entropy is a debated topic. While most authors argue that there is a link between the two, a few argue that they have nothing to do with each other.

The expressions for the two entropies are similar. The information entropy H for equal probabilities $p_i = p = 1/n$ is

$$H = k \log(1/p),$$

where k is a constant which determines the units of entropy. There are many ways of demonstrating the equivalence of "information entropy" and "physics entropy", that is, the equivalence of "Shannon entropy" and "Boltzmann entropy". Nevertheless, some authors argue for dropping the word entropy for the H function of information theory and using Shannon's other term "uncertainty" instead.

Interdisciplinary Applications of Entropy

Although the concept of entropy was originally a thermodynamic construct, it has been adapted in other fields of study, including information theory, psychodynamics, thermoeconomics/ecological economics, and evolution.

Thermodynamic and Statistical Mechanics Concepts

- Entropy unit – a non-S.I. unit of thermodynamic entropy, usually denoted "e.u." and equal to one calorie per Kelvin per mole, or 4.184 Joules per Kelvin per mole.
- Gibbs entropy – the usual statistical mechanical entropy of a thermodynamic system.
- Boltzmann entropy – a type of Gibbs entropy, which neglects internal statistical correlations in the overall particle distribution.
- Tsallis entropy – a generalization of the standard Boltzmann-Gibbs entropy.
- Standard molar entropy – is the entropy content of one mole of substance, under conditions of standard temperature and pressure.
- Residual entropy – the entropy present after a substance is cooled arbitrarily close to absolute zero.
- Entropy of mixing – the change in the entropy when two different chemical substances or components are mixed.
- Loop entropy – is the entropy lost upon bringing together two residues of a polymer within a prescribed distance.
- Conformational entropy – is the entropy associated with the physical arrangement of a polymer chain that assumes a compact or globular state in solution.
- Entropic force – a microscopic force or reaction tendency related to system organization changes, molecular frictional considerations, and statistical variations.
- Free entropy – an entropic thermodynamic potential analogous to the free energy.
- Entropic explosion – an explosion in which the reactants undergo a large change in volume without releasing a large amount of heat.
- Entropy change – a change in entropy dS between two equilibrium states is given by the heat transferred dQ_{rev} divided by the absolute temperature T of the system in this interval.
- Sackur-Tetrode entropy – the entropy of a monatomic classical ideal gas determined via quantum considerations.

The Arrow of Time

Entropy is the only quantity in the physical sciences that seems to imply a particular direction of progress, sometimes called an arrow of time. As time progresses, the second law of thermodynamics states that the entropy of an isolated system never decreases. Hence, from this perspective, entropy measurement is thought of as a kind of clock.

Cosmology

Since a finite universe is an isolated system, the Second Law of Thermodynamics states that its total entropy is constantly increasing. It has been speculated, since the 19th century, that the universe is fated to a heat death in which all the energy ends up as a homogeneous distribution of thermal energy, so that no more work can be extracted from any source.

If the universe can be considered to have generally increasing entropy, then – as Sir Roger Penrose has pointed out – gravity plays an important role in the increase because gravity causes dispersed matter to accumulate into stars, which collapse eventually into black holes. The entropy of a black hole is proportional to the surface area of the black hole's event horizon. Jacob Bekenstein and Stephen Hawking have shown that black holes have the maximum possible entropy of any object of equal size. This makes them likely end points of all entropy-increasing processes, if they are totally effective matter and energy traps. Hawking has, however, recently changed his stance on this aspect.

The role of entropy in cosmology remains a controversial subject. Recent work has cast some doubt on the heat death hypothesis and the applicability of any simple thermodynamic model to the universe in general. Although entropy does increase in the model of an expanding universe, the maximum possible entropy rises much more rapidly, moving the universe further from the heat death with time, not closer. This results in an "entropy gap" pushing the system further away from the posited heat death equilibrium. Other complicating factors, such as the energy density of the vacuum and macroscopic quantum effects, are difficult to reconcile with thermodynamical models, making any predictions of large-scale thermodynamics extremely difficult.

The entropy gap is widely believed to have been originally opened up by the early rapid exponential expansion of the universe.

Thermodynamic Potentials

A thermodynamic potential is a scalar quantity used to represent the thermodynamic state of a system. The concept of thermodynamic potentials was introduced by Pierre Duhem in 1886. Josiah Willard Gibbs in his papers used the term *fundamental functions*. One main thermodynamic potential that has a physical interpretation is the internal energy U. It is the energy of configuration of a given system of conservative forces (that is why it is a potential) and only has meaning with respect to a defined set of references (or data). Expressions for all other thermodynamic energy potentials are derivable via Legendre transforms from an expression for U. In thermodynamics, certain forces, such as gravity, are typically disregarded when formulating expressions for potentials. For example, while all the working fluid in a steam engine may have higher energy due to gravity while sitting on top of Mount Everest than it would at the bottom of the Mariana Trench, the gravitational potential energy term in the formula for the internal energy would usually be ignored because *changes* in gravitational potential within the engine during operation would be negligible.

Description and Interpretation

Five common thermodynamic potentials are:

Name	*Symbol*	*Formula*	*Natural variables*
Internal energy	v	$\int(TdS - pdV + \sum_i \mu_i dN_i)$	$S, V, \{N_i\}$
Helmholtz free energy	F	$U - TS$	$T, V, \{N_i\}$
Enthalpy	H	$U + pV$	$S, p, \{N_i\}$
Gibbs free energy	G	$U + pV - TS$	$T, p, \{N_i\}$
Landau Potential (Grand potential)	Ω, Φ_G	$U - TS - \sum_i \mu_i N_i$	$T, V, \{\mu_i\}$

where T = temperature, S = entropy, p = pressure, V = volume. The Helmholtz free energy is often denoted by the symbol F, but the use of A is preferred by IUPAC. N_i is the number of particles of type i in the system and μ_i is the chemical potential for an i-type particle. For the sake of completeness, the set of all N_i are also included as natural variables, although they are sometimes ignored.

These five common potentials are all energy potentials, but there are also entropy potentials. The thermodynamic square can be used as a tool to recall and derive some of the potentials.

Just as in mechanics, where potential energy is defined as capacity to do work, similarly different potentials have different meanings. Internal energy (U/) is the capacity to do work plus the capacity to release heat. Gibbs energy is the capacity to do non-mechanical work. Enthalpy is the capacity to do non-mechanical work plus the capacity to release heat. Helmholtz free energy is the capacity to do mechanical work (useful work). From these definitions we can say that ΔU is the energy added to the system, ΔF is the total work done on it, ΔG is the non-mechanical work done on it, and ΔH is the sum of non-mechanical work done on the system and the heat given to it. Thermodynamic potentials are very useful when calculating the equilibrium results of a chemical reaction, or when measuring the properties of materials in a chemical reaction. The chemical reactions usually take place under some simple constraints such as constant pressure and temperature, or constant entropy and volume, and when this is true, there is a corresponding thermodynamic potential that comes into play. Just as in mechanics, the system will tend towards lower values of potential and at equilibrium, under these constraints, the potential will take on an unchanging minimum value. The thermodynamic potentials can also be used to estimate the total amount of energy available from a thermodynamic system under the appropriate constraint.

In Particular:

- When the entropy (S/) and "external parametres" (e.g. volume) of a closed system are held constant, the internal energy (U/) decreases and reaches a minimum value at equilibrium. This follows from the first and second laws of thermodynamics and is called the principle of minimum energy. The following three statements are directly derivable from this principle.
- When the temperature (T/) and external parametres of a closed system are held constant, the Helmholtz free energy (F/) decreases and reaches a minimum value at equilibrium.
- When the pressure (p) and external parametres of a closed system are held constant, the enthalpy (H/) decreases and reaches a minimum value at equilibrium.

When the temperature (T/), pressure (p) and external parametres of a closed system are held constant, the Gibbs free energy (G/) decreases and reaches a minimum value at equilibrium.

Natural Variables

The variables that are held constant in this process are termed the natural variables of that potential. The natural variables are important not only for the above mentioned reason, but also because if a thermodynamic potential can be determined as a function of its natural variables, all of the thermodynamic properties of the system can be found by taking partial derivatives of that potential with respect to its natural variables and this is true for no other combination of variables. On the converse, if a thermodynamic potential is not given as a function of its natural variables, it will not, in general, yield all of the thermodynamic properties of the system.

Notice that the set of natural variables for the above four potentials are formed from every combination of the *T-S* and *P-V* variables, excluding any pairs of conjugate variables. There is no reason to ignore the $N_i - \mu_i$ conjugate pairs, and in fact we may define four additional potentials for each species. Using IUPAC notation in which the brackets contain the natural variables (other than the main four), we have:

Formula	***Natural variables***
$U[\mu_j]=U-\mu_j N_j$	$S,V,\{N_{i\neq j}\},\mu_j$
$F[\mu_j]=U-TS-\mu_j N_j$	$T,V,\{N_{i\neq j}\},\mu_j$
$H[\mu_j]=U+pV-\mu_j N_j$	$S,p,\{N_{i\neq j}\},\mu_j$
$G[\mu_j]=U+pV-TS-\mu_j N_j$	$T,p,\{N_{i\neq j}\},\mu_j$

If there is only one species, then we are done. But, if there are, say, two species, then there will be additional potentials such as $U[\mu_1,\mu_2]=U-\mu_1 N_1-\mu_2 N_2$ and so on. If there are *D* dimensions to the thermodynamic space, then there are 2^D unique thermodynamic potentials. For the most simple case, a single phase ideal gas, there will be three dimensions, yielding eight thermodynamic potentials.

In statistical mechanics, the relationship between the Helmholtz free energy and the partition function is fundamental, and is used to calculate the thermodynamic properties of matters.

The Fundamental Equations

The definitions of the thermodynamic potentials may be differentiated and, along with the first and second laws of thermodynamics, a set of differential equations known as the *fundamental equations* follow. (Actually they are all expressions of the same fundamental thermodynamic relation, but are expressed in different variables.) By the first law of thermodynamics, any differential change in the internal energy *U* of a system can be written as the sum of heat flowing into the system and work done by the system on the environment, along with any change due to the addition of new particles to the system:

$$dU = \delta Q - \delta W + \sum_i \mu_i \, dN_i$$

where δQ is the infinitesimal heat flow into the system, and δW is the infinitesimal work done by the system, μ_i is the chemical potential of particle type i and N_i is the number of type i particles. (Note that neither δQ nor δW are exact differentials. Small changes in these variables are, therefore, represented with δ rather than d.)

By the second law of thermodynamics, we can express the internal energy change in terms of state functions and their differentials. In case of reversible changes we have:

$$\delta Q = T\,\mathrm{d}S$$

$$\delta W = p\,\mathrm{d}V$$

where

T is temperature,

S is entropy,

p is pressure,

and V is volume, and the equality holds for reversible processes.

This leads to the standard differential form of the internal energy in case of a quasistatic reversible change:

$$\mathrm{d}U = T\mathrm{d}S - p\mathrm{d}V + \sum_i \mu_i \,\mathrm{d}N_i$$

Since U, S and V are thermodynamic functions of state, the above relation holds also for arbitrary non-reversible changes. If the system has more external variables than just the volume that can change, the fundamental thermodynamic relation generalizes to:

$$dU = T\,dS - \sum_i X_i\,dx_i + \sum_j \mu_j\,dN_j$$

Here the X_i are the generalized forces corresponding to the external variables x_i.

Applying Legendre transforms repeatedly, the following differential relations hold for the four potentials:

$\mathrm{d}U$	=		$T\mathrm{d}S$	−	$p\mathrm{d}V$	$+\sum_i \mu_i \mathrm{d}N_i$
$\mathrm{d}F$	=	−	$S\mathrm{d}T$	−	$p\mathrm{d}V$	$+\sum_i \mu_i \mathrm{d}N_i$
$\mathrm{d}H$	=		$T\mathrm{d}S$	+	$V\mathrm{d}p$	$+\sum_i \mu_i \mathrm{d}N_i$
$\mathrm{d}G$	=	−	$S\mathrm{d}T$	+	$V\mathrm{d}p$	$+\sum_i \mu_i \mathrm{d}N_i$

Note that the infinitesimals on the right-hand side of each of the above equations are of the natural variables of the potential on the left-hand side. Similar equations can be developed for all of the other thermodynamic potentials of the system. There will be one fundamental equation for each thermodynamic potential, resulting in a total of 2^D fundamental equations.

The differences between the four thermodynamic potentials can be summarized as follows:

$$\mathrm{d}(pV) = \mathrm{d}H - \mathrm{d}U = \mathrm{d}G - \mathrm{d}F$$

$$\mathrm{d}(TS) = \mathrm{d}U - \mathrm{d}F = \mathrm{d}H - \mathrm{d}G$$

The Equations of State

We can use the above equations to derive some differential definitions of some thermodynamic parametres. If we define Φ to stand for any of the thermodynamic potentials, then the above equations are of the form:

$$\mathrm{d}\Phi = \sum_i x_i \,\mathrm{d}y_i$$

where x_i and y_i are conjugate pairs, and the y_i are the natural variables of the potential Φ. From the chain rule it follows that:

$$x_j = \left(\frac{\partial \Phi}{\partial y_j}\right)_{\{y_{i\neq j}\}}$$

Where $y_{i\neq j}$ is the set of all natural variables of Φ except y_i. This yields expressions for various thermodynamic parametres in terms of the derivatives of the potentials with respect to their natural variables. These equations are known as *equations of state* since they specify parametres of the thermodynamic state. If we restrict ourselves to the potentials U, F, H and G, then we have:

$$+T = \left(\frac{\partial U}{\partial S}\right)_{V,\{N_i\}} = \left(\frac{\partial H}{\partial S}\right)_{p,\{N_i\}}$$

$$-p = \left(\frac{\partial U}{\partial V}\right)_{S,\{N_i\}} = \left(\frac{\partial F}{\partial V}\right)_{T,\{N_i\}}$$

$$+V = \left(\frac{\partial H}{\partial p}\right)_{S,\{N_i\}} = \left(\frac{\partial G}{\partial p}\right)_{T,\{N_i\}}$$

$$-S = \left(\frac{\partial G}{\partial T}\right)_{p,\{N_i\}} = \left(\frac{\partial F}{\partial T}\right)_{V,\{N_i\}}$$

$$\mu_j = \left(\frac{\partial \phi}{\partial N_j}\right)_{X,Y,\{N_{i\neq j}\}}$$

where, in the last equation, Φ is any of the thermodynamic potentials U, F, H, G and $X, Y, \{N_{j \neq i}\}$ are the set of natural variables for that potential, excluding N_i . If we use all potentials, then we will have more equations of state such as

$$-N_j = \left(\frac{\partial U[\mu_j]}{\partial \mu_j} \right)_{S,V,\{N_{i \neq j}\}}$$

and so on. In all, there will be D equations for each potential, resulting in a total of $D\ 2^D$ equations of state. If the D equations of state for a particular potential are known, then the fundamental equation for that potential can be determined. This means that all thermodynamic information about the system will be known, and that the fundamental equations for any other potential can be found, along with the corresponding equations of state.

The Maxwell Relations

Again, define x_i and y_i to be conjugate pairs, and the y_i to be the natural variables of some potential ϑ. We may take the "cross differentials" of the state equations, which obey the following relationship:

$$\left(\frac{\partial}{\partial y_j} \left(\frac{\partial \Phi}{\partial y_k} \right)_{\{y_{i \neq k}\}} \right)_{\{y_{i \neq j}\}} = \left(\frac{\partial}{\partial y_k} \left(\frac{\partial \Phi}{\partial y_j} \right)_{\{y_{i \neq j}\}} \right)_{\{y_{i \neq k}\}}$$

From these we get the Maxwell relations. There will be $(D-1)2$ of them for each potential giving a total of $D(D-1)2$ equations in all. If we restrict ourselves the U, F, H, G

$$\left(\frac{\partial T}{\partial V} \right)_{S,\{N_i\}} = -\left(\frac{\partial p}{\partial S} \right)_{V,\{N_i\}}$$

$$\left(\frac{\partial T}{\partial p} \right)_{S,\{N_i\}} = +\left(\frac{\partial V}{\partial S} \right)_{p,\{N_i\}}$$

$$\left(\frac{\partial S}{\partial V} \right)_{T,\{N_i\}} = +\left(\frac{\partial p}{\partial T} \right)_{V,\{N_i\}}$$

$$\left(\frac{\partial S}{\partial p} \right)_{T,\{N_i\}} = -\left(\frac{\partial V}{\partial T} \right)_{p,\{N_i\}}$$

Using the equations of state involving the chemical potential we get equations such as:

$$\left(\frac{\partial T}{\partial N_j}\right)_{V,S,\{N_{i\neq j}\}} = \left(\frac{\partial \mu_j}{\partial S}\right)_{V,\{N_i\}}$$

and using the other potentials we can get equations such as:

$$\left(\frac{\partial N_j}{\partial V}\right)_{S,\mu_j,\{N_{i\neq j}\}} = -\left(\frac{\partial p}{\partial \mu_j}\right)_{S,V\{N_{i\neq j}\}}$$

$$\left(\frac{\partial N_j}{\partial N_k}\right)_{S,V,\mu_j,\{N_{i\neq j,k}\}} = -\left(\frac{\partial \mu_k}{\partial \mu_j}\right)_{S,V\{N_{i\neq j}\}}$$

Euler Integrals

Again, define x_i and y_i to be conjugate pairs, and the y_i to be the natural variables of the internal energy. Since all of the natural variables of the internal energy U are extensive quantities

$$U(\{\alpha y_i\}) = \alpha U(\{y_i\})$$

it follows from Euler's homogeneous function theorem that the internal energy can be written as:

$$U(\{y_i\}) = \sum_j y_j \left(\frac{\partial U}{\partial y_j}\right)_{\{y_{i\neq j}\}}$$

From the equations of state, we then have:

$$U = TS - pV + \sum_i \mu_i N_i$$

Substituting into the expressions for the other main potentials we have:

$$F = -pV + \sum_i \mu_i N_i$$

$$H = TS + \sum_i \mu_i N_i$$

$$G = \sum_i \mu_i N_i$$

As in the above sections, this process can be carried out on all of the other thermodynamic potentials. Note that the Euler integrals are sometimes also referred to as fundamental equations.

The Gibbs–Duhem Relation

Deriving the Gibbs–Duhem equation from basic thermodynamic state equations is straightforward. Equating any thermodynamic potential definition with its Euler integral expression yields:

$$U = TS - PV + \sum_i \mu_i N_i$$

Differentiating, and using the second law:

$$dU = TdS - PdV + \sum_i \mu_i \, dN_i$$

yields:

$$0 = SdT - VdP + \sum_i N_i d\mu_i$$

Which is the Gibbs–Duhem relation. The Gibbs–Duhem is a relationship among the intensive parametres of the system. It follows that for a simple system with I components, there will be $I + 1$ independent parametres, or degrees of freedom. For example, a simple system with a single component will have two degrees of freedom, and may be specified by only two parametres, such as pressure and volume for example. The law is named after Josiah Willard Gibbs and Pierre Duhem.

Chemical Reactions

Changes in these quantities are useful for assessing the degree to which a chemical reaction will proceed. The relevant quantity depends on the reaction conditions, as shown in the following table. Δ denotes the change in the potential and at equilibrium the change will be zero.

	Constant V	***Constant p***
Constant S	ΔU	ΔH
Constant T	ΔF	ΔG

Most commonly one considers reactions at constant p and T, so the Gibbs free energy is the most useful potential in studies of chemical reactions.

Gibbs and Helmholtz Functions

The Gibbs–Helmholtz equation is a thermodynamic equation useful for calculating changes in the Gibbs energy of a system as a function of temperature. It is named after Josiah Willard Gibbs and Hermann von Helmholtz.

Equation

The equation is:

$$\left(\frac{\partial(G/T)}{\partial T}\right)_p = -\frac{H}{T^2}$$

where H is the enthalpy, T the absolute temperature and G the Gibbs free energy of the system, all at constant pressure p. The equation states that the change in the G/T ratio at constant pressure as a result of an infinitesimally small change in temperature is a factor H/T^2. The typical applications are to chemical reactions. The equation reads:

$$\left(\frac{\partial(\Delta G^{\ominus}/T)}{\partial T}\right)_p = -\frac{\Delta H}{T^2}$$

with ΔG as the change in Gibbs energy and ΔH as the enthalpy change (considered independent of temperature). The $\ominus$ denotes Standard conditions for temperature and pressure (approximately temperature 298.15 Kelvin and pressure 101,325 pascals).

Integrating with respect to T (again p is constant) it becomes:

$$\frac{\Delta G^{\ominus}(T_2)}{T_2} - \frac{\Delta G^{\ominus}(T_1)}{T_1} = \Delta H^{\ominus}(p)\left(\frac{1}{T_2} - \frac{1}{T_1}\right)$$

This equation quickly enables the calculation of the Gibbs free energy change for a chemical reaction at any temperature T_2 with knowledge of just the Standard Gibbs free energy change of formation and the Standard enthalpy change of formation for the individual components.

Also, using the reaction isotherm equation

$$\frac{\Delta G^{\ominus}}{T} = -R\ln K$$

relating the Gibbs energy to a chemical equilibrium constant, the van 't Hoff equation can be derived.

Derivation

Background: The definition of the Gibbs function is

$$H = G + ST$$

where H is the enthalpy defined by:

$$H = U + pV$$

Taking differentials of each definition to find dH and dG, then using the fundamental thermodynamic relation, aka "master equation" (always true for reversible or irreversibleprocesses):

$$dU = TdS - pdV$$

where$S$ is the entropy, V is volume, (minus sign due to reversibility, in which $dU = 0$: work other than pressure-volume may be done and is equal to $-pV$) leads to the "reversed" form of the initial fundamental relation into a new master equation:

$$dG = -SdT + Vdp$$

This is the Gibbs free energy for a closed system. The Gibbs–Helmholtz equation can be derived by this second master equation, and the chain rule for partial derivatives.

Maxwell Relations

Maxwell's relations are a set of equations in thermodynamics which are derivable from the definitions of the thermodynamic potentials. These relations are named for the nineteenth-century physicist James Clerk Maxwell.

Equation

The Maxwell relations are statements of equality among the second derivatives of the thermodynamic potentials. They follow directly from the fact that the order of differentiation of an analytic function of two variables is irrelevant. If ϑ is a thermodynamic potential and x_i and x_j are two different natural variables for that potential, then the Maxwell relation for that potential and those variables is:

Maxwell relations*(general)*

$$\frac{\partial}{\partial x_j}\left(\frac{\partial \Phi}{\partial x_i}\right) = \frac{\partial}{\partial x_i}\left(\frac{\partial \Phi}{\partial x_j}\right)$$

where the partial derivatives are taken with all other natural variables held constant. It is seen that for every thermodynamic potential there are $n(n - 1)/2$ possible Maxwell relations where n is the number of natural variables for that potential.

The four Most Common Maxwell Relations

The four most common Maxwell relations are the equalities of the second derivatives of each of the four thermodynamic potentials, with respect to their thermal natural variable (temperature T; or entropy$S$) and their *mechanical* natural variable (pressure$P$; or volume$V$):

Maxwell's relations*(common)*

$$+\left(\frac{\partial T}{\partial V}\right)_S = -\left(\frac{\partial P}{\partial S}\right)_V = \frac{\partial^2 U}{\partial S \partial V}$$

$$+\left(\frac{\partial T}{\partial P}\right)_S = +\left(\frac{\partial V}{\partial S}\right)_P = \frac{\partial^2 H}{\partial S \partial P}$$

$$+\left(\frac{\partial S}{\partial V}\right)_T = +\left(\frac{\partial P}{\partial T}\right)_V = -\frac{\partial^2 A}{\partial T \partial V}$$

$$-\left(\frac{\partial S}{\partial P}\right)_T = +\left(\frac{\partial V}{\partial T}\right)_P = \frac{\partial^2 G}{\partial T \partial P}$$

where the potentials as functions of their natural thermal and mechanical variables are the internal energy$U(S, V)$, Enthalpy$H(S, P)$, Helmholtz free energy$A(T, V)$ and Gibbs free energy$G(T, P)$. The thermodynamic square can be used as a mnemonic to recall and derive these relations.

Derivation

Maxwell relations are based on simple partial differentiation rules, in particular the totaldifferential of a function and the symmetry of evaluating second order partial derivatives.

Derivation

Derivation of the Maxwell relations can be deduced from the differential forms of the thermodynamic potentials:

$$\begin{aligned} dU &= TdS - PdV \\ dH &= TdS + VdP \\ dA &= -SdT - PdV \\ dG &= -SdT + VdP \end{aligned}$$

These equations resemble total differentials of the form

$$dz = \left(\frac{\partial z}{\partial x}\right)_y dx + \left(\frac{\partial z}{\partial y}\right)_x dy$$

And indeed, it can be shown for any equation of the form

$$dz = Mdx + Ndy$$

that

$$M = \left(\frac{\partial z}{\partial x}\right)_y, \quad N = \left(\frac{\partial z}{\partial y}\right)_x$$

Consider, as an example, the equation $dH = TdS + VdP$. We can now immediately see that

$$T = \left(\frac{\partial H}{\partial S}\right)_P, \quad V = \left(\frac{\partial H}{\partial P}\right)_S$$

Since we also know that for functions with continuous second derivatives, the mixed partial derivatives are identical (Symmetry of second derivatives), that is, that

$$\frac{\partial}{\partial y}\left(\frac{\partial z}{\partial x}\right)_y = \frac{\partial}{\partial x}\left(\frac{\partial z}{\partial y}\right)_x = \frac{\partial^2 z}{\partial y \partial x} = \frac{\partial^2 z}{\partial x \partial y}$$

we therefore can see that

$$\frac{\partial}{\partial P}\left(\frac{\partial H}{\partial S}\right)_P = \frac{\partial}{\partial S}\left(\frac{\partial H}{\partial P}\right)_S$$

and therefore that

$$\left(\frac{\partial T}{\partial P}\right)_S = \left(\frac{\partial V}{\partial S}\right)_P$$

Each of the four Maxwell relationships given above follows similarly from one of the Gibbs equations.

Extended Derivation

Combined form first and second law of thermodynamics,

$$TdS = dU + PdV \tag{1}$$

U, S, and V are state functions. Let,

$$U = U(x, y)$$

$$S = S(x, y)$$

$$V = V(x, y)$$

$$dU = \left(\frac{\partial U}{\partial x}\right)_y dx + \left(\frac{\partial U}{\partial y}\right)_x dy$$

$$dS = \left(\frac{\partial S}{\partial x}\right)_y dx + \left(\frac{\partial S}{\partial y}\right)_x dy$$

$$dV = \left(\frac{\partial V}{\partial x}\right)_y dx + \left(\frac{\partial V}{\partial y}\right)_x dy$$

Substitute them in Eq.1 and one gets,

$$T\left(\frac{\partial S}{\partial x}\right)_y dx + T\left(\frac{\partial S}{\partial y}\right)_x dy = \left(\frac{\partial U}{\partial x}\right)_y dx + \left(\frac{\partial U}{\partial y}\right)_x dy + P\left(\frac{\partial V}{\partial x}\right)_y dx + P\left(\frac{\partial V}{\partial y}\right)_x dy$$

And also written as,

$$\left(\frac{\partial U}{\partial x}\right)_y dx + \left(\frac{\partial U}{\partial y}\right)_x dy = T\left(\frac{\partial S}{\partial x}\right)_y dx + T\left(\frac{\partial S}{\partial y}\right)_x dy - P\left(\frac{\partial V}{\partial x}\right)_y dx - P\left(\frac{\partial V}{\partial y}\right)_x dy$$

comparing the coefficient of dx and dy, one gets

$$\left(\frac{\partial U}{\partial x}\right)_y = T\left(\frac{\partial S}{\partial x}\right)_y - P\left(\frac{\partial V}{\partial x}\right)_y$$

$$\left(\frac{\partial U}{\partial y}\right)_x = T\left(\frac{\partial S}{\partial y}\right)_x - P\left(\frac{\partial V}{\partial y}\right)_x$$

Differentiating above equations by y, x respectively

$$\left(\frac{\partial^2 U}{\partial y \partial x}\right) = \left(\frac{\partial T}{\partial y}\right)_x \left(\frac{\partial S}{\partial x}\right)_y + T\left(\frac{\partial^2 S}{\partial y \partial x}\right) - \left(\frac{\partial P}{\partial y}\right)_x \left(\frac{\partial V}{\partial x}\right)_y - P\left(\frac{\partial^2 V}{\partial y \partial x}\right) \tag{2}$$

and

$$\left(\frac{\partial^2 U}{\partial x \partial y}\right) = \left(\frac{\partial T}{\partial x}\right)_y \left(\frac{\partial S}{\partial y}\right)_x + T\left(\frac{\partial^2 S}{\partial x \partial y}\right) - \left(\frac{\partial P}{\partial x}\right)_y \left(\frac{\partial V}{\partial y}\right)_x - P\left(\frac{\partial^2 V}{\partial x \partial y}\right) \tag{3}$$

U, S, and V are exact differentials, therefore,

$$\left(\frac{\partial^2 U}{\partial y \partial x}\right) = \left(\frac{\partial^2 U}{\partial x \partial y}\right)$$

$$\left(\frac{\partial^2 S}{\partial y \partial x}\right) = \left(\frac{\partial^2 S}{\partial x \partial y}\right) : \left(\frac{\partial^2 V}{\partial y \partial x}\right) = \left(\frac{\partial^2 V}{\partial x \partial y}\right)$$

Subtract eqn(2) and (3) and one gets

$$\left(\frac{\partial T}{\partial y}\right)_x \left(\frac{\partial S}{\partial x}\right)_y - \left(\frac{\partial P}{\partial y}\right)_x \left(\frac{\partial V}{\partial x}\right)_y = \left(\frac{\partial T}{\partial x}\right)_y \left(\frac{\partial S}{\partial y}\right)_x - \left(\frac{\partial P}{\partial x}\right)_y \left(\frac{\partial V}{\partial y}\right)_x$$

Note: *The above is called the general expression for Maxwell's thermodynamical relation.*

Maxwell's first relation

Allow x = S and y = V and one gets

$$\left(\frac{\partial T}{\partial V}\right)_S = -\left(\frac{\partial P}{\partial S}\right)_V$$

Maxwell's second relation

Allow x = T and y = V and one gets

$$\left(\frac{\partial S}{\partial V}\right)_T = \left(\frac{\partial P}{\partial T}\right)_V$$

Maxwell's third relation

Allow x = S and y = P and one gets

$$\left(\frac{\partial T}{\partial P}\right)_S = \left(\frac{\partial V}{\partial S}\right)_P$$

Maxwell's fourth relation

Allow x = T and y = P and one gets

$$\left(\frac{\partial S}{\partial P}\right)_T = -\left(\frac{\partial V}{\partial T}\right)_P$$

Maxwell's fifth relation

Allow x = P and y = V

$$\left(\frac{\partial T}{\partial P}\right)_V \left(\frac{\partial S}{\partial V}\right)_P - \left(\frac{\partial T}{\partial V}\right)_P \left(\frac{\partial S}{\partial P}\right)_V = 1$$

Maxwell's sixth relation

Allow x = T and y = S and one gets

$$\left(\frac{\partial P}{\partial T}\right)_S \left(\frac{\partial V}{\partial S}\right)_T - \left(\frac{\partial P}{\partial S}\right)_T \left(\frac{\partial V}{\partial T}\right)_S = 1$$

General Maxwell Relationships

The above are not the only Maxwell relationships. When other work terms involving other natural variables besides the volume work are considered or when the number of particles is included as a natural variable, other Maxwell relations become apparent. For example, if we have a single-component gas, then the number of particles N is also a natural variable of the above four thermodynamic potentials. The Maxwell relationship for the enthalpy with respect to pressure and particle number would then be:

$$\left(\frac{\partial \mu}{\partial P}\right)_{S,N} = \left(\frac{\partial V}{\partial N}\right)_{S,P} = \frac{\partial^2 H}{\partial P \partial N}$$

where μ is the chemical potential. In addition, there are other thermodynamic potentials besides the four that are commonly used, and each of these potentials will yield a set of Maxwell relations.

Each equation can be re-expressed using the relationship

$$\left(\frac{\partial y}{\partial x}\right)_z = 1/\left(\frac{\partial x}{\partial y}\right)_z$$

which are sometimes also known as Maxwell relations.

Chapter

3

Pure Substances

PvT Surfaces

PvT Surface for a Substance which Contracts Upon Freezing

The equilibrium states of a simple, compressible substance can be specified in terms of its pressure, volume and temperature. If any two of these state variables is specified, the third is determined. This implies that the states of the substance can be represented as a surface in a three dimensional PvT space. The PvT surface above represents a substance which contracts upon freezing. Most substances do so, the notable exception being water which expands upon freezing. A considerable amount of information about the phases of matter can be illustrated with the PvT surface.

The solid, liquid and gas(vapor) phases can be represented by regions on the surface. Note that there are regions on the surface which represent a single phase, and regions which are combinations of two phases. A point lying on a line between a single-phase and a two-phase region represents a "saturation state". The line between the liquid and the liquid-vapour regions is called the liquid-saturation line and any point on that line represents a saturated-liquid state. A point on the boundary between the vapour and the liquid-vapour regions is called a saturated-vapour state.

Note the critical state where the saturated-liquid and saturated-vapour lines meet. The state variables of this unique point are denoted by P_c, v_c and T_c. If a substance is above the critical temperature T_c, it cannot condense into a liquid, no matter how high the pressure. This merging of the liquid and vapour states above the critical temperature is a characteristic of all known substances. While a pure vapour state can exist at a pressure lower than P_c, at pressures above P_c it is constrained to be a vapour. States with pressures above P_c are described as "supercritical states".

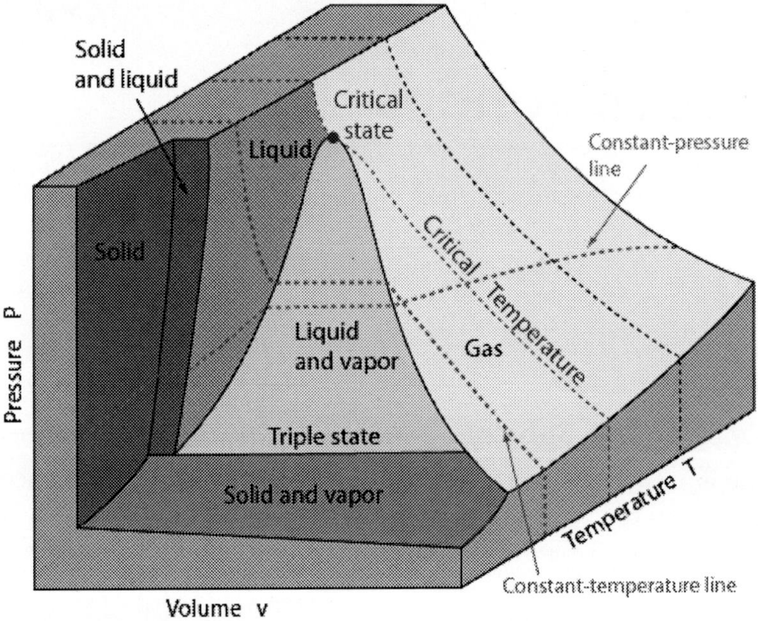

The remarkable "triple state" of matter where solid, liquid and vapour are in equilibrium may be characterized by a temperature called the triple point. The triple state is represented by a line parallel to the Pv plane with a characteristic pressure for the substance but variable volume. The triple point temperature of water is assigned the value 273.16 K and the triple state of water is used as the reference for establishing the Kelvin temperature scale.

T-S and h-s Diagrams

T-S Diagrams (H)

A temperature entropy diagram, or T-s diagram, is used in thermodynamics to visualize changes to temperature and specific entropy during a thermodynamic process or cycle. It is a useful and common tool, particularly because it helps to visualize the heat transfer during a process. For reversible (ideal) processes, the area under the

T-s curve of a process is the heat transferred to the system during that process.

An isentropic process is depicted as a vertical line on a T-s diagram, whereas an isothermal process is a horizontal line.

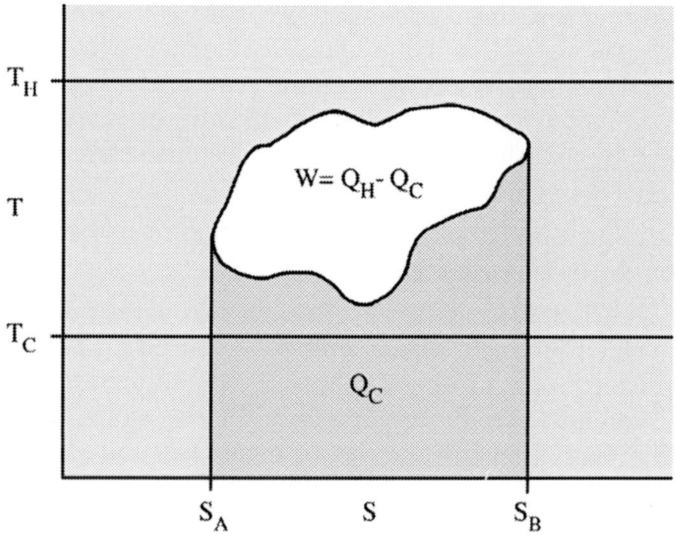

Figure: *This example T-s diagram shows a thermodynamic cycle taking place between a hot reservoir at temperature* T_H *and a cold reservoir at temperature* T_C. *For reversible processes, such as those found in the Carnot cycle, the area in red* Q_C *is the amount of energy exchanged between the system and the cold reservoir. The area in white W is the amount of work energy exchanged by the system with its surroundings. The amount of heat exchanged with the hot reservoir is the sum of the two. The thermal efficiency of the cycle is the ratio of the white area (work) divided by the sum of the white and red areas (total heat).*

h-s Diagrams

An enthalpy–entropy chart, also known as the h–s chart or Mollier diagram, plots the total heat against entropy, describing the enthalpy of a thermodynamic system. A typical chart covers a pressure range of 0.01 - 1000 bar, and temperatures up to 800 degrees Celsius. It shows enthalpy h in terms of internal energy u, pressure P and volume v using the relationship $h = u + Pv$.

History

The diagram was created in 1904, when Richard Mollier plotted the total heat against entropy. At the 1923 Thermodynamics Conference held in Los Angeles it was decided to name, in his honour, as a

"Mollier diagram" any thermodynamic diagram using the Enthalpy h as one of its axes.

Details

On the diagram, lines of constant pressure, constant temperature and volume are plotted, so in a two-phase region, the lines of constant pressure and temperature coincide. Thus, coordinates on the diagram represent entropy and heat. The work done on vapour cycles is represented by length, so it can be measured directly, whereas in a T–s diagram it is shown as an area.

In an isobaric process, the pressure remains constant, so the heat interaction is the change in enthalpy.

In an isenthalpic process, the enthalpy is constant. A vertical line in the h–s chart means an isentropic process and an horizontal line means an isenthalpic process. The process 3-4 in a rankine cycle is isentropic when the steam turbine is said to be an ideal one. So the expansion process in a turbine can be easily calculated using the h–s chart when the process is considered to be ideal (which is the case normally when calculating enthalpies, entropies, etc. Later the deviations from the ideal values can be calculated considering the isentropic efficiency of the steam turbine used.)

In a throttling process, the Joule–Thomson effect means that an adiabatic device uses energy equal to the decrease in the enthalpy of the fluid flowing through the device.

Lines of constant *dryness fraction* (x), sometimes called the *quality* of the steam, are drawn in the wet region and lines of constant temperature are drawn in the superheated region. X gives the fraction (by mass) of gaseous water in the wet region, the remainder being colloidal liquid droplets. Above the heavy line, the temperature is above the boiling point, and the dry (superheated) steam consists of gas only.

In general such charts do not show the values of specific volumes, nor do they show the enthalpies of saturated water at pressures which are of the order of those experienced in condensers in a thermal power station. Hence the chart is only useful for enthalpy changes in the expansion process of the steam cycle.

Applications and Usage

It can be used in practical applications such as malting, to represent the grain-air-moisture system.

The underlying property data for the Mollier diagram is identical to a psychrometric chart. At first inspection, there may appear little resemblance between the charts, but if the user rotates a chart ninety degrees and looks at it in a mirror, the resemblance is apparent. The Mollier diagram coordinates are enthalpy h and humidity ratio x. The enthalpy coordinate is *skewed* and the constant enthalpy lines are parallel and evenly spaced.

The Mollier diagram is preferred by many users in Scandinavia, Eastern Europe, and Russia.

Phase Transformations

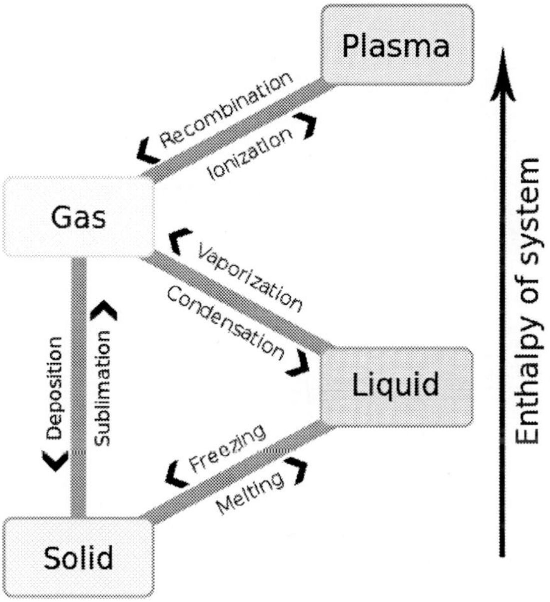

Figure: *This diagram shows the nomenclature for the different phase transitions.*

A phase transition is the transformation of thermodynamic system from one phase or state of matter to another.

A phase of a thermodynamic system and the states of matter have uniform physical properties.

During a phase transition of a given medium certain properties of the medium change, often discontinuously, as a result of some external condition, such as temperature, pressure, and others. For example, a liquid may become gas upon heating to the boiling point, resulting in an abrupt change in volume. The measurement of the external conditions at which the transformation occurs is termed the *phase transition.* Phase transitions are common occurrences observed in nature and many engineering techniques exploit certain types of phase transition.

The term is most commonly used to describe transitions between solid, liquid and gaseousstates of matter, and, in rare cases, plasma.

Types of Phase Transition

Examples of phase transitions include:

- The transitions between the solid, liquid, and gaseous phases of a single component, due to the effects of temperature and/ or pressure:

		To			
		Solid	***Liquid***	***Gas***	***Plasma***
From	**Solid**	Solid-solid transformation	Melting/fusion	Sublimation	—
	Liquid	Freezing	—	Boiling/evaporation	—
	Gas	Deposition	Condensation	—	Ionization
	Plasma	—	—	Recombination/deionization	—

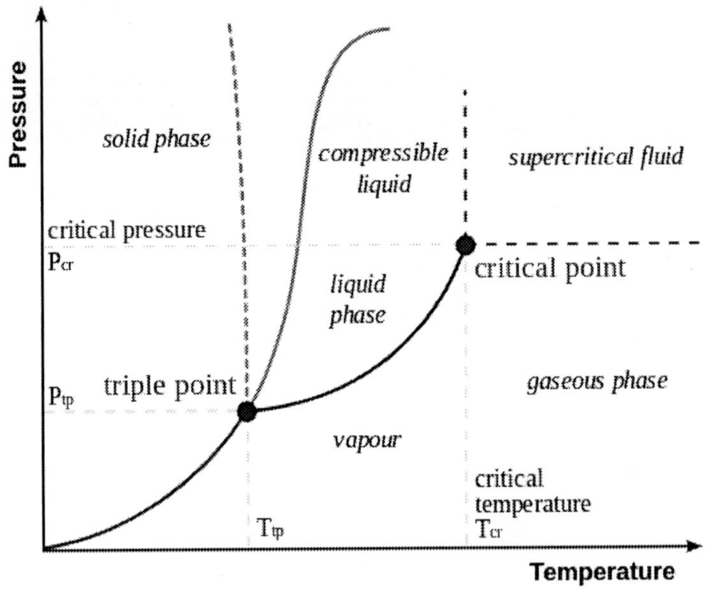

Figure: *A typical phase diagram. The dotted line gives the anomalous behaviour of water.*

- A eutectic transformation, in which a two component single phase liquid is cooled and transforms into two solid phases. The same process, but beginning with a solid instead of a liquid is called a eutectoid transformation.

- A peritectic transformation, in which a two component single phase solid is heated and transforms into a solid phase and a liquid phase.
- A spinodal decomposition, in which a single phase is cooled and separates into two different compositions of that same phase.
- Transition to a mesophase between solid and liquid, such as one of the "liquid crystal" phases.
- The transition between the ferromagnetic and paramagnetic phases of magnetic materials at the Curie point.
- The transition between differently ordered, commensurate or incommensurate, magnetic structures, such as in cerium antimonide.
- The martensitic transformation which occurs as one of the many phase transformations in carbon steel and stands as a model for displacive phase transformations.
- Changes in the crystallographic structure such as between ferrite and austenite of iron.
- Order-disorder transitions such as in alpha-titanium aluminides.
- The dependence of the adsorption geometry on coverage and temperature, such as for hydrogen on iron (110).
- The emergence of superconductivity in certain metals and ceramics when cooled below a critical temperature.
- The transition between different molecular structures (polymorphs, allotropes or polyamorphs), especially of solids, such as between an amorphous structure and a crystal structure, between two different crystal structures, or between two amorphous structures.
- Quantum condensation of bosonic fluids (Bose-Einstein condensation). The superfluid transition in liquid helium is an example of this.
- The breaking of symmetries in the laws of physics during the early history of the universe as its temperature cooled.

Phase transitions occur when the thermodynamic free energy of a system is non-analytic for some choice of thermodynamic variables (cf. phases). This condition generally stems from the interactions of a large number of particles in a system, and does not appear in systems that are too small.

At the phase transition point (for instance, boiling point) the two phases of a substance, liquid and vapour, have identical free energies and therefore are equally likely to exist. Below the boiling point, the liquid is the more stable state of the two, whereas above the gaseous form is preferred.

It is sometimes possible to change the state of a system diabatically (as opposed to adiabatically) in such a way that it can be brought past a phase transition point without undergoing a phase transition. The resulting state is metastable, i.e., less stable than the phase to which the transition would have occurred, but not unstable either. This occurs in superheating, supercooling, and supersaturation, for example.

Classifications

Ehrenfest Classification: Paul Ehrenfest classified phase transitions based on the behaviour of the thermodynamic free energy as a function of other thermodynamic variables. Under this scheme, phase transitions were libelled by the lowest derivative of the free energy that is discontinuous at the transition. *First-order phase transitions* exhibit a discontinuity in the first derivative of the free energy with respect to some thermodynamic variable. The various solid/liquid/gas transitions are classified as first-order transitions because they involve a discontinuous change in density, which is the first derivative of the free energy with respect to chemical potential. *Second-order phase transitions* are continuous in the first derivative (the order parametre, which is the first derivative of the free energy with respect to the external field, is continuous across the transition) but exhibit discontinuity in a second derivative of the free energy. These include the ferromagnetic phase transition in materials such as iron, where the magnetization, which is the first derivative of the free energy with respect to the applied magnetic field strength, increases continuously from zero as the temperature is lowered below the Curie temperature. The magnetic susceptibility, the second derivative of the free energy with the field, changes discontinuously. Under the Ehrenfest classification scheme, there could in principle be third, fourth, and higher-order phase transitions.

Though useful, Ehrenfest's classification has been found to be an inaccurate method of classifying phase transitions, for it does not take into account the case where a derivative of free energy diverges (which is only possible in the thermodynamic limit). For instance, in the ferromagnetic transition, the heat capacity diverges to infinity.

Modern Classifications

In the modern classification scheme, phase transitions are divided into two broad categories, named similarly to the Ehrenfest classes:

First-order phase transitions are those that involve a latent heat. During such a transition, a system either absorbs or releases a fixed (and typically large) amount of energy. During this process, the temperature of the system will stay constant as heat is added: the system is in a “mixed-phase regime” in which some parts of the system have completed the transition and others have not. Familiar examples are the melting of ice or the boiling of water (the water does not instantly turn into vapour, but forms a turbulent mixture of liquid water and vapour bubbles).

Second-order phase transitions are also called *continuous phase transitions*. They are characterized by a divergent susceptibility, an infinite correlation length, and a power-law decay of correlations near criticality. Examples of second-order phase transitions are the ferromagnetic transition, superconducting transition (for a Type-I superconductor the phase transition is second-order at zero external field and for a Type-II superconductor the phase transition is second-order for both normal state-mixed state and mixed state-superconducting state transitions) and the superfluid transition. Lev Landau gave a phenomenological theory of second order phase transitions.

Apart from isolated, simple phase transitions, there exist transition lines as well as multicritical points, when varying external parametres like the magnetic field, composition,...

Several transitions are known as the *infinite-order phase transitions*. They are continuous but break no symmetries. The most famous example is the Kosterlitz–Thouless transition in the two-dimensional XY model. Many quantum phase transitions, e.g., in two-dimensional electron gases, belong to this class.

The liquid-glass transition is observed in many polymers and other liquids that can be supercooled far below the melting point of the crystalline phase. This is atypical in several respects. It is not a transition between thermodynamic ground states: it is widely believed that the true ground state is always crystalline. Glass is a *quenched disorder* state, and its entropy, density, and so on, depend on the thermal history. Therefore, the glass transition is primarily a dynamic phenomenon: on cooling a liquid, internal degrees of freedom successively fall out of equilibrium. Some theoretical methods predict an underlying phase transition in the hypothetical limit of infinitely long relaxation times. No direct experimental evidence supports the existence of these transitions.

Characteristic Properties

Critical Points: In any system containing liquid and gaseous phases, there exists a special combination of pressure and temperature, known as the critical point, at which the transition between liquid and gas becomes a second-order transition. Near the critical point, the fluid is sufficiently hot and compressed that the distinction between the liquid and gaseous phases is almost non-existent. This is associated with the phenomenon of critical opalescence, a milky appearance of the liquid due to density fluctuations at all possible wavelengths (including those of visible light).

Symmetry

Order Parametres: The order parametre is normally a quantity which is zero in one phase (usually above the critical point), and non-zero in the other. It characterises the onset of order at the phase transition. The order parametre susceptibility will usually diverge approaching the critical point. For a ferromagnetic system undergoing a phase transition, the order parametre is the net magnetization. For liquid/gas transitions, the order parametre is the difference of the densities.

When symmetry is broken, one needs to introduce one or more extra variables to describe the state of the system. For example, in the ferromagnetic phase, one must provide the net magnetization, whose direction was spontaneously chosen when the system cooled below the Curie point. Such variables are examples of order parametres. An order parametre is a measure of the degree of order in a system. For example, an order parametre can indicate the degree of order in a liquid crystal. However, note that order parametres can also be defined for non-symmetry-breaking transitions. Some phase transitions, such as superconducting and ferromagnetic, can have order parametres for more than one degree of freedom. In such phases, the order parametre may take the form of a complex number, a vector, or even a tensor, the magnitude of which goes to zero at the phase transition.

There also exist dual descriptions of phase transitions in terms of disorder parametres. These indicate the presence of line-like excitations such as vortex- or defect lines.

Relevance in Cosmology

Symmetry-breaking phase transitions play an important role in cosmology. It has been speculated that, in the hot early universe, the vacuum (i.e. the various quantum fields that fill space) possessed a large number of symmetries. As the universe expanded and cooled, the vacuum underwent a series of symmetry-breaking phase transitions. For example, the electroweak transition broke the SU(2)×U(1) symmetry of the electroweak field into the U(1) symmetry of the present-day electromagnetic field. This transition is important to understanding the asymmetry between the amount of matter and antimatter in the present-day universe.

Progressive phase transitions in an expanding universe are implicated in the development of order in the universe, as is illustrated by the work of Eric Chaisson and David Layzer.

Critical Exponents and Universality Classes

Continuous phase transitions are easier to study than first-order transitions due to the absence of latent heat, and they have been discovered to have many interesting properties. The phenomena associated with continuous phase transitions are called critical phenomena, due to their association with critical points.

It turns out that continuous phase transitions can be characterized by parametres known as critical exponents. The most important one is perhaps the exponent describing the divergence of the thermal correlation length by approaching the transition. For instance, let us examine the behaviour of the heat capacity near such a transition. We vary the temperature T of the system while keeping all the other thermodynamic variables fixed, and find that the transition occurs at some critical temperature T_c. When T is near T_c, the heat capacity C typically has a power law behaviour:

$$C \propto |T_c - T|^{-\alpha}.$$

A similar behaviour, but with the exponent ν instead of α, applies for the correlation length.

The exponent ν is positive. This is different with α. Its actual value depends on the type of phase transition we are considering.

For $-1 < \alpha < 0$, the heat capacity has a "kink" at the transition temperature. This is the behaviour of liquid helium at the lambda transition from a normal state to the superfluid state, for which experiments have found $\alpha = -0.013 \pm 0.003$. At least one experiment was performed in the zero-gravity conditions of an orbiting satellite to minimize pressure differences in the sample. This experimental value of α agrees with theoretical predictions based on variational perturbation theory.

For $0 < \alpha < 1$, the heat capacity diverges at the transition temperature (though, since $\alpha < 1$, the enthalpy stays finite). An example of such behaviour is the 3-dimensional ferromagnetic phase transition. In the three-dimensional Ising model for uniaxial magnets, detailed theoretical studies have yielded the exponent $\alpha \sim +0.110$.

Some model systems do not obey a power-law behaviour. For example, mean field theory predicts a finite discontinuity of the heat capacity at the transition temperature, and the two-dimensional Ising model has a logarithmic divergence. However, these systems are limiting cases and an exception to the rule. Real phase transitions exhibit power-law behaviour.

Several other critical exponents - β, γ, δ, ν, and η - are defined, examining the power law behaviour of a measurable physical quantity near the phase transition. Exponents are related by scaling relations such as $\beta = \gamma/(\delta - 1)$, $\nu = \gamma/(2 - \eta)$. It can be shown that there are only two independent exponents, e.g. ν and η.

It is a remarkable fact that phase transitions arising in different systems often possess the same set of critical exponents. This phenomenon is known as *universality*. For example, the critical exponents at the liquid-gas critical point have been found to be independent of the chemical composition of the fluid. More amazingly, but understandable from above, they are an exact match for the critical exponents of the ferromagnetic phase transition in uniaxial magnets. Such systems are said to be in the same universality class. Universality is a prediction of the renormalization group theory of phase transitions, which states that the thermodynamic properties of a system near a phase transition depend only on a small number of features, such as dimensionality and symmetry, and are insensitive to the underlying microscopic properties of the system. Again, the divergency of the correlation length is the essential point.

Critical Slowing Down and Other Phenomena

There are also other critical phenomena; e.g., besides *static functions* there is also *critical dynamics*. As a consequence, at a phase transition one may observe critical slowing down or *speeding up*. The large *static universality classes* of a continuous phase transition split into smaller *dynamic universality* classes. In addition to the critical exponents, there are also universal relations for certain static or dynamic functions of the magnetic fields and temperature differences from the critical value.

Percolation Theory

Another phenomenon which shows phase transitions and critical exponents is percolation. The simplest example is perhaps percolation in a two dimensional square lattice. Sites are randomly occupied with probability p. For small values of p the occupied sites form only small clusters. At a certain threshold p_c a giant cluster is formed and we have a second order phase transition. The behaviour of P_∞ near p_c is, $P_\infty \sim (p-p_c)^\beta$, where β is a critical exponent.

Clausius Clapeyron Equation

The Clausius–Clapeyron relation, named after Rudolf Clausius and Benoît Paul Émile Clapeyron, is a way of characterizing a discontinuous phase transition between two phases of matter of a single constituent. On a pressure–temperature (P–T) diagram, the line separating the two phases is known as the coexistence curve. The Clausius–Clapeyron relation gives the slope of the tangents to this curve. Mathematically,

$$\frac{\mathrm{d}P}{\mathrm{d}T} = \frac{L}{T\Delta v},$$

where $\mathrm{d}P/\mathrm{d}T$ is the slope of tangent to the coexistence curve at any point, L is the specific latent heat, T is the temperature, and Δv is the specific volume change of the phase transition.

Derivations

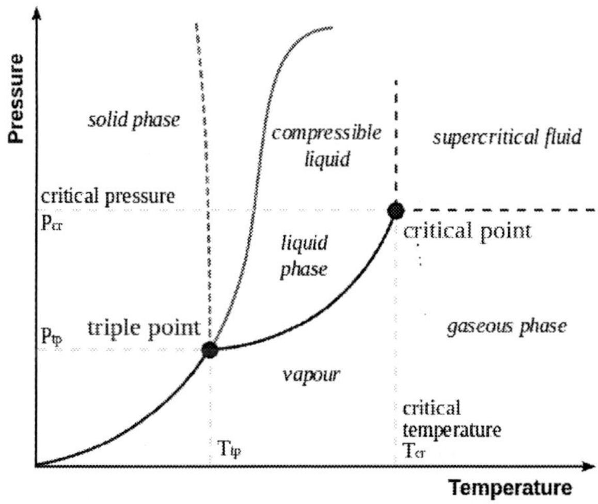

Figure: *A typical phase diagram. The dotted green line gives the anomalous behaviour of water. The Clausius–Clapeyron relation can be used to find the relationship between pressure and temperature along phase boundaries.*

Derivation from State Postulate

Using the state postulate, take the specific entropy, s, for a homogeneous substance to be a function of specific volume v and temperature T.

$$ds = \left(\frac{\partial s}{\partial v}\right)_T dv + \left(\frac{\partial s}{\partial T}\right)_v dT.$$

During a phase change, the temperature is constant, so

$$ds = \left(\frac{\partial s}{\partial v}\right)_T dv.$$

Using the appropriate Maxwell relationgives

$$ds = \left(\frac{\partial P}{\partial T}\right)_v dv.$$

Since temperature and pressure are constant *during a phase transition*, the derivative of pressure with respect to temperature is not a function of the specific volume. Thus the partial derivative may be changed into a total derivative and be factored out when taking an integral from one phase to another,

$$s_\beta - s_\alpha = \frac{dP}{dT}(v_\beta - v_\alpha),$$

$$\frac{dP}{dT} = \frac{s_\beta - s_\alpha}{v_\beta - v_\alpha} = \frac{\Delta s}{\Delta v}.$$

Here $\Delta s \equiv s_\beta - s_\alpha$ and $\Delta v \equiv v_\beta - v_\alpha$ are respectively the change in specific entropy and specific volume from the initial phase α to the final phase β.

For a closed system undergoing an internally reversible process, the first law is

$$du = \delta q - \delta w = T\, ds - P\, dv.$$

Using the definition of specific enthalpy, h, and the fact that the temperature and pressure are constant, we have

$$du + P\, dv = dh = T\, ds \Rightarrow ds = \frac{dh}{T} \Rightarrow \Delta s = \frac{\Delta h}{T} = \frac{L}{T}.$$

After substitution of this result into the derivative of the pressure, one finds

$$\frac{dP}{dT} = \frac{L}{T\Delta v}.$$

This last equation is the Clapeyron equation.

Derivation from Gibbs–Duhem Relation

Suppose two phases, α and β, are in contact and at equilibrium with each other. Then the chemical potentials are related by $\mu_\alpha = \mu_\beta$. Along the coexistence curve, we also have $d\mu_\alpha = d\mu_\beta$. We now use the Gibbs–Duhemrelation $d\mu = M(-s dT + v dP)$, where s and v are, respectively, the specific entropy and specific volume and M is the molar mass, to obtain

$$-(s_\beta - s_\alpha)dT + (v_\beta - v_\alpha)dP = 0.$$

Hence, rearranging, we have

$$\frac{dP}{dT} = \frac{s_\beta - s_\alpha}{v_\beta - v_\alpha}.$$

From which the derivation continues as above. .

Ideal Gas Approximation at Low Temperatures

When the transition is to a gas phase at temperatures much lower than the critical temperature, the specific volume of the gas phase v_g greatly exceeds that of the condensed phase v_c, and so one may approximate $\Delta v = v_g\left(1 - \frac{v_c}{v_g}\right) \approx v_g$. Furthermore, at low pressures, the gas may be approximated by the ideal gas law, so that $v_g = RT / P$, where R is the specific gas constant. Thus,

$$\frac{dP}{dT} = \frac{PL}{T^2 R}.$$

This is the Clausius–Clapeyron equation. In general, L varies along the coexistence curve as a function of temperature. However, if L is a constant, then

$$\frac{dP}{P} = \frac{L}{R}\frac{dT}{T^2},$$

$$\int_{P_1}^{P_2} \frac{dP}{P} = \frac{L}{R}\int \frac{dT}{T^2}$$

$$\ln P\Big|_{P=P_1}^{P_2} = -\frac{L}{R}\cdot\frac{1}{T}\Big|_{T=T_1}^{T_2} \quad \text{or}$$

$$\ln\frac{P_2}{P_1} = \frac{L}{R}\left(\frac{1}{T_1} - \frac{1}{T_2}\right).$$

Here (P_1,T_1) and (P_2,T_2) are two points along the coexistence curve. These last equations are useful because they relate equilibrium or saturation vapour pressure and temperature to the latent heat of the phase change, *without* requiring specific volume data. Note that in this last equation, the subscripts 1 and 2 label different locations on the *P–T* coexistence curve, whereas in previous equations the subscripts α and β label the phases on either side of the coexistence curve.

Applications

Chemistry and Chemical Engineering: For transitions between a gas and a condensed phase with the approximations described above, the expression may be rewritten as

$$\ln P = -\frac{L}{R}\left(\frac{1}{T}\right) + C.$$

For a liquid-gas transition, L is the specific latent heat (or specific enthalpy) of vaporization, whereas for a solid-gas transition, L is the specific latent heat of sublimation. If the latent heat is known, then knowledge of one point on the coexistence curve determines the rest of the curve. Conversely, the relationship between $\ln P$ and $1/T$ is linear, and so linear regression is used to estimate the latent heat.

Meteorology and Climatology

The Clausius–Clapeyron equation for water vapour in typical atmospheric conditions is

$$\frac{\mathrm{d}e_s}{\mathrm{d}T} = \frac{L_v(T)e_s}{R_v T^2}$$

where:

- e_s is saturation water vapour pressure,
- T is a temperature,
- L_v is the specific latent heat of evaporation,
- R_v is water vapour gas constant.

The temperature dependence of the latent heat cannot be neglected in this application. In this context, a very good approximation can usually be made using the August-Roche-Magnus formula (usually called the Magnus or Magnus-Tetens approximation, though this attribution is historically inaccurate):

$$e_s(T) = 6.1094\exp\left(\frac{17.625T}{T + 243.04}\right).$$

$e_s(T)$ is the equilibrium or saturation vapour pressure in hPa as a function of temperature T on the Celsius scale. Since there is only a weak dependence on temperature of the denominator of the exponent, this equation shows that saturation water vapour pressure changes approximately exponentiallywith T.

Example

One of the uses of this equation is to determine if a phase transition will occur in a given situation. Consider the question of how much pressure is needed to melt ice at a temperature ΔT below 0°C. Note that water is unusual in that its change in volume upon melting is negative. We can assume

$$\Delta P = \frac{L}{T\Delta v}\Delta T$$

and substituting in

$L = 3.34\times10^5$ J/kg (latent heat of fusion for water),

$T = 273$ K (absolute temperature), and

$\Delta v = -9.05\times10^{-5}$ m³/kg (change in specific volume from solid to liquid),

we obtain

$$\frac{\Delta P}{\Delta T} = -13.5 \text{ MPa/K}.$$

To provide a rough example of how much pressure this is, to melt ice at –7 °C (the temperature many ice skating rinks are set at) would require balancing a small car (mass = 1000 kg) on a thimble (area = 1 cm²).

Various Thermodynamic Processes and Energy Transfer

A thermodynamic process is the energetic development of a thermodynamic system proceeding from an initial state to a final state. Paths through the space of thermodynamic variables are often specified by holding certain thermodynamic variables constant. A state function is a thermodynamic variable which depends only on the current state of the system, not the path taken to reach that state. Conversely a process function does depend on the path.

Overview

A thermodynamic process can be visualized by graphically plotting the changes to the system's state variables. In the example, four processes are shown. Each process has a well-defined start and end point in the pressure-volume state space. In this particular example, processes 1 and 3 are isothermal, whereas processes 2 and 4 are isochoric. The PV diagram is a particularly useful visualization of a process, because the area under the curve of a process is the amount of work done by the system during that process. Thus work is considered to be a process variable, as its exact value depends on the particular path taken between the start and end points of the process. Similarly, heat may be transferred during a process, and it too is a process variable. In contrast, pressure and volume (as well as numerous other properties) are considered state variables because their values depend only on the position of the start and end points, not the particular path between them.

Conjugate Variable Processes

It is often useful to group processes into pairs, in which each variable held constant is one member of a conjugate pair.

Pressure – Volume

The pressure-volume conjugate pair is concerned with the transfer of mechanical or dynamic energy as the result of work.

- An isobaric process occurs at constant pressure. An example would be to have a movable piston in a cylinder, so that the pressure inside the cylinder is always at atmospheric pressure, although it is isolated from the atmosphere. In other words, the system is dynamically connected, by a movable boundary, to a constant-pressure reservoir.
- An isochoric process is one in which the volume is held constant, meaning that the work done by the system will be zero. It follows that, for the simple system of two dimensions, any heat energy transferred to the system externally will be absorbed as internal energy. An isochoric process is also known as an isometric process or an isovolumetric process. An example would be to place a closed tin can containing only air into a fire. To a first approximation, the can will not expand, and the only change will be that the gas gains internal energy, as evidenced by its increase in temperature and pressure. Mathematically, $\delta Q = dU$. We may say that the system is dynamically insulated, by a rigid boundary, from the environment.

Temperature – Entropy

The temperature-entropy conjugate pair is concerned with the transfer of thermal energy as the result of heating.

- An isothermal process occurs at a constant temperature. An example would be to have a system immersed in a large constant-temperature bath. Any work energy performed by the system will be lost to the bath, but its temperature will remain constant. In other words, the system is thermally connected, by a thermally conductive boundary to a constant-temperature reservoir.
- An adiabatic process is a process in which there is no energy added or subtracted from the system by heating or cooling. For a reversible process, this is identical to an isentropic process. We may say that the system is thermally insulated from its environment and that its boundary is a thermal insulator. If a system has an entropy which has not yet reached its maximum equilibrium value, the entropy will increase even though the system is thermally insulated. Under certain conditions two states of a system may be considered adiabatically accesisble.
- An isentropic process occurs at a constant entropy. For a reversible process this is identical to an adiabatic process. If a system has an entropy which has not yet reached its maximum equilibrium value, a process of cooling may be required to maintain that value of entropy.

Chemical Potential - Particle Number

The processes above have all implicitly assumed that the boundaries are also impermeable to particles. We may assume boundaries that are both rigid and thermally insulating, but are permeable to one or more types of particle. Similar considerations then hold for the chemical potential–particle number conjugate pair, which is concerned with the transfer of energy via this transfer of particles.

- In a constant chemical potential process the system is *particle-transfer connected*, by a particle-permeable boundary, to a constant-μ reservoir.
- In a constant particle number process there is no energy added or subtracted from the system by particle transfer. We may say that the system is *particle-transfer-insulated* from its environment by a boundary that is impermeable to particles.

Thermodynamic Potentials

Any of the thermodynamic potentials may be held constant during a process. For example:

- An isenthalpic process introduces no change in enthalpy in the system.

Polytropic Processes

A polytropic process is a thermodynamic process that obeys the relation:

$$PV^n = C,$$

where P is the pressure, V is volume, n is any real number (the "polytropic index"), and C is a constant. This equation can be used to accurately characterize processes of certain systems, notably the compression or expansion of a gas, but in some cases, liquids and solids.

Quasistatic Process

A quasistatic process is an idealized model of a thermodynamic process that happens infinitely slowly. It is important to note that *no* real process is quasistatic. In practice, such processes can only be approximated by performing them *infinitesimally* slowly. A quasistatic process often ensures that the system will go through a sequence of states that are infinitesimally close to equilibrium (so the system remains in quasistatic equilibrium), in which case the process is typically reversible.

Perfect Gas Laws

The ideal gas law is the equation of state of a hypothetical ideal gas. It is a good approximation to the behaviour of many gases under many conditions, although it has several limitations. It was first stated by Émile Clapeyron in 1834 as a combination of Boyle's law and Charles's law. The ideal gas law is often introduced in its common form:

$$PV = nRT$$

where P is the pressure of the gas, V is the volume of the gas, n is the amount of substance of gas (also known as number of moles), T is the temperature of the gas and R is the ideal, or universal, gas constant.

It can also be derived from kinetic theory, as was achieved (apparently independently) by August Krönig in 1856 and Rudolf Clausius in 1857. Universal gas constant was discovered and first introduced into the ideal gas law instead of a large number of specific gas constants by Dmitri Mendeleev in 1874.

Equation

The state of an amount of gas is determined by its pressure, volume, and temperature. The modern form of the equation relates these simply in two main forms. The temperature used in the equation of state is an absolute temperature: in the SI system of units, kelvin; in the Imperial system, degrees Rankine.

Common Form

The most frequently introduced form is

$$PV = nRT$$

where P is the pressure of the gas, V is the volume of the gas, n is the amount of substance of gas (also known as number of moles), T is the temperature of the gas and R is the ideal, or universal, gas constant, equal to the product of the Boltzmann constant and the Avogadro constant.

In SI units, P is measured in pascals, V is measured in cubic metres, n is measured in moles, and T in kelvin (273.15 Kelvin = 0.00 degrees Celsius). R has the value 8.314 $J \cdot K^{-1} \cdot mol^{-1}$ or 0.08206 $L \cdot atm \cdot mol^{-1} \cdot K^{-1}$ if using pressure in standard atmospheres (atm) instead of pascals, and volume in liters instead of cubic metres.

Molar Form

How much gas is present could be specified by giving the mass instead of the chemical amount of gas. Therefore, an alternative form of the ideal gas law may be useful. The chemical amount (n) (in moles) is equal to the mass (m) (in grams) divided by the molar mass (M) (in grams per mole):

$$n = \frac{m}{M}$$

By replacing n with m / M, and subsequently introducing densityρ = m/V, we get:

$$PV = \frac{m}{M} RT$$

$$P = \rho \frac{R}{M} T$$

Defining the specific gas constantR_{specific} as the ratio R/M,

$$P = \rho R_{\text{specific}} T$$

This form of the ideal gas law is very useful because it links pressure, density, and temperature in a unique formula independent of the quantity of the considered gas. Alternatively, the law may be written in terms of the specific volume$v$, the reciprocal of density, as

$$Pv = R_{\text{specific}}T$$

It is common, especially in engineering applications, to represent the specific gas constant by the symbol *R*. In such cases, the universal gas constant is usually given a different symbol such as *R* to distinguish it. In any case, the context and/or units of the gas constant should make it clear as to whether the universal or specific gas constant is being referred to.

Statistical Mechanics

In statistical mechanics the following molecular equation is derived from first principles:

$$PV = NkT$$

where *P* is the absolute pressure of the gas measured in Pascals; *V* is the volume (in this equation the volume is expressed in metres cubed, as pascal times cubic metre equals one Joule); *N* is the number of particles in the gas; *k* is the Boltzmann constant relating temperature and energy; and *T* is the absolute temperature.

The *actual number* of molecules contrasts to the other formulation, which uses *n*, the *number of moles*. This relation implies that $Nk = nR$, and the consistency of this result with experiment is a good check on the principles of statistical mechanics.

From this we can notice that for an average particle mass of μ times the atomic mass constantm_u (i.e., the mass is μu)

$$N = \frac{m}{\mu m_u}$$

and since $\rho = m/V$, we find that the ideal gas law can be rewritten as:

$$P = \frac{1}{V}\frac{m}{\mu m_u}kT = \frac{k}{\mu m_u}\rho T.$$

In SI units, *P* is measured in pascals; *V* in cubic metres; *N* is a dimensionless number; and *T* in kelvin. *k* has the value $1.38 \cdot 10^{-23} \text{J} \cdot \text{K}^{-1}$ in SI units.

Applications to Thermodynamic Processes

The table below essentially simplifies the ideal gas equation for a particular processes, thus making this equation easier to solve using numerical methods.

A thermodynamic process is defined as a system that moves from state 1 to state 2, where the state number is denoted by subscript. As shown in the first column of the table, basic thermodynamic processes are defined such that one of the gas properties (*P*, *V*, *T*, or *S*) is constant throughout the process.

For a given thermodynamics process, in order to specify the extent of a particular process, one of the properties ratios (listed under the column libelled "known ratio") must be specified (either directly or indirectly). Also, the property for which the ratio is known must be distinct from the property held constant in the previous column (otherwise the ratio would be unity, and not enough information would be available to simplify the gas law equation).

In the final three columns, the properties (*P*, *V*, or *T*) at state 2 can be calculated from the properties at state 1 using the equations listed.

Process	***Constant***	***Known ratio***	***P₂***	***V₂***	***T₂***
Isobaric process	Pressure	V_2/V_1	$P_2 = P_1$	$V_2 = V_1(T_2/T_1)$	$T_2 = T_1(V_2/V_1)$
		T_2/T_1	$P_2 = P_1$	$V_2 = V_1(T_2/T_1)$	$T_2 = T_1(T_2/T_1)$
Isochoric process (Isovolumetric process) (Isometric process)	Volume	P_2/P_1	$P_2 = P_1(T_2/T_1)$	$V_2 = V_1$	$T_2 = T_1(P_2/P_1)$
		T_2/T_1	$P_2 = P_1(T_2/T_1)$	$V_2 = V_1$	$T_2 = T_1(T_2/T_1)$
Isothermal process	Temperature	P_2/P_1	$P_2 = P_1(T_2/T_1)$	$V_2 = V_1/(P_2/P_1)$	$T_2 = T_1$
		V_2/V_1	$P_2 = P_1/(V_2/V_1)$	$V_2 = V_1(V_2/V_1)$	$T_2 = T_1$
Isentropic process (Reversible adiabatic process)	Entropy	P_2/P_1	$P_2 = P_1(V_2/V_1)$	$V_2 = V_1(P_2/P_1)^{(-1/\gamma)}$	$T_2 = T_1(P_2/P_1)^{(1-1/\gamma)}$
		V_2/V_1	$P_2 = P_1(V_2/V_1)^{-\gamma}$	$V_2 = V_1(P_2/P_1)$	$T_2 = T_1(V_2/V_1)^{(1-\gamma)}$
		T_2/T_1	$P_2 = P_1(T_2/T_1)^{\gamma/(\gamma-1)}$	$V_2 = V_1(T_2/T_1)^{1/(1-\gamma)}$	$T_2 = T_1(T_2/T_1)$
Polytropic process	$P V^n$	P_2/P_1	$P_2 = P_1(P_2/P_1)$	$V_2 = V_1(P_2/P_1)^{(-1/n)}$	$T_2 = T_1(P_2/P_1)^{(1-1/n)}$
		V_2/V_1	$P_2 = P_1(V_2/V_1)^{-n}$	$V_2 = V_1(V_2/V_1)$	$T_2 = T_1(V_2/V_1)^{(1-n)}$
		T_2/T_1	$P_2 = P_1(T_2/T_1)^{n/(n-1)}$	$V_2 = V_1(T_2/T_1)^{1/(1-n)}$	$T_2 = T_1(T_2/T_1)$

In an isentropic process, system entropy (*S*) is constant. Under these conditions, $P_1V_1^{\gamma} = P_2V_2^{\gamma}$, where γ is defined as the heat capacity ratio, which is constant for an ideal gas. The value used for γ is typically 1.4 for diatomic gases like nitrogen (N_2) and oxygen (O_2), (and air, which is 99% diatomic). Also γ is typically 1.6 for monatomic gases like the noble gaseshelium (He), and argon (Ar). In internal combustion engines γ varies between 1.35 and 1.15, depending on constitution gases and temperature.

Deviations From Ideal Behaviour of Real Gases

The equation of state given here applies only to an ideal gas, or as an approximation to a real gas that behaves sufficiently like an ideal gas. There are in fact many different forms of the equation of state. Since the ideal gas law neglects both molecular size and intermolecular attractions, it is most accurate for monatomic gases at high temperatures and low pressures. The neglect of molecular size becomes less important for lower densities, i.e. for larger volumes at lower pressures, because the average distance between adjacent molecules becomes much larger than the molecular size. The relative importance of intermolecular attractions diminishes with increasing thermal kinetic energy, i.e., with increasing temperatures. More detailed *equations of state*, such as the van der Waals equation, account for deviations from ideality caused by molecular size and intermolecular forces.

A residual property is defined as the difference between a real gas property and an ideal gas property, both considered at the same pressure, temperature, and composition.

Derivations

Theoretical: Kinetic theory The ideal gas law can also be derived from first principles using the kinetic theory of gases, in which several simplifying assumptions are made, chief among which are that the molecules, or atoms, of the gas are point masses, possessing mass but no significant volume, and undergo only elastic collisions with each other and the sides of the container in which both linear momentum and kinetic energy are conserved.

Statistical Mechanics

Let $q = (q_x, q_y, q_z)$ and $p = (p_x, p_y, p_z)$ denote the position vector and momentum vector of a particle of an ideal gas, respectively. Let F denote the net force on that particle. Then the time average momentum of the particle is:

$$\begin{aligned}\langle q \cdot F \rangle &= \left\langle q_x \frac{dp_x}{dt} \right\rangle + \left\langle q_y \frac{dp_y}{dt} \right\rangle + \left\langle q_z \frac{dp_z}{dt} \right\rangle \\ &= -\left\langle q_x \frac{\partial H}{\partial q_x} \right\rangle - \left\langle q_y \frac{\partial H}{\partial q_y} \right\rangle - \left\langle q_z \frac{\partial H}{\partial q_z} \right\rangle = -3k_B T,\end{aligned}$$

where the first equality is Newton's second law, and the second line uses Hamilton's equations and the equipartition theorem. Summing over a system of N particles yields

$$3Nk_BT = -\left\langle \sum_{k=1}^{N} \mathrm{q}_k \cdot \mathrm{F}_k \right\rangle.$$

By Newton's third law and the ideal gas assumption, the net force on the system is the force applied by the walls of their container, and this force is given by the pressure P of the gas. Hence

$$-\left\langle \sum_{k=1}^{N} \mathrm{q}_k \cdot \mathrm{F}_k \right\rangle = P\oint_{\text{surface}} \mathrm{q} \cdot d\mathrm{S},$$

where dS is the infinitesimal area element along the walls of the container. Since the divergence of the position vector q is

$$\nabla \cdot \mathrm{q} = \frac{\partial q_x}{\partial q_x} + \frac{\partial q_y}{\partial q_y} + \frac{\partial q_z}{\partial q_z} = 3,$$

thedivergence theorem implies that

$$P\oint_{\text{surface}} \mathrm{q} \cdot d\mathrm{S} = P\int_{\text{volume}} (\nabla \cdot \mathrm{q})\, dV = 3PV,$$

where dV is an infinitesimal volume within the container and V is the total volume of the container.

Putting these equalities together yields

$$3Nk_BT = -\left\langle \sum_{k=1}^{N} \mathrm{q}_k \cdot \mathrm{F}_k \right\rangle = 3PV,$$

which immediately implies the ideal gas law for N particles:

$$PV = Nk_BT = nRT,$$

where $n = N/N_A$ is the number of moles of gas and $R = N_A k_B$ is the gas constant.

Changes in Internal Energy

In thermodynamics, the internal energy is the total energy contained by a thermodynamic system. It is the energy needed to create the system but excludes the energy to displace the system's surroundings, any energy associated with a move as a whole, or due to external force fields. Internal energy has two major components, kinetic energy and potential energy. The kinetic energy is due to the motion of the system's particles (translations, rotations, vibrations), and the potential energy is associated with the static rest mass energy of the constituents of matter, static electric energy of atoms within molecules or crystals, and the static energy of chemical bonds. The internal energy of a system can be changed by heating the system or by doing work on it; the first law of thermodynamics states that the increase in internal energy is equal to the total heat added and work done by the surroundings. If the system is isolated from its surroundings, its internal energy cannot change.

For practical considerations in thermodynamics and engineering it is rarely necessary or convenient to consider all energies belonging to the total intrinsic energy of a sample system, such as the energy given by the equivalence of mass. Typically, descriptions only include components relevant to the system under study. Thermodynamics is chiefly concerned only with *changes* in the internal energy.

The internal energy is a state function of a system, because its value depends only on the current state of the system and not on the path taken or process undergone to arrive at this state. It is an extensive quantity. The SI unit of energy is the joule (J). Some authors use a corresponding intensivethermodynamic property called specific internal energy which is internal energy per unit of mass (kilogram) of the system in question. The SI unit of specific internal energy is J/kg. If intensive internal energy is expressed relative to units of amount of substance (mol), then it is referred to as *molar internal energy* and the unit is J/mol.

From the standpoint of statistical mechanics, the internal energy is equal to the ensemble average of the total energy of the system. It is also called intrinsic energy.

Description and Definition

The internal energy (U) is the sum of all forms of energy (E_i) intrinsic to a thermodynamic system:

$$U = \sum_i E_i$$

It is the energy needed to create the system. It may be divided into potential energy (U_{pot}) and kinetic energy (U_{kin}) components:

$$U = U_{pot} + U_{kin}$$

The kinetic energy of a system arises as the sum of the motions of all the system's particles, whether it be the motion of atoms, molecules, atomic nuclei, electrons, or other particles. The potential energy includes all energies given by the mass of particles, by the chemical composition, i.e. the chemical energy stored in chemical bonds having the potential to undergo chemical reactions, the nuclear energy stored by the configuration of protons, neutrons, and other elementary particles in atomic nuclei, and the physical force fields within the system, such as due to internal induced electric or magnetic dipolemoment, as well as the energy of deformation of solids (stress-strain).

Internal energy does not include the energy due to motion of a system as a whole. It further excludes any kinetic or potential energy the body may have because of its location in external gravitational, electrostatic, or electomagnetic fields. It does, however, include the contribution to the energy due to the coupling of the internal degrees of freedom of the object to such the field. In such a case, the field is included in the thermodynamic description of the object in the form of an additional external parametre.

For practical considerations in thermodynamics or engineering, it is rarely necessary, convenient, nor even possible, to consider all energies belonging to the total intrinsic energy of a sample system, such as the energy given by the equivalence of mass. Typically, descriptions only include components relevant to the system under study. Indeed in most systems under consideration, especially through thermodynamics, it is impossible to calculate the total internal energy. Therefore, a convenient null reference point may be chosen for the internal energy.

The internal energy is an extensive property: it depends on the size of the system, or on the amount of substance it contains.

At any temperature greater than absolute zero, potential energy and kinetic energy are constantly converted into one another, but the sum remains constant in an isolated system. In the classical picture of thermodynamics, kinetic energy vanishes at zero temperature and the internal energy is purely potential energy. However, quantum mechanics has demonstrated that even at zero temperature particles maintain a residual energy of motion, the zero point energy. A system at absolute zero is merely in its quantum-mechanical ground state, the lowest energy state available. At absolute zero a system has attained its minimum attainable entropy.

The kinetic energy portion of the internal energy gives rise to the temperature of the system. Statistical mechanics relates the pseudo-random kinetic energy of individual particles to the mean kinetic energy of the entire ensemble of particles comprising a system. Furthermore it relates the mean kinetic energy to the macroscopically observed empirical property that is expressed as temperature of the system. This energy is often referred to as the *thermal energy* of a system, relating this energy, like the temperature, to the human experience of hot and cold.

Statistical mechanics considers any system to be statistically distributed across an ensemble of Nmicrostates. Each microstate has an energyE_i and is associated with a probability p_i. The internal energy is the mean value of the system's total energy, i.e., the sum of all microstate energies, each weighted by their probability of occurrence:

$$U = \sum_{i=1}^{N} p_i E_i .$$

This is the statistical expression of the first law of thermodynamics.

Internal Energy Changes

Thermodynamics is chiefly concerned only with the *changes*, ΔU, in internal energy. The most important parametres in thermodynamics when considering the changes in total energy are the changes due to the flow of heat Q and due to mechanical work, i.e. from changes in volume of the system under an external pressure. Accordingly, the internal energy change ΔU for a process may be written more specifically as

$$\Delta U = Q + W_{mech} + W_{extra}$$

where Q is heat added to a system and W_{mech} is the mechanical work performed by the surroundings due to pressure or volume changes in the system. All other perturbations and energies added by other processes, such as an electric current introduced into an electronic circuit, is summarized as the term W_{extra}.

When a system is heated, it receives energy in form of heat. This energy increases the internal energy. However, it may be extremely difficult to determine how this extra energy is stored. In general, except in an ideal gas, it is redistributed between kinetic and potential energy. The net increase in kinetic energy is measurable by an increase in the temperature of the system. The equipartition theorem states that increase in thermal energy is distributed between the available degrees of freedom of the fundamental oscillators in the system. In an ideal gas all of the extra energy results in a temperature increase, as it is stored solely as kinetic energy. The heat introduced to a system while the temperature changes is often called sensible heat.

Another method to change the internal energy of a system is by performing work on the system, either in mechanical form by changing pressure or volume, or by other perturbations, such as directing an electrical current through the system. Finally, the internal energy increases when additional mass is transferred into the system. If a system undergoes certain phase transformations while being heated, such as melting and vaporization, it may be observed that the temperature of the system does not change until the entire sample has completed the transformation. The energy introduced into the system while the temperature did not change is called a *latent energy*, or latent heat, in contrast to sensible heat. It increases only the potential energy of the system, but not its thermal energy component.

Internal Energy of the Ideal Gas

Thermodynamics often uses the concept of the ideal gas for teaching purposes, and as an approximation for working systems. The ideal gas is a gas of particles considered as point objects that interact only by elastic collisions and fill a volume such that their free mean path between collisions is much larger than their diametre. Such systems are approximated by the monoatomic gases, helium and the other noble gases. Here the kinetic energy consists only of the translational energy of the individual atoms. Monoatomic particles do not rotate or vibrate, and are not electronically excited to higher energies except at very high temperatures.

Therefore practical internal energy changes in an ideal gas may be described solely by changes in its kinetic energy. Kinetic energy is simply the internal energy of the perfect gas and depends entirely on its pressure, volume and thermodynamic temperature. The internal energy of an ideal gas is proportional to its mass (number of moles) N and to its temperature T

$$U = cNT,$$

where c is the heat capacity (at constant volume) of the gas. The internal energy may be written as a function of the three extensive properties S, V, N (entropy, volume, mass) in the following way

$$U(S,V,T) = const \cdot e^{\frac{S}{cN}} V^{\frac{-R}{c}} N^{\frac{R+c}{c}},$$

where *const* is an arbitrary positive constant and where R is the Universal Gas Constant. It is easily seen that U is a linearly homogeneous function of the three variables and that it is weakly convex. Knowing temperature and pressure to be the derivatives $T = \frac{\partial U}{\partial S}$, $p = -\frac{\partial U}{\partial V}$, the Ideal Gas Law $pV = RNT$ immediately follows.

Internal Energy of a Closed Thermodynamic System

This above summation of all components of change in internal energy assume that a positive energy denotes heat added to the system or work done on the system, while a negative energy denotes work of the system on the environment. Typically this relationship is expressed in infinitesimal terms using the differentials of each term. Only the internal energy is an exact differential. For a system undergoing only thermodynamics processes, i.e. a closed system that can exchange only heat and work, the change in the internal energy is

$$dU = \delta Q + \delta W$$

which constitutes the first law of thermodynamics. It may be expressed in terms of other thermodynamic parametres. Each term is composed of an intensive variable (a generalized force) and its conjugate infinitesimal extensive variable (a generalized displacement). For example, for a non-viscous fluid, the mechanical work done on the system may be related to the pressure$p$ and volume$V$. The pressure is the intensive generalized force, while the volume is the extensive generalized displacement:

$$\delta W = -p\mathrm{d}V\ .$$

This defines the direction of work, W, to be energy flow from the working system to the surroundings, indicated by a negative term. Taking the direction of heat transfer Q to be into the working fluid and assuming a reversible process, the heat is

$$\delta Q = T\mathrm{d}S\ .$$

T istemperature

S isentropy

and the change in internal energy becomes

$$\mathrm{d}U = T\mathrm{d}S - p\mathrm{d}V$$

Changes Due to Temperature and Volume

The expression relating changes in internal energy to changes in temperature and volume is

$$dU = C_V dT + \left[T\left(\frac{\partial p}{\partial T}\right)_V - p\right]dV \qquad (1)$$

This is useful if the equation of state is known.

In case of an ideal gas, we can derive that $dU = C_v dT$, i.e. the internal energy of an ideal gas can be written as a function that depends only on the temperature.

Changes Due to Temperature and Pressure

When dealing with fluids or solids, an expression in terms of the temperature and pressure is usually more useful:

$$dU = \left(C_p - \alpha pV\right)dT + \left(\beta_T p - \alpha T\right)Vdp$$

where it is assumed that the heat capacity at constant pressure is related to the heat capacity at constant volume according to:

$$C_p = C_V + VT\frac{\alpha^2}{\beta_T}$$

Derivation of dU in terms of dT and dP.

Changes Due to Volume at Constant Pressure

The internal pressure is defined as a partial derivative of the internal energy with respect to the volume at constant temperature:

$$\pi_T = \left(\frac{\partial U}{\partial V}\right)_T$$

Internal Energy of Multi-component Systems

In addition to including the entropy S and volume V terms in the internal energy, a system is often described also in terms of the number of particles or chemical species it contains:

$$U = U(S, V, N_1, \ldots, N_n)$$

where N_j are the molar amounts of constituents of type j in the system. The internal energy is an extensive function of the extensive variables S, V, and the amounts N_j, the internal energy may be written as a linearly homogeneous function of first degree:

$$U(\alpha S, \alpha V, \alpha N_1, \alpha N_2, \ldots) = \alpha U(S, V, N_1, N_2, \ldots)$$

where α is a factor describing the growth of the system. The differential internal energy may be written as

$$dU = \frac{\partial U}{\partial S}dS + \frac{\partial U}{\partial V}dV + \sum_i \frac{\partial U}{\partial N_i}dN_i = T\,dS - p\,dV + \sum_i \mu_i dN_i$$

which shows temperature T to be the partial derivative of U with respect to entropy S and pressure p to be the negative of the similar derivative with respect to volume V

$$T = \frac{\partial U}{\partial S},$$

$$p = -\frac{\partial U}{\partial V},$$

and where the coefficients μ_i are the chemical potentials for the components of type i in the system. The chemical potentials are defined as the partial derivatives of the energy with respect to the variations in composition:

$$\mu_i = \left(\frac{\partial U}{\partial N_i}\right)_{S,V,N_{j\neq i}}$$

As conjugate variables to the composition $\{N_j\}$, the chemical potentials are intensive properties, intrinsically characteristic of the system, and not dependent on its extent. Because of the extensive nature of U and its variables, the differential dU may be integrated and yields an expression for the internal energy:

$$U = TS - pV + \sum_i \mu_i N_i \, .$$

The sum over the composition of the system is the Gibbs energy:

$$G = \sum_i \mu_i N_i$$

that arises from changing the composition of the system at constant temperature and pressure. For a single component system, the chemical potential equals the Gibbs energy per amount of substance, i.e. particles or moles according to the original definition of the unit for $\{N_j\}$.

Internal Energy in an Elastic Medium

For an elastic medium the mechanical energy term of the internal energy must be replaced by the more general expression involving the stress σ_{ij} and strain ε_{ij}. The infinitesimal statement is:

$$\mathrm{d}U = T\mathrm{d}S + V\sigma_{ij}\mathrm{d}\varepsilon_{ij}$$

where E instein notation has been used for the tensors, in which there is a summation over all repeated indices in the product term. The Euler theorem yields for the internal energy:

$$U = TS + \frac{1}{2}\sigma_{ij}\varepsilon_{ij}$$

For a linearly elastic material, the stress can be related to the strain by:

$$\sigma_{ij} = C_{ijkl}\varepsilon_{kl}$$

Where C_{ijkl} is an element of the 4th-rank elastic constant tensor of the medium.

Computational Methods

The path integral Monte Carlo method is a numerical approach for determining the values of the internal energy, based on quantum dynamical principles.

History

James Joule studied the relationship between heat, work, and temperature. He observed that if he did mechanical work on a fluid, such as water, by agitating the fluid, its temperature increased. He proposed that the mechanical work he was doing on the system was converted to *thermal energy*. Specifically, he found that 4185.5 joules of energy were needed to raise the temperature of a kilogram of water by one degree Celsius.

Throttling and Free Expansion Processes

In thermodynamics, the Joule–Thomson effect or Joule–Kelvin effect or Kelvin–Joule effect or Joule–Thomson expansion describes the temperature change of a gas or liquid when it is forced through a valve or porous plug while kept insulated so that no heat is exchanged with the environment. This procedure is called a *throttling process* or *Joule–Thomson process.* At room temperature, all gases except hydrogen, helium and neon cool upon expansion by the Joule–Thomson process. The effect is named for James Prescott Joule and William Thomson, 1st Baron Kelvin, who discovered it in 1852 following earlier work by Joule on Joule expansion, in which a gas undergoes free expansion in a vacuum.

In the Joule experiment, the gas expands in a vacuum and the temperature drop of the system is zero, if the gas were ideal. The throttling process is of the highest technical importance. It is at the heart of thermal machines such as refrigerators, air conditioners, heat pumps, and liquefiers. Furthermore, throttling is a fundamentally irreversible process. The throttling due to the flow resistance in supply lines, heat exchangers, regenerators, and other components of (thermal) machines is a source of losses that limits the performance.

Description

The *adiabatic* (no heat exchanged) expansion of a gas may be carried out in a number of ways. The change in temperature experienced by the gas during expansion depends not only on the initial and final pressure, but also on the manner in which the expansion is carried out.

- If the expansion process is reversible, meaning that the gas is in thermodynamic equilibrium at all times, it is called an *isentropic* expansion. In this scenario, the gas does positive work during the expansion, and its temperature decreases.
- In a free expansion, on the other hand, the gas does no work and absorbs no heat, so the internal energy is conserved. Expanded in this manner, the temperature of an ideal gas would remain constant, but the temperature of a real gas may either increase or decrease, depending on the initial temperature and pressure.
- The method of expansion discussed in this article, in which a gas or liquid at pressure P_1 flows into a region of lower pressure P_2 via a valve or porous plug under steady state conditions and without change in kinetic energy, is called the Joule–Thomson process. During this process, enthalpy remains unchanged.

A throttling process proceeds along a constant-enthalpy line in the direction of decreasing pressure, which means that the process occurs from right to left on a T-P diagram. As we proceed along a constant-enthalpy line from high enough pressures the temperature increases, until the inversion temperature. After this, as the fluid continues its expansion, the temperature drops. If we do this for several constant enthalpies and join the inversion points a line called the inversion line is obtained. This line intersects the T-axis at some temperature, named the maximum inversion temperature. For hydrogen this temperature is -68°C. In Vapour-compression refrigeration we need to throttle the gas and cool it at the same time. This poses a problem for substances whose maximum inversion temperature is well below room temperature. Thus hydrogen must be cooled below its inversion temperature if any cooling is achieved by throttling.

Thermodynamic Interpretation of the experiment:

If we consider that certain amount of gas has passed through the porous plug then the pressure and temperature on the left side of the porous plug are P_1 and T_1 and a certain volume of V_1.

On the right a similar amount of the gas will be at a pressure of P_2, the temperature T_2 and will occupy a volume V_2.

As the gas is compressed the work done on the gas is P_1V_1 and the resultant work done by the gas during this expansion is P_2V_2. This gives a relation for work done as $W = P_2V_2 - P_1V_1$

Physical Mechanism

As a gas expands, the average distance between molecules grows. Because of intermolecular attractive forces, expansion causes an increase in the potential energy of the gas. If no external work is extracted in the process and no heat is transferred, the total energy of the gas remains the same because of the conservation of energy. The increase in potential energy thus implies a decrease in kinetic energy and therefore in temperature.

A second mechanism has the opposite effect. During gas molecule collisions, kinetic energy is temporarily converted into potential energy. As the average intermolecular distance increases, there is a drop in the number of collisions per time unit, which causes a decrease in average potential energy. Again, total energy is conserved, so this leads to an increase in kinetic energy (temperature). Below the Joule–Thomson inversion temperature, the former effect (work done internally against intermolecular attractive forces) dominates, and free expansion causes a decrease in temperature. Above the inversion temperature, gas molecules move faster and so collide more often, and the latter effect (reduced collisions causing a decrease in the average potential energy) dominates: Joule–Thomson expansion causes a temperature increase.

The Joule–Thomson (Kelvin) Coefficient

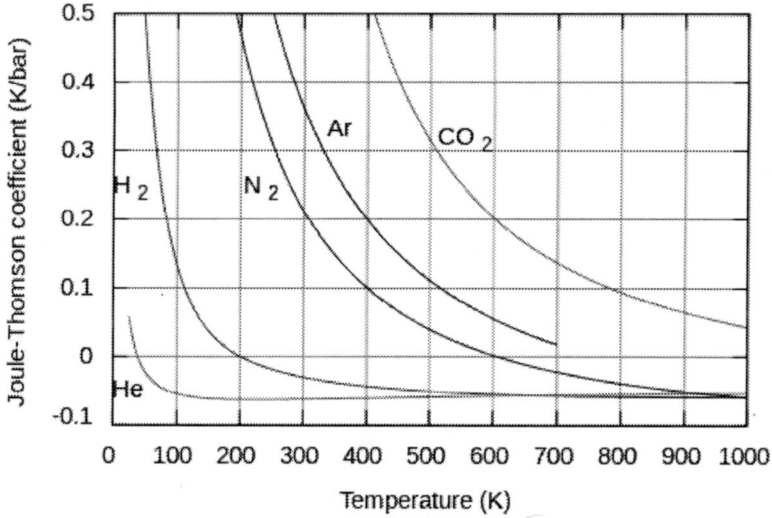

Figure: *Joule-Thomson coefficients for various gases at atmospheric pressure.*

The rate of change of temperature T with respect to pressure P in a Joule–Thomson process (that is, at constant enthalpy H) is the *Joule–Thomson (Kelvin) coefficient* μ_{JT}. This coefficient can be expressed in terms of the gas's volume V, its heat capacity at constant pressure C_p, and its coefficient of thermal expansion α as:

$$\mu_{JT} = \left(\frac{\partial T}{\partial P}\right)_H = \frac{V}{C_p}(\alpha T - 1)$$

See the Derivation of the Joule–Thomson (Kelvin) coefficient below for the proof of this relation. The value of μ_{JT} is typically expressed in °C/bar (SI units: K/Pa) and depends on the type of gas and on the temperature and pressure of the gas before expansion. Its pressure dependence is usually only a few percent for pressures up to 100 bar.

All real gases have an *inversion point* at which the value of μ_{JT} changes sign. The temperature of this point, the *Joule–Thomson inversion temperature*, depends on the pressure of the gas before expansion.

In a gas expansion the pressure decreases, so the sign of ∂T is negative by definition. With that in mind, the following table explains when the Joule–Thomson effect cools or warms a real gas:

If the gas temperature is	*then μ_{JT} is*	*since ∂P is*	*thus ∂T must be*	*so the gas*
below the inversion temperature	positive	always negative	negative	cools
above the inversion temperature	negative	always negative	positive	warms

Helium and hydrogen are two gases whose Joule–Thomson inversion temperatures at a pressure of one atmosphere are very low (e.g., about 51 K (–222 °C) for helium). Thus, helium and hydrogen warm up when expanded at constant enthalpy at typical room temperatures. On the other hand nitrogen and oxygen, the two most abundant gases in air, have inversion temperatures of 621 K (348 °C) and 764 K (491 °C) respectively: these gases can be cooled from room temperature by the Joule–Thomson effect.

For an ideal gas, μ_{JT} is always equal to zero: ideal gases neither warm nor cool upon being expanded at constant enthalpy.

Applications

In practice, the Joule–Thomson effect is achieved by allowing the gas to expand through a throttling device (usually a valve) which must be very well insulated to prevent any heat transfer to or from the gas. No external work is extracted from the gas during the expansion (the gas must not be expanded through a turbine, for example). The effect is applied in the Linde technique as a standard process in the petrochemical industry, where the cooling effect is used to liquefy gases, and also in many cryogenic applications (e.g. for the production of liquid oxygen, nitrogen, and argon). Only when the Joule–Thomson coefficient for the given gas at the given temperature is greater than zero can the gas be liquefied at that temperature by the Linde cycle. In other words, a gas must be below its inversion temperature to be liquefied by the Linde cycle. For this reason, simple Linde cycle liquefiers cannot normally be used to liquefy helium, hydrogen, or neon.

Proof That the Specific Enthalpy Remains Constant

In thermodynamics so-called "specific" quantities are quantities per kilogram and are denoted by lower-case characters. So h, u, and v are the enthalpy, internal energy, and volume per kilogram, respectively. In a Joule–Thomson process the specific enthalpy$h$ remains constant. To prove this, the first step is to compute the net work done when a mass m of the gas moves through the plug. This amount of gas has a volume of $V_1 = mv_1$ in the region at pressure P_1 (region 1) and a volume $V_2 = mv_2$ when in the region at pressure P_2 (region 2). Then the work done on the gas by the rest of the gas in region 1 is $= mP_1v_1$. In region 2 the amount of work done by the gas is mP_2v_2. So, the total work done by the gas is

$$mP_2v_2 - mP_1v_1.$$

The change in internal energy plus the work done by the gas is, by the first law of thermodynamics, the total amount of heat absorbed by the gas (here it is assumed that there is no change in bulk kinetic energy). In the Joule–Thomson process the gas is insulated, so no heat is absorbed. This means that

$$mu_2 - mu_1 + mP_2v_2 - mP_1v_1 = 0$$

where u_1 and u_2 denote the specific internal energies of the gas in regions 1 and 2, respectively. Using the definition of the specific enthalpy $h = u + Pv$, the above equation implies that

$$h_1 = h_2$$

where h_1 and h_2 denote the specific enthalpies of the gas in regions 1 and 2, respectively.

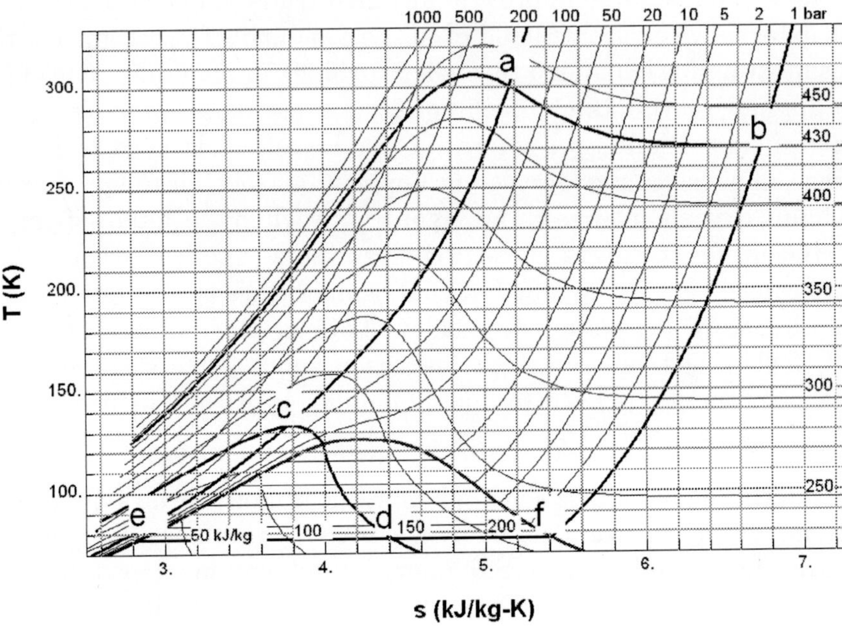

Figure: *Ts diagram of nitrogen. The red dome represents the two-phase region with the low-entropy side the saturated liquid and the high-entropy side the saturated gas. The black curves give the Ts relation along isobars. The pressures are indicated in bar. The blue curves are isenthalps (curves of constant specific enthalpy). The values are indicated in blue in kJ/kg. The specific points a, b, etc., are treated in the main text.*

Throttling in the Ts Diagram

A very convenient way to get a quantitative understanding of the throttling process is by using diagrams. There are many types of diagrams (hT, hP, etc.). Commonly used are the so-called Ts-diagrams. Fig.2 is the Ts-diagram of nitrogen as an example. In Fig.2 various points are indicated as follows:

aT = 300 K, p = 200 bar, s = 5.16 kg/(kgK), h = 430 kJ/kg;

bT = 270 K, p = 1 bar, s = 6.79 kg/(kgK), h = 430 kJ/kg;

cT = 133 K, p = 200 bar, s = 3.75 kg/(kgK), h = 150 kJ/kg;

dT = 77.2 K, p = 1 bar, s = 4.40 kg/(kgK), h = 150 kJ/kg;

eT = 77.2 K, p = 1 bar, s = 2.83 kg/(kgK), h = 28 kJ/kg (saturated liquid at 1 bar);

fT = 77.2 K, p = 1 bar, s = 5.41 kg/(kgK), h =230 kJ/kg (saturated gas at 1 bar);

As shown before, throttling keeps h constant. E.g. throttling from 200 bar and 300 K (point a in Fig.2) follows the isenthalp (line of constant specific enthalpy) of 430 kJ/kg. At 1 bar it results in point b which has a temperature of 270 K. So throttling from 200 to 1 bar gives a cooling from room temperature to below the freezing point of water. Throttling from 200 bar and an initial temperature of 133 K (point c in Fig.2) to 1 bar results in point d, which is in the two-phase region of nitrogen at a temperature of 72.2 K. Since the enthalpy is an extensive parametre the enthalpy in d (h_d) is equal to the enthalpy in e (h_e) multiplied with the liquid fraction in d (x_d) plus the enthalpy in f (h_f) multiplied with the gas fraction in d ($1-x_d$). So

$$h_d = x_d h_e + (1 - x_d) h_f .$$

With Numbers: $150 = x_d\, 28 + (1 - x_d)\, 230$ so x_d is about 0.40. This means that the mass fraction of the liquid in the liquid-gas mixture leaving the throttling valve is 40%.

Derivation of the Joule–Thomson (Kelvin) Coefficient

In this Section a derivation of the formula

$$\mu_{JT} \equiv \left(\frac{\partial T}{\partial P}\right)_H = \frac{V}{C_p}(\alpha T - 1)$$

for the Joule–Thomson (Kelvin) coefficient is given.

The partial derivative of T with respect to P at constant H can be computed by expressing the differential of the enthalpy, dH, in terms of dT and dP, and solving for the ratio of dT and dP with dH = 0. The differential of the enthalpy is given by:

$$\mathrm{d}H = T\mathrm{d}S + V\mathrm{d}P.$$

Here, S is the entropy of the gas. Expressing dS in terms of dT and dP gives:

$$\mathrm{d}H = T\left(\frac{\partial S}{\partial T}\right)_P \mathrm{d}T + \left[V + T\left(\frac{\partial S}{\partial P}\right)_T\right]\mathrm{d}P.$$

Using

$$C_p = T\left(\frac{\partial S}{\partial T}\right)_P ,$$

we can write:

$$\mathrm{d}H = C_p \mathrm{d}T + \left[V + T\left(\frac{\partial S}{\partial P}\right)_T\right]\mathrm{d}P.$$

The remaining partial derivative of S can be expressed in terms of the coefficient of thermal expansion via a Maxwell relation as follows. The differential of the Gibbs energy is given by:

$$dG = -SdT + VdP.$$

The symmetry of partial derivatives of G with respect to T and P implies that:

$$\left(\frac{\partial S}{\partial P}\right)_T = -\left(\frac{\partial V}{\partial T}\right)_P = -V\alpha$$

whereα is the cubic coefficient of thermal expansion. Using this relation dH can be expressed as

$$dH = C_p dT + V(1 - T\alpha)dP.$$

Solving for $(\partial T/\partial P)_H$ by equating dH to zero gives:

$$\left(\frac{\partial T}{\partial P}\right)_H = \frac{V}{C_p}(\alpha T - 1).$$

The cooling of a gas by pure isentropic expansion (a reversibleadiabatic process where work, done by gas-expansion, causes it to cool) is *not* Joule-Thomson cooling.

Joule's Second Law

It is easy to verify that for an ideal gas$\alpha T = 1$, so the temperature change of an ideal gas at a Joule–Thomson expansion is zero. This implies that the internal energy of the ideal gas depends only on its temperature (not pressure or volume), a result known as Joule's second law.

Chapter

4

Deviations From Perfect Gas Model

Vander Waals Equation of State

The van der Waals equation is an equation of state for a fluid composed of particles that have a non-zero volume and a pairwise attractive inter-particle force (such as the van der Waals force). It was derived in 1873 by Johannes Diderik van der Waals, who received the Nobel prize in 1910 for "*his work on the equation of state for gases and liquids*". The equation is based on a modification of the ideal gas law and approximates the behaviour of real fluids, taking into account the nonzero size of molecules and the attraction between them.

Equation

The equation uses the following state variables: the pressure of the fluid p, total volume of the container containing the fluid V, number of moles n, and absolute temperature of the system T.

One form of the equation is

$$\left(p+\frac{a'}{v^2}\right)(v-b')=kT$$

where

$$v=V/N$$

is the volume of the container shared between each particle (not the velocity of a particle),

$$N = N_A n$$

is the total number of particles, and

$$k = R / N_A$$

is Boltzmann's constant, given by the universal gas constant R and Avogadro's constant N_A.

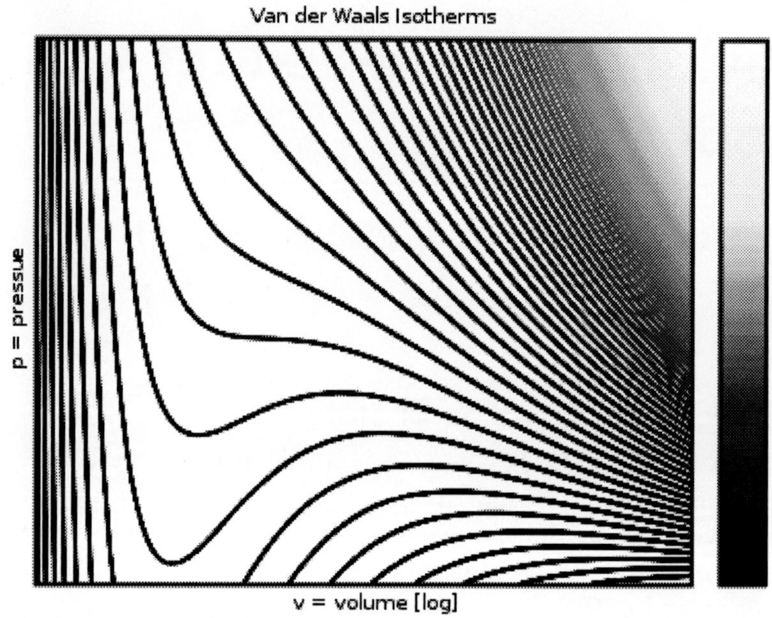

Figure: *The van der Waals isotherms: the model correctly predicts a mostly incompressible liquid phase, but the oscillations in the phase transition zone do not fit experimental data.*

Extra parametres are introduced: a is a measure for the attraction between the particles, and b is the average volume excluded from v by a particle.

The equation can be cast into the better known form

$$\left(p + \frac{n^2 a}{V^2}\right)(V - nb) = nRT$$

where

$$a = N_A^2 a'$$

is a measure of the attraction between the particles,

$$b = N_A b'$$

is the volume excluded by a mole of particles.

A careful distinction must be drawn between the volume *available to* a particle and the volume *of* a particle. In particular, in the first equation v refers to the empty space available per particle. That is, v is the volume V of the container divided by the total number nN_A of particles.

The parametre b, on the other hand, is proportional to the proper volume of a single particle - the volume bounded by the atomic radius. This is the volume to be subtracted from v because of the space taken up by one particle.

In van der Waals' original derivation, given below, b is four times the proper volume of the particle. Observe further that the pressure p goes to infinity when the container is completely filled with particles so that there is no void space left for the particles to move. This occurs when $V = nb$.

Validity

Above the critical temperature the van der Waals equation is an improvement of the ideal gas law, and for lower temperatures the equation is also qualitatively reasonable for the liquid state and the low-pressure gaseous state. However, the van der Waals model is not appropriate for rigorous quantitative calculations, remaining useful only for teaching and qualitative purposes.

In the first-order phase transition the range of (P, V, T), where the liquidphase and the gas phase are in equilibrium, it does not exhibit the empirical fact that p is constant as a function of V for a given temperature: although this behaviour can be easily inserted into the van der Waals model, the result is no longer a simple analytical model, and others (such as those based on the principle of corresponding states) achieve a better fit with roughly the same work.

Derivation

Most textbooks give two different derivations. One is the conventional derivation that goes back to van der Waals and the other is a statistical mechanics derivation. The latter has the major advantage that it makes explicit the intermolecular potential, which is neglected in the first derivation. The conventional van der Waals equation is a mechanical equation of state, which cannot be used to specify all thermodynamic functions, while the statistical mechanical derivation yields the partition function for the system, which does allow all thermodynamic functions to be specified, including the mechanical equation of state.

Conventional Derivation

Consider first one mole of gas which is composed of non-interacting point particles that satisfy the ideal gas law

$$p = \frac{RT}{V_m}.$$

Next assume that all particles are hard spheres of the same finite radius r (the van der Waals radius). The effect of the finite volume of the particles is to decrease the available void space in which the particles are free to move. We must replace V by $V - b$, where b is called the *excluded volume*. The corrected equation becomes

$$p = \frac{RT}{V_m - b}.$$

The excluded volume b is not just equal to the volume occupied by the solid, finite-sized particles, but actually four times that volume. To see this, we must realize that a particle is surrounded by a sphere of radius $r = 2r$ (two times the original radius) that is forbidden for the centres of the other particles. If the distance between two particle centres were to be smaller than $2r$, it would mean that the two particles penetrate each other, which, by definition, hard spheres are unable to do.

The excluded volume per particle (of average diametre d or radius r) is

$$b' = 4\pi d^3 / 3 = 8 \times (4\pi r^3 / 3) \quad \rightarrow \quad b' = 4 \times (4\pi r^3 / 3),$$

which was divided by 2 to prevent overcounting. So $b2$ is four times the proper volume of the particle. It was a point of concern to van der Waals that the factor four yields an upper bound; empirical values for $b2$ are usually lower. Of course, molecules are not infinitely hard, as van der Waals thought, and are often fairly soft.

Next, we introduce a pairwise attractive force between the particles. Van der Waals assumed that, not withstanding the existence of this force, the density of the fluid is homogeneous. Further he assumed that the range of the attractive force is so small that the great majority of the particles do not feel that the container is of finite size. That is, the bulk of them have more attracting particles to their right than to their left when they are relatively close to the left-hand wall of the container. The same statement holds with left and right interchanged. Given the homogeneity of the fluid, the bulk of the particles do not experience a net force pulling them to the right or to the left. This is different for the particles in surface layers directly adjacent to the walls. They feel a net force from the bulk particles pulling them into the container, because this force is not compensated by particles on the side where the wall is (another assumption here is that there is no interaction between walls and particles, which is not true as can be seen from the phenomenon of droplet formation; most types of liquid show adhesion). This net force decreases the force exerted onto the wall by the particles in the surface layer. The net force on a surface particle, pulling it into the container, is proportional to the number density

$$C = N_{\mathrm{A}} / V_{\mathrm{m}}.$$

The number of particles in the surface layers is, again by assuming homogeneity, also proportional to the density. In total, the force on the walls is decreased by a factor proportional to the square of the density, and the pressure (force per unit surface) is decreased by

$$a'C^2 = a'\left(\frac{N_{\mathrm{A}}}{V_{\mathrm{m}}}\right)^2 = \frac{a}{V_{\mathrm{m}}^2},$$

so that

$$p = \frac{RT}{V_{\mathrm{m}} - b} - \frac{a}{V_{\mathrm{m}}^2} \Rightarrow \left(p + \frac{a}{V_{\mathrm{m}}^2}\right)(V_{\mathrm{m}} - b) = RT.$$

Upon writing n for the number of moles and $nV_{\mathrm{m}} = V$, the equation obtains the second form given above,

$$\left(p + \frac{n^2 a}{V^2}\right)(V - nb) = nRT.$$

It is of some historical interest to point out that van der Waals in his Nobel prize lecture gave credit to Laplace for the argument that pressure is reduced proportional to the square of the density.

Statistical Thermodynamics Derivation

The canonical partition function$Q$ of an ideal gas consisting of $N = nN_{\mathrm{A}}$ identical particles, is

$$Q = \frac{q^N}{N!} \quad \text{with} \quad q = \frac{V}{\Lambda^3}$$

where Λ is the thermal de Broglie wavelength,

$$\Lambda = \sqrt{\frac{h^2}{2\pi mkT}}$$

with the usual definitions: h is Planck's constant, m the mass of a particle, kBoltzmann's constant and T the absolute temperature. In an ideal gas q is the partition function of a single particle in a container of volume V. In order to derive the van der Waals equation we assume now that each particle moves independently in an average potential field offered by the other particles. The averaging over the particles is easy because we will assume that the particle density of the van der Waals fluid is homogeneous. The interaction between a pair of particles, which are hard spheres, is taken to be

$$u(r) = \begin{cases} \infty & \text{when} \quad r < d, \\ -\epsilon\left(\dfrac{d}{r}\right)^6 & \text{when} \quad r \geq d, \end{cases}$$

r is the distance between the centres of the spheres and d is the distance where the hard spheres touch each other (twice the van der Waals radius). The depth of the van der Waals well is ϵ.

Because the particles are independent, the total partition function still factorizes,

$$Q = q^N / N!,$$

but the intermolecular potential necessitates two modifications to q. First, because of the finite size of the particles, not all of V is available, but only $V - Nb'$, where (just as in the conventional derivation above)

$$b' = 2\pi d^3 / 3 \cdot$$

Second, we insert a Boltzmann factor exp[- $\Phi/2kT$] to take care of the average intermolecular potential. We divide here the potential by two because this interaction energy is shared between two particles. Thus

$$q = \frac{(V - Nb')e^{-\phi/(2kT)}}{\Lambda^3}.$$

The total attraction felt by a single particle is

$$\phi = \int_d^\infty u(r)\frac{N}{V}4\pi r^2 dr,$$

where we assumed that in a shell of thickness dr there are $N/V\ 4\pi r^2 dr$ particles. This is a mean field approximation; the position of the particles is averaged. In reality the density close to the particle is different than far away as can be described by a pair correlation function. Furthermore it is neglected that the fluid is enclosed between walls. Performing the integral we get

$$\phi = -2a'\frac{N}{V} \quad \text{with} \quad a' = \epsilon\frac{2\pi d^3}{3} = \epsilon b'.$$

Hence, we obtain,

$$\ln Q = N\ln(V - Nb') + \frac{N^2 a'}{VkT} - N\ln(\Lambda^3) - \ln N!$$

From statistical thermodynamics we know that

$$p = kT\frac{\partial \ln Q}{\partial V},$$

so that we only have to differentiate the terms containing V. We get

$$p = \frac{NkT}{V - Nb'} - \frac{N^2 a'}{V^2} \Rightarrow \left(p + \frac{N^2 a'}{V^2}\right)(V - Nb') = NkT \Rightarrow \left(p + \frac{n^2 a}{V^2}\right)(V - nb) = nRT.$$

Other Thermodynamic Parametres

We reiterate that the extensive volume V is related to the volume per particle $v = V/N$ where $N = nN_A$ is the number of particles in the system.

The equation of state does not give us all the thermodynamic parametres of the system. We can take the equation for the Helmholtz energy A

$$A = -kT\ln Q.$$

From the equation derived above for ln$Q$, we find

$$A(T,V,N) = -NkT\left[1 + \ln\left(\frac{(V - Nb')T^{3/2}}{N\Phi}\right)\right] - \frac{a'N^2}{V}.$$

Where Φ is an undetermined constant, which may be taken from the Sackur-Tetrode equation for an ideal gas to be:

$$\Phi = T^{3/2}\Lambda^3 = \frac{h}{\sqrt{2\pi mk}}$$

This equation expresses A in terms of its natural variables V and T, and therefore gives us all thermodynamic information about the system. The mechanical equation of state was already derived above

$$p = -\left(\frac{\partial A}{\partial V}\right)_T = \frac{NkT}{V - Nb'} - \frac{a'N^2}{V^2}.$$

The entropy equation of state yields the entropy (S)

$$S = -\left(\frac{\partial A}{\partial T}\right)_V = Nk\left[\ln\left(\frac{(V - Nb')T^{3/2}}{N\Phi}\right) + \frac{5}{2}\right]$$

from which we can calculate the internal energy

$$U = A + TS = \frac{3}{2}NkT - \frac{a'N^2}{V}.$$

Similar equations can be written for the other thermodynamic potential and the chemical potential, but expressing any potential as a function of pressure p will require the solution of a third-order polynomial, which yields a complicated expression. Therefore, expressing the enthalpy and the Gibbs energy as functions of their natural variables will be complicated.

Reduced Form

Although the material constants a and b in the usual form of the van der Waals equation differs for every single fluid considered, the equation can be recast into an invariant form applicable to *all* fluids.

Defining the following reduced variables (f_R, f_C are the reduced and critical variable versions of f, respectively),

$$p_R = \frac{p}{p_C}, \qquad v_R = \frac{v}{v_C}, \quad \text{and} \quad T_R = \frac{T}{T_C},$$

where

$$p_C = \frac{a'}{27b'^2}, \qquad v_C = 3b', \quad \text{and} \quad kT_C = \frac{8a'}{27b'}$$

as shown by Salzman.

The first form of the van der Waals equation of state given above can be recast in the following reduced form:

$$\left(p_R + \frac{3}{v_R^2}\right)(v_R - 1/3) = (8/3)T_R$$

This equation is *invariant* for all fluids; that is, the same reduced form equation of state applies, no matter what a andb may be for the particular fluid.

This invariance may also be understood in terms of the principle of corresponding states. If two fluids have the same reduced pressure, reduced volume, and reduced temperature, we say that their states are corresponding. The states of two fluids may be corresponding even if their measured pressure, volume, and temperature are very different. If the two fluids' states are corresponding, they exist in the same regime of the reduced form equation of state. Therefore, they will respond to changes in roughly the same way, even though their measurable physical characteristics may differ significantly.

Cubic Equation

The van der Waals equation is a cubic equation of state. That is we can write the equation into a cubic form of the volume. In the reduced formulation the cubic equation is:

$$v_R^3 - \frac{1}{3}\left(1 + \frac{8T_R}{p_R}\right)v_R^2 + \frac{3}{p_R}v_R - \frac{1}{p_R} = 0$$

At the critical temperature, where $T_R = p_R = 1$ we get as expected

$$v_R^3 - 3v_R^2 + 3v_R - 1 = (v_R - 1)^3 = 0 \quad \Leftrightarrow \quad v_R = 1$$

For $T_R < 1$, there are 3 values for v_R. For $T_R > 1$, there is 1 real value for v_R.

Application to Compressible Fluids

The equation is also usable as a PVT equation for compressible fluids (e.g. polymers). In this case specific volume changes are small and it can be written in a simplified form:

$$(p + A)(V - B) = CT,$$

where p is the pressure, V is specific volume, T is the temperature and A, B, C are parametres.

Maxwell Equal Area Rule

Below the critical temperature $T_R < 1$ an isotherm of the van der Waals equation oscillates as shown.

Along the red portion of the isotherm $(\partial P / \partial V)_{T,N} > 0$ which is unstable; the van der Waals equation fails to describe real substances in this region because the equation always assumes that the fluid is uniform while between a and c on the isotherm it becomes more stable to be a coexistence of two different phases, a denser phase which we normally call *liquid* and a sparser phase which we normally call *gas*. To fix this problem James Clerk Maxwell (1875) replaced the isotherm between a and$c$ with a horizontal line positioned so that the areas of the two hatched regions are equal.

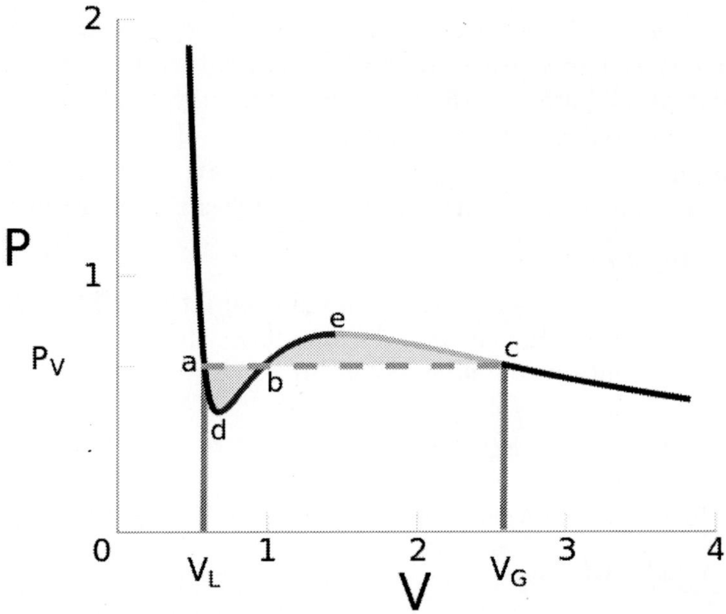

Figure: *Maxwell's rule eliminates the oscillating behaviour of the isotherm in the phase transition zone by defining it as a certain isobar in that zone. The above isotherm is for a reduced temperature of* $T_R = 0.9$. *The Maxwell correction is at a vapour pressure of* $P_V \approx 0.64700$ *between the reduced volume of the pure liquid* $V_L \approx 0.60340$ *and the pure gas* $V_G \approx 2.3488$ *at the vapour pressure.*

The flat line portion of the isotherm now corresponds to liquid-vapour equilibrium. The portions *a–d* and *c–e* are interpreted as metastable states of super-heated liquid and super-cooled vapour respectively. The equal area rule can be expressed as:

$$P_V(V_G - V_L) = \int_{V_L}^{V_G} P\,dV$$

where P_V is the vapour pressure (flat portion of the curve), V_L is the volume of the pure liquid phase at point *a* on the diagram, and V_G is the volume of the pure gas phase at point *c* on the diagram. The sum of these two volumes will equal the total volume *V*.

Maxwell justified the rule by saying that work done on the system in going from *c* to *b* should equal work released on going from *a* to *b*. (The area on a *PV* diagram corresponds to mechanical work). That's because the change in the free energy function $A(T,V)$ equals the work done during a reversible process, and the free energy function - being a state variable - should take on a unique value regardless of path.

In particular, the value of A at point b should calculate the same regardless of whether the path came from left or right, or went straight across the horizontal isotherm or around the original van der Waals isotherm. Maxwell's argument is not totally convincing since it requires a reversible path through a region of thermodynamic instability. Nevertheless, more subtle arguments based on modern theories of phase equilibrium seem to confirm the Maxwell Equal Area construction and it remains a valid modification of the van der Waals equation of state.

The Maxwell equal area rule can also be derived from an assumption of equal chemical potentialμ of coexisting liquid and vapour phases. On the isotherm shown in the above plot, points a and$c$ are the only pair of points which fulfill the equilibrium condition of having equal pressure, temperature and chemical potential. It follows that systems with volumes intermediate between these two points will consist of a mixture of the pure liquid and gas with specific volumes equal to the pure liquid and gas phases at points a and c.

The van der Waals equation may be solved for V_G and V_L as functions of the temperature and the vapour pressure P_V. Since:

$$P = \left(\frac{\partial A}{\partial V}\right)_{T,N}$$

where A is the Helmholtz free energy, it follows that the equal area rule can be expressed as:

$$P_V = \frac{A(V_G, T, N) - A(V_L, T, N)}{V_G - V_L}$$

Since the gas and liquid volumes are functions of P_V and T only, this equation is then solved numerically to obtain P_V as a function of temperature (and number of particles N), which may then be used to determine the gas and liquid volumes.

Compressibility Charts

The compressibility factor (Z), also known as the compression factor, is the ratio of the molar volume of a gas to the molar volume of an ideal gas at the same temperature and pressure. It is a useful thermodynamic property for modifying the ideal gas law to account for the real gas behaviour. In general, deviation from ideal behaviour becomes more significant the closer a gas is to a phase change, the lower the temperature or the larger the pressure. Compressibility factor values are usually obtained by calculation from equations of state (EOS), such as the virial equation which take compound specific empirical constants as input. For a gas that is a mixture of two or more pure gases (air or natural gas, for example), a gas composition is required before compressibility can be calculated.

Alternatively, the compressibility factor for specific gases can be read from generalized compressibility charts that plot Z as a function of pressure at constant temperature.

Definition and Physical Significance

The compressibility factor is defined as

$$Z = \frac{V_{\mathrm{m}}}{(V_{\mathrm{m}})_{\text{ideal gas}}} = \frac{pV_{\mathrm{m}}}{RT},$$

where V_{m} is the molar volume, $(V_{\mathrm{m}})_{\text{ideal gas}} = RT / p$ is the molar volume of the corresponding ideal gas, p is the pressure, T is the temperature, and R is the gas constant. For engineering applications, it is frequently expressed as

$$Z = \frac{p}{\rho R_{\text{specific}} T},$$

where ρ is the density of the gas and $R_{\text{specific}} = R / M$ is the specific gas constant, M being the molar mass.

For an ideal gas the compressibility factor is $Z = 1$ per definition. In many real world applications requirements for accuracy demand that deviations from ideal gas behaviour, i.e., real gas behaviour, is taken into account. The value of Z generally increases with pressure and decreases with temperature. At high pressures molecules are colliding more often. This allows repulsive forces between molecules to have a noticeable effect, making the molar volume of the real gas (V_{m}) greater than the molar volume of the corresponding ideal gas ($(V_{\mathrm{m}})_{\text{ideal gas}} = RT / p$), which causes Z to exceed one. When pressures are lower, the molecules are free to move. In this case attractive forces dominate, making $Z < 1$. The closer the gas is to its critical point or its boiling point, the more Z deviates from the ideal case.

Generalized Compressibility Factor Graphs for Pure Gases

The unique relationship between the compressibility factor and the reduced temperature, T_r, and the reduced pressure, P_r, was first recognized by Johannes Diderik van der Waals in 1873 and is known as the two-parametre principle of corresponding states. The principle of corresponding states expresses the generalization that the properties of a gas which are dependent on intermolecular forces are related to the critical properties of the gas in a universal way. That provides a most important basis for developing correlations of molecular properties.

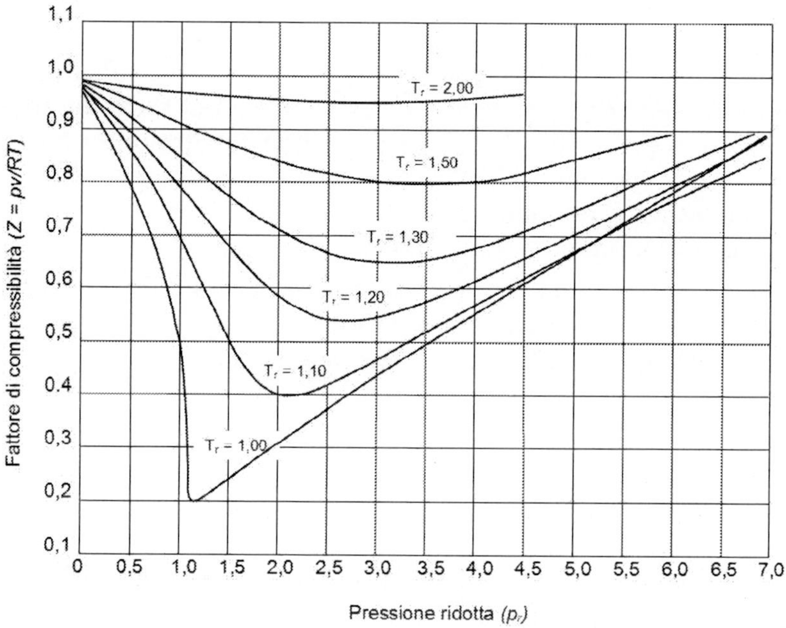

Figure: *Generalized compressibility factor diagram.*

As for the compressibility of gases, the principle of corresponding states indicates that any pure gas at the same reduced temperature, T_r, and reduced pressure, P_r, should have the same compressibility factor.

The reduced temperature and pressure are defined by

$$T_r = \frac{T}{T_c} \text{ and } P_r = \frac{P}{P_c}.$$

Here T_c and P_c are known as the critical temperature and critical pressure of a gas. They are characteristics of each specific gas with T_c being the temperature above which it is not possible to liquify a given gas and P_c is the minimum pressure required to liquify a given gas at its critical temperature. Together they define the critical point of a fluid above which distinct liquid and gas phases of a given fluid do not exist.

The pressure-volume-temperature (PVT) data for real gases varies from one pure gas to another.

However, when the compressibility factors of various single-component gases are graphed versus pressure along with temperature isotherms many of the graphs exhibit similar isotherm shapes.

In order to obtain a generalized graph that can be used for many different gases, the reduced pressure and temperature, P_r and T_r, are used to normalize the compressibility factor data. Above figure is an example of a generalized compressibility factor graph derived from hundreds of experimental PVT data points of 10 pure gases, namely methane, ethane, ethylene, propane, n-butane, i-pentane, n-hexane, nitrogen, carbon dioxide and steam.

There are more detailed generalized compressibility factor graphs based on as many as 25 or more different pure gases, such as the Nelson-Obert graphs. Such graphs are said to have an accuracy within 1-2 percent for Z values greater than 0.6 and within 4-6 percent for Z values of 0.3-0.6.

The generalized compressibility factor graphs may be considerably in error for strongly polar gases which are gases for which the centres of positive and negative charge do not coincide.

In such cases the estimate for Z may be in error by as much as 15-20 percent.

The quantum gases hydrogen, helium, and neon do not conform to the corresponding-states behaviour and the reduced pressure and temperature for those three gases should be redefined in the following manner to improve the accuracy of predicting their compressibility factors when using the generalized graphs:

$$T_r = \frac{T}{T_c + 8} \text{ and } P_r = \frac{P}{P_c + 8}$$

where the temperatures are in kelvin and the pressures are in atmospheres.

Theoretical Models

The virial equation is especially useful to describe the causes of non-ideality at a molecular level (very few gases are mono-atomic) as it is derived directly from statistical mechanics:

$$\frac{PV_{\mathrm{m}}}{RT} = 1 + \frac{B}{V_{\mathrm{m}}} + \frac{C}{V_{\mathrm{m}}^2} + \frac{D}{V_{\mathrm{m}}^3} + \ldots$$

Where the coefficients in the numerator are known as virial coefficients and are functions of temperature.

The virial coefficients account for interactions between successively larger groups of molecules. For example, B accounts for interactions between pairs, C for interactions between three gas molecules, and so on. Because interactions between large numbers of molecules are rare, the virial equation is usually truncated after the third term.

The Real gas article features more theoretical methods to compute compressibility factors

Experimental Values

It is extremely difficult to generalize at what pressures or temperatures the deviation from the ideal gas becomes important. As a rule of thumb, the ideal gas law is reasonably accurate up to a pressure of about 2 atm, and even higher for small non-associating molecules. For example methyl chloride, a highly polar molecule and therefore with significant intermolecular forces, the experimental value for the compressibility factor is $Z = 0.9152$ at a pressure of 10 atm and temperature of 100 °C. For air (small non-polar molecules) at approximately the same conditions, the compressibility factor is only $Z = 1.0025$.

Compressibility of Air

Normal air comprises in crude numbers 80 percent nitrogen N_2 and 20 percent oxygen O_2. Both molecules are small and non-polar (and therefore non-associating). We can therefore expect that the behaviour of air within broad temperature and pressure ranges can be approximated as an ideal gas with reasonable accuracy. Experimental values for the compressibility factor confirm this.

Compressibility factor for air (experimental values)														
	Pressure, bar (absolute)													
Temp, K	1	5	10	20	40	60	80	100	150	200	250	300	400	500
75	0.0052	0.0260	0.0519	0.1036	0.2063	0.3082	0.4094	0.5099	0.7581	1.0125				
80		0.0250	0.0499	0.0995	0.1981	0.2958	0.3927	0.4887	0.7258	0.9588	1.1931	1.4139		
90	0.9764	0.0236	0.0453	0.0940	0.1866	0.2781	0.3686	0.4681	0.6779	0.8929	1.1098	1.3110	1.7161	2.1105
100	0.9797	0.8872	0.0453	0.0900	0.1782	0.2635	0.3498	0.4337	0.6386	0.8377	1.0395	1.2227	1.5937	1.9536
120	0.9880	0.9373	0.8860	0.6730	0.1778	0.2557	0.3371	0.4132	0.5964	0.7720	0.9530	1.1076	1.5091	1.7366
140	0.9927	0.9614	0.9205	0.8297	0.5856	0.3313	0.3737	0.4340	0.5909	0.7699	0.9114	1.0393	1.3202	1.5903
160	0.9951	0.9748	0.9489	0.8954	0.7803	0.6603	0.5696	0.5489	0.6340	0.7564	0.8840	1.0105	1.2585	1.4970
180	0.9967	0.9832	0.9660	0.9314	0.8625	0.7977	0.7432	0.7084	0.7180	0.7986	0.9000	1.0068	1.2232	1.4361
200	0.9978	0.9886	0.9767	0.9539	0.9100	0.8701	0.8374	0.8142	0.8061	0.8549	0.9311	1.0185	1.2054	1.3944
250	0.9992	0.9957	0.9911	0.9822	0.9671	0.9549	0.9463	0.9411	0.9450	0.9713	1.0152	1.0702	1.1990	1.3392
300	0.9999	0.9987	0.9974	0.9950	0.9917	0.9901	0.9903	0.9930	1.0074	1.0326	1.0669	1.1089	1.2073	1.3163
350	1.0000	1.0002	1.0004	1.0014	1.0038	1.0075	1.0121	1.0183	1.0377	1.0635	1.0947	1.1303	1.2116	1.3015
400	1.0002	1.0012	1.0025	1.0046	1.0100	1.0159	1.0229	1.0312	1.0533	1.0795	1.1087	1.1411	1.2117	1.2890
450	1.0003	1.0016	1.0034	1.0063	1.0133	1.0210	1.0287	1.0374	1.0614	1.0913	1.1183	1.1463	1.2090	1.2778
500	1.0003	1.0020	1.0034	1.0074	1.0151	1.0234	1.0323	1.0410	1.0650	1.0913	1.1183	1.1463	1.2051	1.2667
600	1.0004	1.0022	1.0039	1.0081	1.0164	1.0253	1.0340	1.0434	1.0678	1.0920	1.1172	1.1427	1.1947	1.2475
800	1.0004	1.0020	1.0038	1.0077	1.0157	1.0240	1.0321	1.0408	1.0621	1.0844	1.1061	1.1283	1.1720	1.2150
1000	1.0004	1.0018	1.0037	1.0068	1.0142	1.0215	1.0290	1.0365	1.0556	1.0744	1.0948	1.1131	1.1515	1.1889

Mole Fraction

In chemistry, the mole fraction x_i is defined as the amount of a constituent n_i divided by the total amount of all constituents in a mixture n_{tot}:

$$x_i = \frac{n_i}{n_{tot}}$$

The sum of all the mole fractions is equal to 1:

$$\sum_{i=1}^{N} n_i = n_{tot}\,; \sum_{i=1}^{N} x_i = 1$$

The mole fraction is also called the amount fraction. It is identical to the number fraction, which is defined as the number of molecules of a constituent N_i divided by the total number of all molecules N_{tot}. It is one way of expressing the composition of a mixture with a dimensionless quantity (mass fraction is another). The mole fraction is sometimes denoted by the lowercase Greek letter χ (*chi*) instead of a Roman *x*. For mixtures of gases, IUPAC recommends the letter *y*.

Properties

Mole fraction is used very frequently in the construction of phase diagrams. It has a number of advantages:

- it is not temperature dependent (such as molar concentration) and does not require knowledge of the densities of the phase(s) involved
- a mixture of known mole fraction can be prepared by weighing off the appropriate masses of the constituents
- the measure is *symmetric*: in the mole fractions x=0.1 and x=0.9, the roles of 'solvent' and 'solute' are reversed.
- In a mixture of ideal gases, the mole fraction can be expressed as the ratio of partial pressure to total pressure of the mixture.

Related Quantities

Mass Fraction: The mass fraction w_i can be calculated using the formula

$$w_i = x_i \cdot \frac{M_i}{M}$$

where M_i is the molar mass of the component *i* and M is the average molar mass of the mixture.

Replacing the expression of the molar mass:

$$w_i = x_i \cdot \frac{M_i}{\sum_i x_i M_i}$$

Mole Percentage

Multiplying mole fraction by 100 gives the mole percentage, also referred as amount/amount percent (abbreviated as n/n%).

Mass Concentration

The conversion to and from mass concentration ρ_i is given by:

$$x_i = \frac{\rho_i}{\rho} \cdot \frac{M}{M_i}$$

where M is the average molar mass of the mixture.

Molar Concentration

The conversion to molar concentration c_i is given by:

$$c_i = \frac{x_i \cdot \rho}{M} = x_i c$$

or

$$c_i = \frac{x_i \cdot \rho}{\sum_i x_i M_i}$$

where M is the average molar mass of the solution, c total molar concentration and ρ is the density of the solution .

Mass and Molar Mass

The mole fraction can be calculated from the masses m_i and molar masses M_i of the components:

$$x_i = \frac{\frac{m_i}{M_i}}{\sum_i \frac{m_i}{M_i}}$$

Mass Fraction

In chemistry, the mass fraction w_i is the ratio of one substance with mass m_i to the mass of the total mixture m_{tot}, defined as

$$w_i = \frac{m_i}{m_{tot}}$$

The sum of all the mass fractions is equal to 1:

$$\sum_{i=1}^{N} m_i = m_{tot}; \sum_{i=1}^{N} w_i = 1$$

Mass fraction is one way of expressing the composition of a mixture in a dimensionless size (mole fraction is another).

For elemental analysis, mass fraction (or "mass percent composition") can also refer to the ratio of the mass of one element to the total mass of a compound. It can be calculated for any compound using its empirical formula. or its chemical formula

Terminology

"Percent concentration" does not refer to this quantity. This improper name persists, especially in elementary textbooks. In biology, the unit "%" is sometimes (incorrectly) used to denote mass concentration, also called "mass/volume percentage." A solution with 1 g of solute dissolved in a final volume of 100 mL of solution would be libelled as "1 %" or "1 % m/v" (mass/volume). The notation is mathematically flawed because the unit "%" can only be used for dimensionless quantities. "Percent solution" or "percentage solution" are thus terms best reserved for "weight percent solutions" (mass/w = m% = mass solute/mass total solution after mixing), or "volume percent solutions" (v/v = v% = volume solute per volume of total solution after mixing). The very ambiguous terms "percent solution" and "percentage solutions" with no other qualifiers, continue to occasionally be encountered.

In thermal engineeringvapor quality is used for the mass fraction of vapour in the steam.

In alloys, especially those of noble metals, the term fineness is used for the mass fraction of the noble metal in the alloy.

Properties

The mass fraction is independent of temperature.

Related Quantities

Mass Concentration: The mass fraction of a component in a solution is the ratio of the mass concentration of that component ρ_i (density of that component in the mixture) to the density of solution ρ.

$$w_i = \frac{\rho_i}{\rho}$$

Molar Concentration

The relation to molar concentration is like that from above substituting the relation between mass and molar concentration.

$$w_i = \frac{c_i M_i}{\rho}$$

Mass Percentage

Multiplying mass fraction by 100 gives the mass percentage. It is sometimes called weight percent (wt%) or weight-weight percentage. However, since mass and weight are different quantities, this is incorrect.

The mole fraction x_i can be calculated using the formula

$$x_i = w_i \cdot \frac{M}{M_i}$$

where M_i is the molar mass of the component i and M is the average molar mass of the mixture.

Replacing the expression of the molar mass produces:

$$x_i = \frac{\frac{w_i}{M_i}}{\sum_i \frac{w_i}{M_i}}$$

Spatial Variation and Gradient

In a spatially non-uniform mixture, the mass fraction gradient triggers the phenomenon of diffusion.

Gravimetric Analysis

Gravimetric analysis describes a set of methods in analytical chemistry for the quantitative determination of an analyte based on the mass of a solid. A simple example is the measurement of solids suspended in a water sample: A known volume of water is filtered, and the collected solids are weighed.

In most cases, the analyte must first be converted to a solid by precipitation with an appropriate reagent. The precipitate can then be collected by filtration, washed, dried to remove traces of moisture from the solution, and weighed. The amount of analyte in the original sample can then be calculated from the mass of the precipitate and its chemical composition.

In other cases, it may be easier to remove the analyte by vaporization. The analyte might be collected—perhaps in a cryogenic trap or on some absorbent material such as activated carbon — and measured directly. Or, the sample can be weighed before and after it is dried; the difference between the two masses gives the mass of analyte lost. This is especially useful in determining the water content of complex materials such as foodstuffs.

Procedure

1. The sample is dissolved, if it is not already insoluble.
2. The solution may be treated to adjust the pH (so that the proper precipitate is formed, or to suppress the formation of other precipitates). If it is known that species are present which interfere (by also forming precipitates under the same conditions as the analyte), the sample might require treatment with a different reagent to remove these interferents.
3. The precipitating reagent is added at a concentration that favours the formation of a "good" precipitate. This may require low concentration, extensive heating (often described as "digestion"), or careful control of the pH. Digestion can help reduce the amount of coprecipitation.
4. After the precipitate has formed and been allowed to "digest", the solution is carefully filtered. The filter is chosen to trap the precipitate; smaller particles are more difficult to filter.
   - Depending on the procedure followed, the filter might be a piece of ashless filter paper in a fluted funnel, or a filter crucible. Filter paper is convenient because it does not typically require cleaning before use; however, filter paper can be chemically attacked by some solutions (such as concentrated acid or base), and may tear during the filtration of large volumes of solution.
   - The alternative is a crucible whose bottom is made of some porous material, such as sintered glass, porcelain or sometimes metal. These are chemically inert and mechanically stable, even at elevated temperatures. However, they must be carefully cleaned to minimize contamination or carryover(cross-contamination). Crucibles are often used with a mat of glass or asbestos fibres to trap small particles.
   - After the solution has been filtered, it should be tested to make sure that the analyte has been completely

precipitated. This is easily done by adding a few drops of the precipitating reagent; if a precipitate is observed, the precipitation is incomplete.

5. After filtration, the precipitate – including the filter paper or crucible – is heated. This achieves three purposes:
   - o The remaining moisture is removed (drying).
   - o Secondly, the precipitate is converted to a more chemically stable form. For instance, calcium ion might be precipitated using oxalate ion, to produce calcium oxalate (CaC_2O_4); it might then be heated to convert it into the oxide (CaO). It is vital that the empirical formula of the weighed precipitate be known, and that the precipitate be pure; if two forms are present, the results will be inaccurate.
   - o The precipitate cannot be weighed with the necessary accuracy in place on the filter paper; nor can the precipitate be completely removed from the filter paper in order to weigh it. The precipitate can be carefully heated in a crucible until the filter paper has burned away; this leaves only the precipitate. (As the name suggests, "ashless" paper is used so that the precipitate is not contaminated with ash.)
6. After the precipitate is allowed to cool (preferably in a desiccator to keep it from absorbing moisture), it is weighed (in the crucible). The mass of the crucible is subtracted from the combined mass, giving the mass of the precipitated analyte. Since the composition of the precipitate is known, it is simple to calculate the mass of analyte in the original sample.

Example

A chunk of ore is treated with concentrated nitric acid and potassium chlorate to convert all of the sulfur to sulfate (SO_4^{2-}). The nitrate and chlorate are removed by treating the solution with concentrated HCl. The sulfate is precipitated with barium (Ba^{2+}) and weighed as $BaSO_4$.

Advantages

Gravimetric analysis, if methods are followed carefully, provides for exceedingly precise analysis. In fact, gravimetric analysis was used to determine the atomic masses of many elements to six figure accuracy. Gravimetry provides very little room for instrumental error and does not require a series of standards for calculation of an unknown. Also, methods often do not require expensive equipment. Gravimetric analysis, due to its high degree of accuracy, when performed correctly, can also be used to calibrate other instruments in lieu of reference standards.

Disadvantages

Gravimetric analysis usually only provides for the analysis of a single element, or a limited group of elements, at a time. Comparing modern dynamic flash combustion coupled with gas chromatography with traditional combustion analysis will show that the former is both faster and allows for simultaneous determination of multiple elements while traditional determination allowed only for the determination of carbon and hydrogen. Methods are often convoluted and a slight mis-step in a procedure can often mean disaster for the analysis (colloid formation in precipitation gravimetry, for example). Compare this with hardy methods such as spectrophotometry and one will find that analysis by these methods is much more efficient.

Dalton's Law of Partial Pressures

In chemistry and physics, Dalton's law (also called Dalton's law of partial pressures) states that the total pressure exerted by the mixture of non-reactive gases is equal to the sum of the partial pressures of individual gases. This empirical law was observed by John Dalton in 1801 and is related to the idealgas laws.

Mathematically, the pressure of a mixture of gases can be defined as the summation

$$P_{\text{total}} = \sum_{i=1}^{n} p_i \text{ or } P_{\text{total}} = p_1 + p_2 + \cdots + p_n$$

where $p_1, p_2, \ldots, p_n$ represent the partial pressure of each component.

It is assumed that the gases do not react with each other

$$p_i = P_{\text{total}} y_i$$

where y_i is the mole fraction of the i-th component in the total mixture of n components .

The relationship below provides a way to determine the volume based concentration of any individual gaseous component

$$P_i = \frac{P_{\text{total}} C_i}{1{,}000{,}000}$$

where C_i is the concentration of the i-th component expressed in ppm.

Dalton's law is not exactly followed by real gases. Those deviations are considerably large at high pressures. In such conditions, the volume occupied by the molecules can become significant compared to the free space between them. In particular, the short average distances between molecules raises the intensity of intermolecular forces between gas molecules enough to substantially change the pressure exerted by them. Neither of those effects are considered by the ideal gas model.

Avogadro's Laws of Additive Volumes

Avogadro's law (sometimes referred to as Avogadro's hypothesis or Avogadro's principle) is an experimental gas law relating volume of a gas to the amount of substance of gas present. A modern statement of Avogadro's law is:

For a given mass of an ideal gas, the volume and amount (moles) of the gas are directly proportional if the temperature and pressure are constant.

which can be written as:

$$V \propto n$$

or

$$\frac{V}{n} = k$$

where:

V is the volume of the gas

n is the amount of substance of the gas (measured in moles).

k is a constant.

This law explains how, under the same condition of temperature and pressure, equal volumes of all gases contain the same number of molecules. For comparing the same substance under two different sets of conditions, the law can be usefully expressed as follows:

$$\frac{V_1}{n_1} = \frac{V_2}{n_2}$$

The equation shows that, as the number of moles of gas increases, the volume of the gas also increases in proportion. Similarly, if the number of moles of gas is decreased, then the volume also decreases. Thus, the number of molecules or atoms in a specific volume of ideal gas is independent of their size or the molar mass of the gas.

The law is named after Amedeo Avogadro who, in 1811, hypothesized that two given samples of an ideal gas, of the same volume and at the same temperature and pressure, contain the same number of molecules. As an example, equal volumes of molecular hydrogen and nitrogen contain the same number of molecules when they are at the same temperature and pressure, and observe ideal gas behaviour. In practice, real gases show small deviations from the ideal behaviour and the law holds only approximately, but is still a useful approximation for scientists.

Mathematical Definition

Avogadro's law is stated mathematically as:

$$\frac{V}{n} = k$$

Where:

V is the volume of the gas(es).

n is the amount of substance of the gas.

k is a proportionality constant.

The most significant consequence of Avogadro's law is that the ideal gas constant has the same value for all gases. This means that:

$$\frac{p_1 \cdot V_1}{T_1 \cdot n_1} = \frac{p_2 \cdot V_2}{T_2 \cdot n_2} = constant$$

Where:

p is the pressure of the gas in the cell

T is the temperature in kelvin of the gas

Ideal Gas Law

A common rearrangement of this equation is by letting R be the proportionality constant, and rearranging as follows:

$$pV = nRT$$

This equation is known as the ideal gas law.

Molar Volume

Taking STP to be 101.325 kPa and 273.15 K, we can find the volume of one mole of a gas:

$$V_{\mathrm{m}} = \frac{V}{n} = \frac{RT}{p} = \frac{(8.314\mathrm{Jmol^{-1}K^{-1}})(273.15\mathrm{K})}{101325\mathrm{Pa}} = 22.41\mathrm{dm^3mol^{-1}}$$

For 100.000 kPa and 273.15 K, the molar volume of an ideal gas is 22.712 dm^3mol^{-1}.

Volume Fraction

In chemistry, the volume fraction φ_i is defined as the volume of a constituent V_i divided by the volume of all constituents of the mixture V prior to mixing:

$$\phi_i = \frac{V_i}{V}$$

The SI-unit is m^3/m^3.

Volume fraction is also called volume concentration in ideal solutions where the volumes of the constituents are additive (the volume of the solution is equal to the sum of the volumes of its ingredients).

The sum of all volume fractions of a mixture is equal to 1:

$$\sum_{i=1}^{N} V_i = V; \qquad \sum_{i=1}^{N} \phi_i = 1$$

Partial Pressure

In a mixture of gases, each gas has a partial pressure which is the hypothetical pressure of that gas if it alone occupied the volume of the mixture at the same temperature. The total pressure of an ideal gas mixture is the sum of the partial pressures of each individual gas in the mixture.

It relies on the following isotherm relation:

$$V_x \times p_{tot} = V_{tot} \times p_x$$

- V_x is the partial volume of any individual gas component (X)
- V_{tot} is the total volume in gas mixture
- p_x is the partial pressure of gas X
- p_{tot} is the total pressure of gas mixture
- n_x is the amount of substance of a gas (X)
- n_{tot} is the total amount of substance in gas mixture

The partial pressure of a gas is a measure of thermodynamic activity of the gas's molecules. Gases dissolve, diffuse, and react according to their partial pressures, and not according to their concentrations in gas mixtures or liquids.

This general property of gases is also true of chemical reactions of gases in biology. For example, the necessary amount of oxygen for human respiration, and the amount that is toxic, is set by the partial pressure of oxygen alone. This is true across a very wide range of different concentrations of oxygen present in various inhaled breathing gases, or dissolved in blood.

Partial Volume (Amagat's Law of Additive Volume)

The partial volume of a particular gas in a mixture is the volume of one component of the gas mixture. It is useful in gas mixtures, e.g. air, to focus on one particular gas component, e.g. oxygen.

It can be approximated both from partial pressure and molar fraction:

$$V_x = V_{tot} \times \frac{p_x}{p_{tot}} = V_{tot} \times \frac{n_x}{n_{tot}}$$

- V_x is the partial volume of any individual gas component (X)
- V_{tot} is the total volume in gas mixture
- p_x is the partial pressure of gas X
- p_{tot} is the total pressure of gas mixture
- n_x is the amount of substance of a gas (X)
- n_{tot} is the total amount of substance in gas mixture

Vapour Pressure

Vapour pressure is the pressure of a vapour in equilibrium with its non-vapour phases (i.e., liquid or solid). Most often the term is used to describe a liquid's tendency to evaporate. It is a measure of the tendency of molecules and atoms to escape from a liquid or a solid. A liquid's atmospheric pressure boiling point corresponds to the temperature at which its vapour pressure is equal to the surrounding atmospheric pressure and it is often called the normal boiling point. The higher the vapour pressure of a liquid at a given temperature, the lower the normal boiling point of the liquid. The vapour pressure chart to the right has graphs of the vapour pressures versus temperatures for a variety of liquids. As can be seen in the chart, the liquids with the highest vapour pressures have the lowest normal boiling points.

For example, at any given temperature, methyl chloride has the highest vapour pressure of any of the liquids in the chart. It also has the lowest normal boiling point (-24.2 °C), which is where the vapour pressure curve of methyl chloride (the blue line) intersects the horizontal pressure line of one atmosphere (atm) of absolute vapour pressure.

Equilibrium Constants of Reactions Involving Gas Mixtures

It is possible to work out the equilibrium constant for a chemical reaction involving a mixture of gases given the partial pressure of each gas and the overall reaction formula. For a reversible reaction involving gas reactants and gas products, such as:

$$aA + bB \leftrightarrow cC + dD$$

the equilibrium constant of the reaction would be:

$$K_p = \frac{p_C^c \, p_D^d}{p_A^a \, p_B^b}$$

where:

K_p	=	the equilibrium constant of the reaction
a	=	coefficient of reactant A
b	=	coefficient of reactant B
c	=	coefficient of product C
d	=	coefficient of product D
p_C^c	=	the partial pressure of C raised to the power of c
p_D^d	=	the partial pressure of D raised to the power of d
p_A^a	=	the partial pressure of A raised to the power of a
p_B^b	=	the partial pressure of B raised to the power of b

For reversible reactions, changes in the total pressure, temperature or reactant concentrations will shift the equilibrium so as to favour either the right or left side of the reaction in accordance with Le Chatelier's Principle. However, the reaction kinetics may either oppose or enhance the equilibrium shift. In some cases, the reaction kinetics may be the over-riding factor to consider.

Henry's Law and the Solubility of Gases

Gases will dissolve in liquids to an extent that is determined by the equilibrium between the undissolved gas and the gas that has dissolved in the liquid (called the *solvent*). The equilibrium constant for that equilibrium is:

$$k = \frac{p_x}{C_x} \qquad (1)$$

where:		
k	=	the equilibrium constant for the solvation process
p_x	=	partial pressure of gas x in equilibrium with a solution containing some of the gas
C_x	=	the concentration of gas x in the liquid solution

The form of the equilibrium constant shows that the concentration of a solute gas in a solution is directly proportional to the partial pressure of that gas above the solution. This statement is known as Henry's Law and the equilibrium constant is quite often referred to as the Henry's Law constant.

Henry's Law is sometimes written as:

$$k' = \frac{C_x}{p_x} \quad (2)$$

where k' is also referred to as the Henry's Law constant. As can be seen by comparing equations (1) and (2) above, k' is the reciprocal of k. Since both may be referred to as the Henry's Law constant, readers of the technical literature must be quite careful to note which version of the Henry's Law equation is being used.

Henry's Law is an approximation that only applies for dilute, ideal solutions and for solutions where the liquid solvent does not react chemically with the gas being dissolved.

Partial Pressure in Diving Breathing Gases

In recreational diving and professional diving the richness of individual component gases of breathing gases is expressed by partial pressure.

Using diving terms, partial pressure is calculated as:

partial pressure = (total absolute pressure) × (volume fraction of gas component)

For the component gas "i":

ppi = P × Fi

For example, at 50 metres (165 feet), the total absolute pressure is 6 bar (600 kPa) (i.e., 1 bar of atmospheric pressure + 5 bar of water pressure) and the partial pressures of the main components of air, oxygen 21% by volume and nitrogen 79% by volume are:

ppN2 = 6 bar × 0.79 = 4.7 bar absolute

ppO2 = 6 bar × 0.21 = 1.3 bar absolute

where:		
ppi	=	partial pressure of gas component i = P_i in the terms used in this article
P	=	total pressure = P in the terms used in this article
Fi	=	volume fraction of gas component i = mole fraction, x_i, in the terms used in this article
ppN2	=	partial pressure of nitrogen = P_{N_2} in the terms used in this article
ppO2	=	partial pressure of oxygen = P_{O_2} in the terms used in this article

The minimum safe lower limit for the partial pressures of oxygen in a gas mixture is 0.16 bar (16 kPa) absolute. Hypoxia and sudden unconsciousness becomes a problem with an oxygen partial pressure of less than 0.16 bar absolute.

Oxygen toxicity, involving convulsions, becomes a problem when oxygen partial pressure is too high. The NOAA Diving Manual recommends a maximum single exposure of 45 minutes at 1.6 bar absolute, of 120 minutes at 1.5 bar absolute, of 150 minutes at 1.4 bar absolute, of 180 minutes at 1.3 bar absolute and of 210 minutes at 1.2 bar absolute.

Oxygen toxicity becomes a risk when these oxygen partial pressures and exposures are exceeded. The partial pressure of oxygen determines the maximum operating depth of a gas mixture.

Nitrogen narcosis is a problem when breathing gases at high pressure. Typically, the maximum total partial pressure of narcotic gases used when planning for technical diving is 4.5 bar absolute, based on an equivalent narcotic depth of 35 metres (115 ft).

In Medicine

The partial pressures of particularly oxygen (p_{O_2}) and carbon dioxide (p_{CO_2}) are important parametres in tests of arterial blood gases, but can also be measured in, for example, cerebrospinal fluid.

Reference ranges for p_{O_2} and p_{CO_2}						
	Unit	*Arterial blood gas*	*Venous blood gas*	*Cerebrospinal fluid*	*Alveolar pulmonary gas pressures*
p_{O_2}	kPa	11–13	4.0–5.3	5.3–5.9	14.2
	mmHg	75–100	30–40	40–44	107
p_{CO_2}	kPa	4.7–6.0	5.5–6.8	5.9–6.7	4.8
	mmHg	35–45	41–51	44–50	36

Equivalent Gas Constant

Values of R	Units($V\ P\ T^{-1}\ n^{-1}$)
8.3144621(75)	$J\ K^{-1}\ mol^{-1}$
5.189×10^{19}	$eV\ K^{-1}\ mol^{-1}$
0.08205746(14)	$L\ atm\ K^{-1}\ mol^{-1}$
1.9858775(34)	$cal\ K^{-1}\ mol^{-1}$
$1.9858775(34)\times10^{-3}$	$kcal\ K^{-1}\ mol^{-1}$
$8.3144621(75)\times10^{7}$	$erg\ K^{-1}\ mol^{-1}$
8.3144621(75)	$L\ kPa\ K^{-1}\ mol^{-1}$
$8.3144621(75)\times10^{3}$	$cm^3 kPa\ K^{-1}\ mol^{-1}$
8.3144621(75)	$m^3\ Pa\ K^{-1}\ mol^{-1}$
8.3144621(75)	$cm^3\ MPa\ K^{-1}\ mol^{-1}$
$8.3144621(75)\times10^{-5}$	$m^3\ bar\ K^{-1}\ mol^{-1}$
8.205746×10^{-5}	$m^3\ atm\ K^{-1}\ mol^{-1}$
8.205746×10^{-2}	$m^3\ atm\ K^{-1}\ kg\text{-}mol^{-1}$
82.05746	$cm^3\ atm\ K^{-1}\ mol^{-1}$
84.78402×10^{-6}	$m^3\ kgf/cm^2\ K^{-1}\ mol^{-1}$
$8.3144621(75)\times10^{-2}$	$L\ bar\ K^{-1}\ mol^{-1}$
$62.36367(11)\times10^{-3}$	$m^3\ mmHg\ K^{-1}\ mol^{-1}$
62.36367(11)	$L\ mmHg\ K^{-1}\ mol^{-1}$
62.36367(11)	$L\ Torr\ K^{-1}\ mol^{-1}$
6.132440(10)	$ft\ lbf\ K^{-1}\ g\text{-}mol^{-1}$
1,545.34896(3)	$ft\ lbf\ R^{-1}\ lb\text{-}mol^{-1}$
10.73159(2)	$ft^3\ psi\ R^{-1}\ lb\text{-}mol^{-1}$
0.7302413(12)	$ft^3\ atm\ R^{-1}\ lb\text{-}mol^{-1}$
1.31443	$ft^3\ atm\ K^{-1}\ lb\text{-}mol^{-1}$
998.9701(17)	$ft^3\ mmHg\ K^{-1}\ lb\text{-}mol^{-1}$
1.986	$Btu\ lb\text{-}mol^{-1}\ R^{-1}$

The gas constant (also known as the molar, universal, or ideal gas constant, denoted by the symbol R or R) is a physical constant which is featured in many fundamental equations in the physical sciences, such as the ideal gas law and the Nernst equation.

It is equivalent to the Boltzmann constant, but expressed in units of energy (i.e. the pressure-volume product) per temperature increment per *mole* (rather than energy per temperature increment per *particle*). The constant is also a combination of the constants from Boyle's law, Charles's law, Avogadro's law, and Gay-Lussac's law.

Physically, the gas constant is the constant of proportionality that happens to relate the energy scale in physics to the temperature scale, when a mole of particles at the stated temperature is being considered. Thus, the value of the gas constant ultimately derives from historical decisions and accidents in the setting of the energy and temperature scales, plus similar historical setting of the value of the molar scale used for the counting of particles. The last factor is not a consideration in the value of the Boltzmann constant, which does a similar job of equating linear energy and temperature scales.

The gas constant value is

$$R = 8.3144621(75)\frac{\text{J}}{\text{mol K}}$$

The two digits in parentheses are the uncertainty (standard deviation) in the last two digits of the value. The relative uncertainty is 9.1×10^{-7}. Some have suggested that it might be appropriate to name the symbol R the Regnault constant in honour of the FrenchchemistHenri Victor Regnault, whose accurate experimental data was used to calculate the early value of the constant; however, the exact reason for the original representation of the constant by the letter R is elusive.

The gas constant occurs in the ideal gas law, as follows:

$$PV = nRT = mR_{\text{specific}}T$$

where P is the absolute pressure (SI unit pascals), V is the volume of gas (SI unit cubic metres), n is the chemical amount of gas (SI unit moles), m is the mass (SI unit kilograms) contained in V, and T is the thermodynamic temperature (SI unit kelvins). The gas constant is expressed in the same physical units as molar entropy and molar heat capacity.

Dimensions of R

From the general equation $PV = nRT$ we get

$R = PV/nT$ or (pressure × volume)/ (amount × temperature).

As pressure is defined as force per unit area, we can also write the gas equation as

R = [(force/area) × volume] / (amount × temperature).

Again, area is nothing but $(\text{length})^2$ and volume is equal to $(\text{length})^3$. Therefore,

R = [force/ $(\text{length})^2$] $(\text{length})^3$ / (amount × temperature).

Since force × length = work,

R = (work) / (amount × temperature).

The physical significance of R is work per degree per mole. It may be expressed in any set of units representing work or energy (such as joules), other units representing degrees of temperature (such as degrees celsius or Fahrenheit), and any system of units designating a mole or a similar pure number that allows an equation of macroscopic mass and fundamental particle numbers in a system, such as an ideal gas.

Relationship with the Boltzmann Constant

The Boltzmann constantk_B (often abbreviated k) may be used in place of the gas constant by working in pure particle count, N, rather than amount of substance, n, since

$$R = N_A k_B,$$

where N_A is the Avogadro constant. For example, the ideal gas law in terms of Boltzmann's constant is

$$PV = Nk_BT.$$

where N is the number of particles (molecules in this case).

Measurement

As of 2006, the most precise measurement of R is obtained by measuring the speed of sound $c_a(p, T)$ in argon at the temperature T of the triple point of water (used to define the kelvin) at different pressures p, and extrapolating to the zero-pressure limit $c_a(0, T)$. The value of R is then obtained from the relation

$$c_a^2(0,T) = \frac{\gamma_0 RT}{A_r(\mathrm{Ar})M_u},$$

where:

- γ_0 is the heat capacity ratio (5/3 for monatomic gases such as argon);
- T is the temperature, T_{TPW} = 273.16 K by definition of the kelvin;
- A_r(Ar) is the relative atomic mass of argon and $M_u = 10^{-3}$ kg mol^{-1}.

Specific Gas Constant

$R_{specific}$ for dry air	Units
287.058	J kg^{-1} K^{-1}
53.3533	ft lbf lb^{-1} °R^{-1}
1716.49	ft lbf slug^{-1} °R^{-1}

Based on a mean molar massfor dry air of 28.9645 g/mol.

The specific gas constant of a gas or a mixture of gases ($R_{specific}$) is given by the molar gas constant, divided by the molar mass (M) of the gas/mixture.

$$R_{\text{specific}} = \frac{R}{M}$$

Just as the ideal gas constant can be related to the Boltzmann constant, so can the specific gas constant by dividing the Boltzmann constant by the molecular mass of the gas.

$$R_{\text{specific}} = \frac{k_{\text{B}}}{m}$$

Another important relationship comes from thermodynamics. Mayer's relation relates the specific gas constant to the specific heats for a calorically perfect gas and a thermally perfect gas.

$$R_{\text{specific}} = c_{\text{p}} - c_{\text{v}}$$

where c_p is the specific heat for a constant pressure and c_v is the specific heat for a constant volume.

It is common, especially in engineering applications, to represent the specific gas constant by the symbol R. In such cases, the universal gas constant is usually given a different symbol such as R to distinguish it. In any case, the context and/or units of the gas constant should make it clear as to whether the universal or specific gas constant is being referred to.

U.S. Standard Atmosphere

The U.S. Standard Atmosphere, 1976 (USSA1976) defines the gas constant R^* as:

$$R^* = 8.31432 \times 10^3 \frac{\text{Nm}}{\text{kmolK}}.$$

The USSA1976 does recognize, however, that this value is not consistent with the cited values for the Avogadro constant and the Boltzmann constant. This disparity is not a significant departure from accuracy, and USSA1976 uses this value of R^* for all the calculations of the standard atmosphere. When using the ISO value of R, the calculated pressure increases by only 0.62 pascal at 11 kilometres (the equivalent of a difference of only 0.174 metre or 6.8 inches) and an increase of 0.292 Pa at 20 km (the equivalent of a difference of only 0.338 m or 13.2 in).

Individual Gas Constants

Individual gas constants in base SI units of J/kg.K can also be derived for any gas species, by making use of their molar mass. An average value could also be derived for gas mixtures. Use of individual gas constants may make it more difficult to follow the workings of a calculation, as the relevant values tend to be less well known, and less intuitive, than the fixed value of the universal gas constants, and the well-known values of gas molecular masses.

Enthalpy

Enthalpy is a measure of the total energy of a thermodynamic system. It includes the system's internal energy or thermodynamic potential (a state function), as well as its volume and pressure (the energy required to "make room for it" by displacing its environment, which is an extensive quantity). The unit of measurement for enthalpy in the International System of Units (SI) is the joule, but other historical, conventional units are still in use, such as the British thermal unit and the calorie.

The enthalpy is the preferred expression of system energy changes in many chemical, biological, and physical measurements, because it simplifies certain descriptions of energy transfer. Enthalpy change accounts for energy transferred to the environment at constant pressure through expansion or heating.

The total enthalpy, H, of a system cannot be measured directly. Thus, change in enthalpy, ΔH, is a more useful quantity than its absolute value. The change ΔH is positive in endothermic reactions, and negative in heat-releasing exothermic processes. ΔH of a system is equal to the sum of non-mechanical work done on it and the heat supplied to it.

For processes under constant pressure, ΔH is equal to the change in the internal energy of the system, plus the work that the system has done on its surroundings. This means that the change in enthalpy under such conditions is the heat absorbed (or released) by the material through a chemical reaction or by external heat transfer. Enthalpies for chemical substances at constant pressure assume standard state: most commonly 1 bar pressure. Standard state does not, strictly speaking, specify a temperature, but expressions for enthalpy generally reference the standard heat of formation at 25 °C.

Enthalpies of ideal gases and incompressible solids and liquids do not depend on pressure, unlike entropy and Gibbs energy. Real materials at common temperatures and pressures usually closely approximate this behaviour, which greatly simplifies enthalpy calculation and use in practical designs and analyses.

Origins

The word *enthalpy* is based on the Greek noun*enthalpos* (—νèαλπïò), which means *heating.* It comes from the Classical Greek prefix ν-, *en-*, meaning *to put into*, and the verb èÜλπåév, *thalpein*, meaning "to heat". The word *enthalpy* is often incorrectly attributed to Benoit Paul Émile Clapeyron and Rudolf Clausius through the 1850 publication of their Clausius-Clapeyron relation. This misconception was popularized by the 1927 publication of *The Mollier Steam Tables and Diagrams*. However, neither the concept, the word, nor the symbol for enthalpy existed until well after Clapeyron's death.

The earliest writings to contain the concept of enthalpy did not appear until 1875, when Josiah Willard Gibbs introduced "a heat function for constant pressure". However, Gibbs did not use the word "enthalpy" in his writings. The actual word first appears in the scientific literature in a 1909 publication by J. P. Dalton. According to that publication, Heike Kamerlingh Onnes (1853-1926) actually coined the word.

Over the years, many different symbols were used to denote enthalpy. It was not until 1922 that Alfred W. Porter proposed the symbol "H" as the accepted standard, thus finalizing the terminology still in use today.

Formal Definition

The enthalpy of a homogeneous system is defined as:

$$H = U + pV$$

where

H is the enthalpy of the system

U is the internal energy of the system

p is the pressure of the system

V is the volume of the system.

The enthalpy is an extensive property. This means that, for homogeneous systems, the enthalpy is proportional to the size of the system. It is convenient to introduce the specific enthalpy $h = H/m$ where m is the mass of the system, or the molar enthalpy $H_m = H/n$, where n is the number of moles (h and H_m are intensive properties).

For inhomogeneous systems the enthalpy is the sum of the enthalpies of the composing subsystems

$$H = \Sigma_k H_k$$

where the label k refers to the various subsystems. In case of continuously varying p, T, and/or composition the summation becomes an integral:

$$H = \int \rho h \mathrm{d}V,$$

where ρ is the density.

The enthalpy $H(S,p)$ of homogeneous systems can be derived as a characteristic function of the entropy S and the pressure p as follows: we start from the first law of thermodynamics for closed systems for an infinitesimal process

$$\mathrm{d}U = \delta Q - \delta W.$$

Here, δQ is a small amount of heat added to the system and δW a small amount of work performed by the system. In a homogeneous system only reversible processes can take place so the second law of thermodynamics gives $\delta Q = T\mathrm{d}S$ with T the absolute temperature of the system. Furthermore, if only pV work is done, $\delta W = p\mathrm{d}V$. As a result

$$\mathrm{d}U = T\mathrm{d}S - p\mathrm{d}V.$$

Adding $\mathrm{d}(pV)$ to both sides of this expression gives

$$\mathrm{d}U + \mathrm{d}(pV) = T\mathrm{d}S - p\mathrm{d}V + \mathrm{d}(pV)$$

or

$$\mathrm{d}(U + pV) = T\mathrm{d}S + V\mathrm{d}p.$$

So

$$\mathrm{d}H(S,p) = T\mathrm{d}S + V\mathrm{d}p.$$

Other Expressions

The expression of $\mathrm{d}H$ in terms of entropy and pressure may be unfamiliar to many readers. However, there are expressions in terms of more familiar variables such as temperature and pressure

$$\mathrm{d}H = C_p \mathrm{d}T + V(1 - \alpha T)\mathrm{d}p.$$

Here C_p is the heat capacity at constant pressure and α is the coefficient of (cubic) thermal expansion

$$\alpha = \frac{1}{V}\left(\frac{\partial V}{\partial T}\right)_p.$$

With this expression one can, in principle, determine the enthalpy if C_p and V are known as functions of p and T.

In a more general form, the first law describes the internal energy with additional terms involving the chemical potential and the number of particles of various types. The differential statement for dH then becomes:

$$dH = TdS + Vdp + \sum_i \mu_i dN_i$$

where μ_i is the chemical potential per particle for an i-type particle, and N_i is the number of such particles. The last term can also be written as $\mu_i dn_i$ (with dn_i the number of moles of component i added to the system and, in this case, μ_i the molar chemical potential) or as $\mu_i dm_i$ (with dm_i the mass of component i added to the system and, in this case, μ_i the specific chemical potential).

Enthalpy Versus Internal Energy

The U term can be interpreted as the energy required to create the system, and the pV term as the energy that would be required to "make room" for the system if the pressure of the environment remained constant. When a system, for example, $n$moles of a gas of volume$V$ at pressure$p$ and temperature T, is created or brought to its present state from absolute zero, energy must be supplied equal to its internal energy U plus pV, where pV is the work done in pushing against the ambient (atmospheric) pressure.

In basic physics and statistical mechanics it may be more interesting to study the internal properties of the system and therefore the internal energy is used. In basic chemistry, experiments are often conducted at atmospheric pressure and H is therefore more useful for reaction energy calculations. Furthermore the enthalpy is the workhorse of engineering thermodynamics as we will see later.

Relationship to Heat

In order to discuss the relation between the enthalpy increase and heat supply we return to the first law for closed systems: $dU = \delta Q - \delta W$. We apply it to the special case that the pressure at the surface is uniform. In this case the work term can be split in two contributions, the so-called pV work, given by $-pdV$ (where here p is the pressure at the surface, dV is the increase of the volume of the system) and other types of work $\delta W'$ such as by a shaft or by electromagnetic interaction. So we write $\delta W = -pdV + \delta W'$. In this case the first law reads

$$dU = \delta Q - p dV - \delta W'$$

or

$$dH = \delta Q + V dp - \delta W'.$$

From this relation we see that the increase in enthalpy of a system is equal to the added heat ·

$$dH = \delta Q$$

provided that the system is under constant pressure ($dp = 0$) and that the only work done by the system is expansion work ($\delta W' = 0$)

Applications

In thermodynamics, one can calculate enthalpy by determining the requirements for creating a system from "nothingness"; the mechanical work required, pV, differs based upon the constancy of conditions present at the creation of the thermodynamic system. Internal energy, U, must be supplied to remove particles from a surrounding in order to allow space for the creation of a system, providing that environmental variables, such as pressure (p) remain constant. This internal energy also includes the energy required for activation and the breaking of bonded compounds into gaseous species.

This process is calculated within enthalpy calculations as $U + pV$, to label the amount of energy or work required to "set aside space for" and "create" the system; describing the work done by both the reaction or formation of systems, and the surroundings. For systems at constant pressure, the change in enthalpy is the heat received by the system. Therefore, the change in enthalpy can be devised or represented without the need for compressive or expansive mechanics; for a simple system, with a constant number of particles, the difference in enthalpy is the maximum amount of thermal energy derivable from a thermodynamic process in which the pressure is held constant.

The term pV is the work required to displace the surrounding atmosphere in order to vacate the space to be occupied by the system.

Heat of Reaction

The total enthalpy of a system cannot be measured directly; the *enthalpy change* of a system is measured instead. Enthalpy change is defined by the following equation:

$$\Delta H = H_f - H_i$$

where

ΔH is the "enthalpy Change"

H_f is the final enthalpy of the system, expressed in joules. In a chemical reaction, H_f is the enthalpy of the products.

H_i is the initial enthalpy of the system, expressed in joules. In a chemical reaction, H_i is the enthalpy of the reactants.

For an exothermicreaction at constant pressure, the system's change in enthalpy equals the energy released in the reaction, including the energy retained in the system and lost through expansion against its surroundings. In a similar manner, for an endothermic reaction, the system's change in enthalpy is equal to the energy *absorbed* in the reaction, including the energy *lost by* the system and *gained* from compression from its surroundings. A relatively easy way to determine whether or not a reaction is exothermic or endothermic is to determine the sign of ΔH. If ΔH is positive, the reaction is endothermic, that is heat is absorbed by the system due to the products of the reaction having a greater enthalpy than the reactants. On the other hand if ΔH is negative, the reaction is exothermic, that is the overall decrease in enthalpy is achieved by the generation of heat.

Although enthalpy is commonly used in engineering and science, it is impossible to measure directly, as enthalpy has no datum (reference point). Therefore enthalpy can only accurately be used in a closed system. However, few real-world applications exist in closed isolation, and it is for this reason that two or more closed systems cannot correctly be compared using enthalpy as a basis.

Specific Enthalpy

As noted before, the specific enthalpy of a uniform system is defined as $h = H/m$ where m is the mass of the system. The SI unit for specific enthalpy is joule per kilogram. It can be expressed in other specific quantities by $h = u + pv$, where u is the specific internal energy, p is the pressure, and v is specific volume, which is equal to $1/\rho$, where ρ is the density.

Enthalpy Changes

An enthalpy change describes the change in enthalpy observed in the constituents of a thermodynamic system when undergoing a transformation or chemical reaction. It is the difference between the enthalpy after the process has completed, i.e. the enthalpy of the products, and the initial enthalpy of the system, i.e. the reactants. These processes are reversible and the enthalpy for the reverse process is the negative value of the forward change.

A common standard enthalpy change is the enthalpy of formation, which has been determined for a large number of substances. Enthalpy changes are routinely measured and compiled in chemical and physical reference works, such as the CRC Handbook of Chemistry and Physics. The following is a selection of enthalpy changes commonly recognized in thermodynamics. When used in these recognized terms the qualifier *change* is usually dropped and the property is simply termed *enthalpy of 'process'*. Since these properties are often used as reference values it is very common to quote them for a standardized set of environmental parametres, or standard conditions, including:

- A temperature of 25 °C or 298 K,
- A pressure of one atmosphere (1 atm or 101.325 kPa),
- A concentration of 1.0 M when the element or compound is present in solution,
- Elements or compounds in their normal physical states, i.e. standard state.

For such standardized values the name of the enthalpy is commonly prefixed with the term *standard*, e.g. *standard enthalpy of formation.*

Chemical properties:

- Enthalpy of reaction, defined as the enthalpy change observed in a constituent of a thermodynamic system when one mole of substance reacts completely.
- Enthalpy of formation, defined as the enthalpy change observed in a constituent of a thermodynamic system when one mole of a compound is formed from its elementary antecedents.
- Enthalpy of combustion, defined as the enthalpy change observed in a constituent of a thermodynamic system when one mole of a substance burns completely with oxygen.
- Enthalpy of hydrogenation, defined as the enthalpy change observed in a constituent of a thermodynamic system when one mole of an unsaturated compound reacts completely with an excess of hydrogen to form a saturated compound.
- Enthalpy of atomization, defined as the enthalpy change required to atomize one mole of compound completely.
- Enthalpy of neutralization, defined as the enthalpy change observed in a constituent of a thermodynamic system when one mole of water is formed when an acid and a base react.
- Standard Enthalpy of solution, defined as the enthalpy change observed in a constituent of a thermodynamic system when one mole of a solute is dissolved completely in an excess of solvent, so that the solution is at infinite dilution.
- Standard enthalpy of Denaturation (biochemistry), defined as the enthalpy change required to denature one mole of compound.

- Enthalpy of hydration, defined as the enthalpy change observed when one mole of gaseous ions are completely dissolved in water forming one mole of aqueous ions.

Physical properties:

- Enthalpy of fusion, defined as the enthalpy change required to completely change the state of one mole of substance between solid and liquid states.
- Enthalpy of vaporization, defined as the enthalpy change required to completely change the state of one mole of substance between liquid and gaseous states.
- Enthalpy of sublimation, defined as the enthalpy change required to completely change the state of one mole of substance between solid and gaseous states.
- Lattice enthalpy, defined as the energy required to separate one mole of an ionic compound into separated gaseous ions to an infinite distance apart (meaning no force of attraction).

Open Systems

In thermodynamicopen systems, matter may flow in and out of the system boundaries. The first law of thermodynamics for open systems states: The increase in the internal energy of a system is equal to the amount of energy added to the system by matter flowing in and by heating, minus the amount lost by matter flowing out and in the form of work done by the system. The first law for open systems is given by:

$$dU = \delta Q + dU_{in} - dU_{out} - \delta W$$

where U_{in} is the average internal energy entering the system and U_{out} is the average internal energy leaving the system.

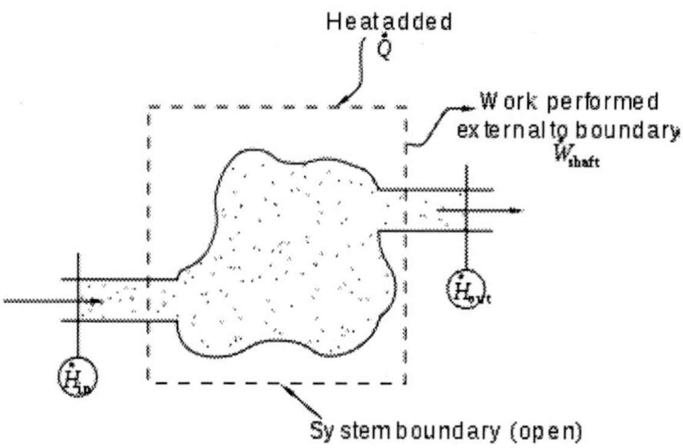

Figure: *During steady, continuous operation, an energy balance applied to an open system equates shaft work performed by the system to heat added plus net enthalpy added*

The region of space enclosed by open system boundaries is usually called a control volume, and it may or may not correspond to physical walls. If we choose the shape of the control volume such that all flow in or out occurs perpendicular to its surface, then the flow of matter into the system performs work as if it were a piston of fluid pushing mass into the system, and the system performs work on the flow of matter out as if it were driving a piston of fluid. There are then two types of work performed: *flow work* described above, which is performed on the fluid (this is also often called *pV work*), and *shaft work*, which may be performed on some mechanical device.

These two types of work are expressed in the equation:

$$\delta W = \mathrm{d}(p_{out}V_{out}) - \mathrm{d}(p_{in}V_{in}) + \delta W_{shaft}$$

Substitution into the equation above for the control volume cv yields:

$$\mathrm{d}U_{cv} = \delta Q + \mathrm{d}U_{in} + \mathrm{d}(p_{in}V_{in}) - \mathrm{d}U_{out} - \mathrm{d}(p_{out}V_{out}) - \delta W_{shaft}\,.$$

The definition of enthalpy, H, permits us to use this thermodynamic potential to account for both internal energy and pV work in fluids for open systems:

$$\mathrm{d}U_{cv} = \delta Q + \mathrm{d}H_{in} - \mathrm{d}H_{out} - \delta W_{shaft}\,.$$

This expression is described by Fig.1. If we allow also the system boundary to move (e.g. due to moving pistons) we get a rather general form of the first law for open systems. In terms of time derivatives it reads

$$\frac{\mathrm{d}U}{\mathrm{d}t} = \Sigma_k \dot{Q}_k + \Sigma_k \dot{H}_k - \Sigma_k p_k \frac{\mathrm{d}V_k}{\mathrm{d}t} - P,$$

where Σ represent algebraic sums and the indices k refer to the various places where heat is supplied, matter flows into the system, and boundaries are moving. The $\dot{H}_k$ terms represent enthalpy flows, which can be written as

$$\dot{H}_k = h_k \dot{m}_k = H_m \dot{n}_k$$

with $\dot{m}_k$ the mass flow and $\dot{n}_k$ the molar flow at position k respectively. The term $\mathrm{d}V_k/\mathrm{d}t$ represents the rate of change of the system volume at position k that results in pV power done by the system. The parametre P represents all other forms of power done by the system such as shaft power, but it can also be e.g. electric power produced by an electrical power plant. Note that the previous expression holds true only if the kinetic energy flow rate is conserved between system inlet and outlet. Otherwise, it has to be included in the enthalpy balance. During steady-state operation of a device, the average dU/dt may be set equal to zero. This yields a useful expression for the average power generation for these devices in the absence of chemical reactions

$$P = \Sigma_k \left\langle \dot{Q}_k \right\rangle + \Sigma_k \left\langle \dot{H}_k \right\rangle - \Sigma_k \left\langle p_k \frac{\mathrm{d}V_k}{\mathrm{d}t} \right\rangle$$

where the angle brackets denote time averages. The technical importance of the enthalpy is directly related to its presence in the first law for open systems, as formulated above.

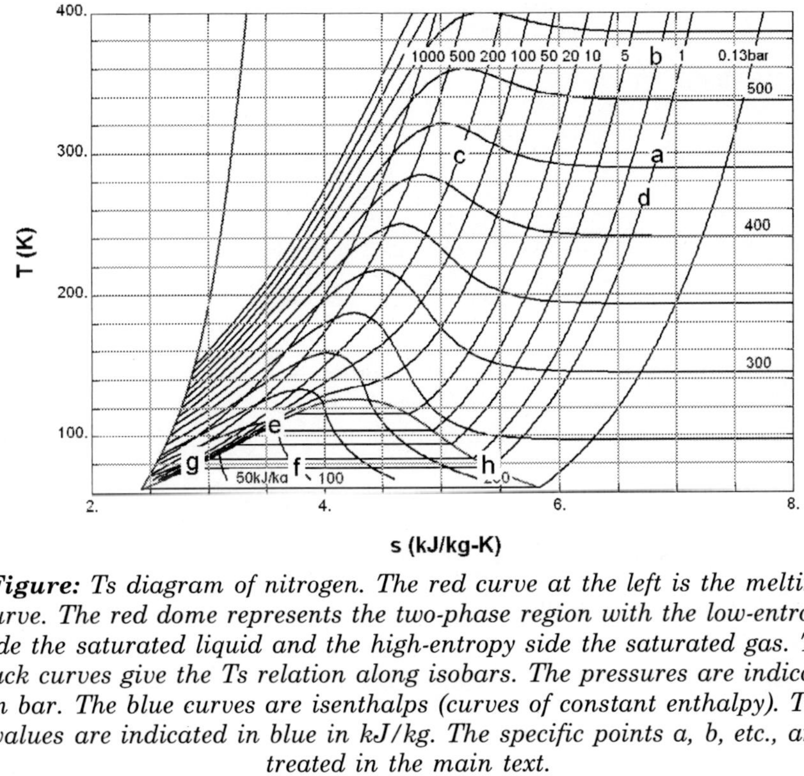

__Figure:__ Ts diagram of nitrogen. The red curve at the left is the melting curve. The red dome represents the two-phase region with the low-entropy side the saturated liquid and the high-entropy side the saturated gas. The black curves give the Ts relation along isobars. The pressures are indicated in bar. The blue curves are isenthalps (curves of constant enthalpy). The values are indicated in blue in kJ/kg. The specific points a, b, etc., are treated in the main text.

Diagrams

Nowadays the enthalpy values of important substances can be obtained via commercial software. Practically all relevant material properties can be obtained either in tabular or in graphical form. There are many types of diagrams, such as hT diagrams, which give the specific enthalpy as function of temperature for various pressures and hp diagrams, which give h as function of p for various T. One of the most common diagrams is the temperature-entropy diagram

(Ts-diagram). An example is *Fig.2*, which is the Ts-diagram of nitrogen. It gives the melting curve and saturated liquid and vapour values together with isobars and isenthalps. These diagrams are powerful tools in the hands of the thermal engineer.

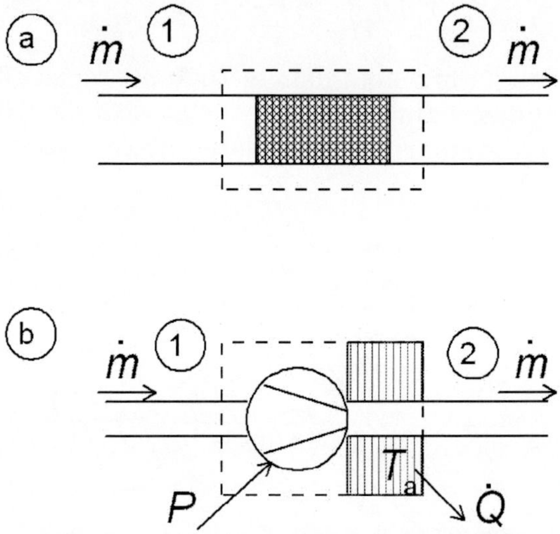

Figure: *Two open systems in the steady state. Fluid enters the system (dotted rectangle) at point 1 and leaves it at point 2. The mass flow is* $\dot{m}$. *a: schematic diagram of the throttling process. b: schematic diagram of a compressor. A power* P *is applied and a heat flow* $\dot{Q}$ *is released to the surroundings at ambient temperature* T_a.

Some Basic Applications

The points a through h in Fig.2 play a role in the discussion in this Section.

a T = 300 K, p = 1 bar, s = 6.85 kJ/(kgK), h = 461 kJ/kg;

b T = 380 K, p = 2 bar, s = 6.85 kJ/(kgK), h = 530 kJ/kg;

c T = 300 K, p = 200 bar, s = 5.16 kJ/(kgK), h = 430 kJ/kg;

d T = 270 K, p = 1 bar, s = 6.79 kJ/(kgK), h = 430 kJ/kg;

e T = 108 K, p = 13 bar, s = 3.55 kJ/(kgK), h = 100 kJ/kg (saturated liquid at 13 bar);

f T = 77.2 K, p = 1 bar, s = 3.75 kJ/(kgK), h = 100 kJ/kg;

g T = 77.2 K, p = 1 bar, s = 2.83 kJ/(kgK), h = 28 kJ/kg (saturated liquid at 1 bar);

h T = 77.2 K, p = 1 bar, s = 5.41 kJ/(kgK), h =230 kJ/kg (saturated gas at 1 bar);

Throttling

One of the simple applications of the concept of enthalpy is the so-called throttling process, also known as Joule-Thomson expansion. It concerns a steady adiabatic flow of a fluid through a flow resistance (valve, porous plug, or any other type of flow resistance) as shown in Fig.3a. This process is very important since it is at the heart of domestic refrigerators where it is responsible for the temperature drop between ambient temperature and the interior of the fridge. It is also the final stage in many types of liquefiers.

In the first law for open systems, applied to the system in Fig.3a, all terms are zero except the terms for the enthalpy flow. Hence

$$0 = \dot{m}h_1 - \dot{m}h_2.$$

Since the mass flow is constant the specific enthalpies at the two sides of the flow resistance are the same

$$h_1 = h_2.$$

The consequences of this relation can be demonstrated using Fig.2. Point c in Fig.2 is at 200 bar and room temperature (300 K). A Joule-Thomson expansion from 200 to 1 bar follows a curve (not shown in Fig.2) in between the 400 and 450 kJ/kg isenthalps and ends in point d, which is at a temperature of about 270 K. Hence the expansion from 200 bar to 1 bar cools nitrogen from 300 K to 270 K. In the valve there is a lot of friction and a lot of entropy is produced, but still the final temperature is below the starting value!

Point e is chosen so that it is on the saturated liquid line with h = 100 kJ/kg. It corresponds roughly with p = 13 bar and T = 108 K. Throttling from this point to a pressure of one bar ends in the two-phase region (point f). This means that a mixture of gas and liquid leaves the throttling valve. Since the enthalpy is an extensive parametre the enthalpy in f (h_f) is equal to the enthalpy in g (h_g) multiplied with the liquid fraction in f (x_f) plus the enthalpy in h (h_h) multiplied with the gas fraction in f (1-x_f). So

$$h_f = x_f h_g + (1 - x_f)h_h.$$

With numbers: 100 = x_f 28 + (1 - x_f)230 so x_f = 0.64. This means that the mass fraction of the liquid in the liquid-gas mixture that leaves the throttling valve is 64%.

Compressors

Figure above is a schematic drawing of a compressor. A power P is applied e.g. as electrical power. If the compression is adiabatic the gas temperature goes up. In the reversible case it would be at constant entropy, which corresponds with a vertical line in figure E.g. compressing nitrogen from 1 bar (point a) to 2 bar (point b) would result in a temperature increase from 300 K to 380 K.

In order to let the compressed gas exit at ambient temperature T_a heat exchange, e.g. by cooling water, is necessary. In the ideal case the compression is isothermal.

The average heat flow to the surroundings is $\dot{Q}$. Since the system is in the steady state the first law gives

$$0 = -\dot{Q} + \dot{m}h_1 - \dot{m}h_2 + P.$$

The minimum power, needed for the compression is realized if the compression is reversible. In that case the second law of thermodynamics for open systems gives

$$0 = -\frac{\dot{Q}}{T_a} + \dot{m}s_1 - \dot{m}s_2.$$

Eliminating $\dot{Q}$ gives for the minimum power

$$\frac{P_{\min}}{\dot{m}} = h_2 - h_1 - T_a(s_2 - s_1).$$

E.g. compressing 1 kg of nitrogen from 1 bar to 200 bar costs at least $(h_c - h_a) - T_a(s_c - s_a)$. With the data, obtained with Fig.2, we find a value of (430-461) - 300(5.16 - 6.85) = 476 kJ/kg.

The relation for the power can be further simplified by writing it as

$$\frac{P_{\min}}{\dot{m}} = \int_1^2 (\mathrm{d}h - T_a \mathrm{d}s).$$

With $\mathrm{d}h = T\mathrm{d}s + v\mathrm{d}p$ this results in the final relation

$$\frac{P_{\min}}{\dot{m}} = \int_1^2 v\mathrm{d}p.$$

Atmospheric Air

Air, mixture of gases comprising the Earth's atmosphere. The mixture contains a group of gases of nearly constant concentrations and a group with concentrations that are variable in both space and time. The atmospheric gases of steady concentration (and their proportions in percentage by volume) are as follows:

nitrogen (N_2)	78.084
oxygen (O_2)	20.946
argon (Ar)	0.934
neon (Ne)	0.0018
helium (He)	0.000524
methane (CH_4)	0.0002
krypton (Kr)	0.000114
hydrogen (H_2)	0.00005
nitrous oxide (N_2O)	0.00005
xenon (Xe)	0.0000087

The uniformity of composition is maintained by mixing associated with atmospheric motions; but, above a height of about 90 km (55 miles), diffusional processes become more important than mixing, and the lighter gases (hydrogen and helium, in particular) are more abundant above that level.

Of the gases present in variable concentrations, water vapour, ozone, carbon dioxide, sulfur dioxide, and nitrogen dioxide are of principal importance. The typical concentration ranges of these gases (in percentage by volume) are as follows:

water vapour (H_2O)	0 to 7
carbon dioxide (CO_2)	0.01 to 0.1 (average about 0.032)
ozone (O_3)	0 to 0.01
sulfur dioxide (SO_2)	0 to 0.0001
nitrogen dioxide (NO_2)	0 to 0.000002

Although present in relatively small amounts, these variable constituents may be very important for maintaining life on Earth's surface. Water vapour is the source for all forms of precipitation and is an important absorber and emitter of infrared radiation. Carbon dioxide, besides being involved in the process of photosynthesis, is also an important absorber and emitter of infrared radiation. Ozone, which is present mainly in the atmospheric region 10 to 50 km (6 to 30 miles) above the Earth's surface, is an effective absorber of ultraviolet radiation from the Sun and effectively shields the Earth from all radiation of wavelengths less than 3,000 angstroms.

Psychometric Properties

Psychometrics is the field of study concerned with the theory and technique of psychologicalmeasurement, which includes the measurement of knowledge, abilities, attitudes, personality traits, and educational measurement. The field is primarily concerned with the construction and validation of measurement instruments such as questionnaires, tests, and personality assessments.

It involves two major research tasks, namely: (i) the construction of instruments and procedures for measurement; and (ii) the development and refinement of theoretical approaches to measurement. Those who practice psychometrics are known as psychometricians. All psychometricians possess a specific psychometric qualification, and while many are clinical psychologists, others work as human resources or learning and development professionals.

19th Century Foundation

Psychological testing has come from two streams of thought: one, from Darwin, Galton, and Cattell on the measurement of individual differences, and the second, from Herbart, Weber, Fechner, and Wundt and their psychophysical measurements of a similar construct. The second set of individuals and their research is what has led to the development of experimental psychology, and standardized testing.

Victorian Stream

Charles Darwin was the inspiration behind Sir Francis Galton who led to the creation of psychometrics. In 1859, Charles Darwin published his book "The Origin of Species", which pertained to individual differences in animals. This book discussed how individual members in a species differ and how they possess characteristics that are more adaptive and successful or less adaptive and less successful. Those who are adaptive and successful are the ones that survive and give way to the next generation, who would be just as or more adaptive and successful. This idea, studied previously in animals, led to Galton's interest and study of human beings and how they differ one from another, and more importantly, how to measure those differences.

Galton wrote a book entitled "Hereditary Genius" about different characteristics that people possess and how those characteristics make them more "fit" than others. Today these differences, such as sensory and motor functioning (reaction time, visual acuity, and physical strength) are important domains of scientific psychology. Much of the early theoretical and applied work in psychometrics was undertaken in an attempt to measure intelligence. Francis Galton, often referred to as "the father of psychometrics," devised and included mental tests among his anthropometric measures. James McKeen Cattell, who is considered a pioneer of psychometrics went on to extend Galton's work. Cattell also coined the term *mental test*, and is responsible for the research and knowledge which ultimately led to the development of modern tests. (Kaplan & Saccuzzo, 2010)

German Stream

The origin of psychometrics also has connections to the related field of psychophysics. Around the same time that Darwin, Galton, and Cattell were making their discoveries, J.E. Herbart was also interested in "unlocking the mysteries of human consciousness" through the scientific method. (Kaplan & Saccuzzo, 2010) Herbart was responsible for creating mathematical models of the mind, which were influential in educational practices in years to come.

Following Herbart, E.H. Weber built upon Herbart's work and tried to prove the existence of a psychological threshold saying that a minimum stimulus was necessary to activate a sensory system. After Weber, G.T. Fechner expanded upon the knowledge he gleaned from Herbart and Weber, to devise the law that the strength of a sensation grows as the logarithm of the stimulus intensity. A follower of Weber and Fechner, Wilhelm Wundt is credited with founding the science of psychology. It is Wundt's influence that paved the way for others to develop psychological testing.

20th Century

The psychometrician L. L. Thurstone, founder and first president of the Psychometric Society in 1936, developed and applied a theoretical approach to measurement referred to as the law of comparative judgement, an approach that has close connections to the psychophysical theory of Ernst Heinrich Weber and Gustav Fechner. In addition, Spearman and Thurstone both made important contributions to the theory and application of factor analysis, a statistical method developed and used extensively in psychometrics. In the late 1950s, Leopold Szondi made an historical and epistemological assessment of the impact of statistical thinking onto psychology during previous few decades: "in the last decades, the specifically psychological thinking has been almost completely suppressed and removed, and replaced by a statistical thinking. Precisely here we see the cancer of testology and testomania of today."

More recently, psychometric theory has been applied in the measurement of personality, attitudes, and beliefs, and academic achievement. Measurement of these unobservable phenomena is difficult, and much of the research and accumulated science in this discipline has been developed in an attempt to properly define and quantify such phenomena. Critics, including practitioners in the physical sciences and social activists, have argued that such definition and quantification is impossibly difficult, and that such measurements are often misused, such as with psychometric personality tests used in employment procedures:

"For example, an employer wanting someone for a role requiring consistent attention to repetitive detail will probably not want to give that job to someone who is very creative and gets bored easily."

Figures who made significant contributions to psychometrics include Karl Pearson, Henry F. Kaiser, Carl Brigham, L. L. Thurstone, Georg Rasch, Eugene Galanter, Johnson O'Connor, Frederic M. Lord, Ledyard R Tucker, Arthur Jensen, and David Andrich.

Psychometric, psychometrician and psychometrist appreciation week is the first week in November.

Definition of Measurement in the Social Sciences

The definition of measurement in the social sciences has a long history. A currently widespread definition, proposed by Stanley Smith Stevens (1946), is that measurement is "the assignment of numerals to objects or events according to some rule." This definition was introduced in the paper in which Stevens proposed four levels of measurement. Although widely adopted, this definition differs in important respects from the more classical definition of measurement adopted in the physical sciences, which is that *measurement is the numerical estimation and expression of the magnitude of one quantity relative to another* (Michell, 1997).

Indeed, Stevens's definition of measurement was put forward in response to the British Ferguson Committee, whose chair, A. Ferguson, was a physicist. The committee was appointed in 1932 by the British Association for the Advancement of Science to investigate the possibility of quantitatively estimating sensory events. Although its chair and other members were physicists, the committee also included several psychologists. The committee's report highlighted the importance of the definition of measurement. While Stevens's response was to propose a new definition, which has had considerable influence in the field, this was by no means the only response to the report. Another, notably different, response was to accept the classical definition, as reflected in the following statement:

Measurement in psychology and physics are in no sense different. Physicists can measure when they can find the operations by which they may meet the necessary criteria; psychologists have but to do the same. They need not worry about the mysterious differences between the meaning of measurement in the two sciences. (Reese, 1943, p. 49)

These divergent responses are reflected in alternative approaches to measurement. For example, methods based on covariance matrices are typically employed on the premise that numbers, such as raw scores derived from assessments, are measurements. Such approaches implicitly entail Stevens's definition of measurement, which requires only that numbers are *assigned* according to some rule. The main research task, then, is generally considered to be the discovery of associations between scores, and of factors posited to underlie such associations.

On the other hand, when measurement models such as the Rasch model are employed, numbers are not assigned based on a rule. Instead, in keeping with Reese's statement above, specific criteria for measurement are stated, and the goal is to construct procedures or operations that provide data that meet the relevant criteria. Measurements are estimated based on the models, and tests are conducted to ascertain whether the relevant criteria have been met.

Instruments and Procedures

The first psychometric instruments were designed to measure the concept of intelligence. The best known historical approach involved the Stanford-Binet IQ test, developed originally by the French psychologist Alfred Binet. Intelligence tests are useful tools for various purposes. An alternative conception of intelligence is that cognitive capacities within individuals are a manifestation of a general component, or general intelligence factor, as well as cognitive capacity specific to a given domain.

Psychometrics is applied widely in educational assessment to measure abilities in domains such as reading, writing, and mathematics. The main approaches in applying tests in these domains have been Classical Test Theory and the more recent Item Response Theory and Rasch measurement models. These latter approaches permit joint scaling of persons and assessment items, which provides a basis for mapping of developmental continua by allowing descriptions of the skills displayed at various points along a continuum. Such approaches provide powerful information regarding the nature of developmental growth within various domains.

Another major focus in psychometrics has been on personality testing. There have been a range of theoretical approaches to conceptualizing and measuring personality. Some of the better known instruments include the Minnesota Multiphasic Personality Inventory, the Five-Factor Model (or "Big 5") and tools such as Personality and Preference Inventory and the Myers-Briggs Type Indicator. Attitudes have also been studied extensively using psychometric approaches. A common method in the measurement of attitudes is the use of the Likert scale. An alternative method involves the application of unfolding measurement models, the most general being the Hyperbolic Cosine Model (Andrich & Luo, 1993).

Theoretical Approaches

Psychometricians have developed a number of different measurement theories. These include classical test theory (CTT) and item response theory (IRT). An approach which seems mathematically to be similar to IRT but also quite distinctive, in terms of its origins and features, is represented by the Rasch model for measurement. The development of the Rasch model, and the broader class of models to which it belongs, was explicitly founded on requirements of measurement in the physical sciences.

Psychometricians have also developed methods for working with large matrices of correlations and covariances. Techniques in this general tradition include: factor analysis, a method of determining the underlying dimensions of data; multidimensional scaling, a method for finding a simple representation for data with a large number of latent dimensions; and data clustering, an approach to finding objects that are like each other. All these multivariate descriptive methods try to distill large amounts of data into simpler structures. More recently, structural equation modelling and path analysis represent more sophisticated approaches to working with large covariance matrices. These methods allow statistically sophisticated models to be fitted to data and tested to determine if they are adequate fits.

One of the main deficiencies in various factor analyses is a lack of consensus in cutting points for determining the number of latent factors. A usual procedure is to stop factoring when eigenvalues drop below one because the original sphere shrinks. The lack of the cutting points concerns other multivariate methods, also.

Key Concepts

Key concepts in classical test theory are reliability and validity. A reliable measure is one that measures a construct consistently across time, individuals, and situations. A valid measure is one that measures what it is intended to measure. Reliability is necessary, but not sufficient, for validity.

Both reliability and validity can be assessed statistically. Consistency over repeated measures of the same test can be assessed with the Pearson correlation coefficient, and is often called *test-retest reliability*. Similarly, the equivalence of different versions of the same measure can be indexed by a Pearson correlation, and is called *equivalent forms reliability* or a similar term.

Internal consistency, which addresses the homogeneity of a single test form, may be assessed by correlating performance on two halves of a test, which is termed *split-half reliability*; the value of this Pearson product-moment correlation coefficient for two half-tests is adjusted with the Spearman–Brown prediction formula to correspond to the correlation between two full-length tests. Perhaps the most commonly used index of reliability is Cronbach's α, which is equivalent to the mean of all possible split-half coefficients. Other approaches include the intra-class correlation, which is the ratio of variance of measurements of a given target to the variance of all targets.

There are a number of different forms of validity. Criterion-related validity can be assessed by correlating a measure with a criterion measure known to be valid. When the criterion measure is collected at the same time as the measure being validated the goal is to establish *concurrent validity*; when the criterion is collected later the goal is to establish *predictive validity*. A measure has *construct validity* if it is related to measures of other constructs as required by theory. *Content validity* is a demonstration that the items of a test are drawn from the domain being measured. In a personnel selection example, test content is based on a defined statement or set of statements of knowledge, skill, ability, or other characteristics obtained from a *job analysis*.

Item response theory models the relationship between latent traits and responses to test items. Among other advantages, IRT provides a basis for obtaining an estimate of the location of a test-taker on a given latent trait as well as the standard error of measurement of that location. For example, a university student's knowledge of history can be deduced from his or her score on a university test and then be compared reliably with a high school student's knowledge deduced from a less difficult test. Scores derived by classical test theory do not have this characteristic, and assessment of actual ability (rather than ability relative to other test-takers) must be assessed by comparing scores to those of a "norm group" randomly selected from the population.

In fact, all measures derived from classical test theory are dependent on the sample tested, while, in principle, those derived from item response theory are not.

Standards of Quality

The considerations of validity and reliability typically are viewed as essential elements for determining the quality of any test. However, professional and practitioner associations frequently have placed these concerns within broader contexts when developing standards and making overall judgements about the quality of any test as a whole within a given context. A consideration of concern in many applied research settings is whether or not the metric of a given psychological inventory is meaningful or arbitrary.

Testing Standards

In this field, the *Standards for Educational and Psychological Testing* place standards about validity and reliability, along with errors of measurement and related considerations under the general topic of test construction, evaluation and documentation. The second major topic covers standards related to fairness in testing, including fairness in testing and test use, the rights and responsibilities of test takers, testing individuals of diverselinguistic backgrounds, and testing individuals with disabilities. The third and final major topic covers standards related to testing applications, including the responsibilities of test users, psychological testing and assessment, educational testing and assessment, testing in employment and credentialing, plus testing in programme evaluation and public policy.

Evaluation Standards

In the field of evaluation, and in particular educational evaluation, the Joint Committee on Standards for Educational Evaluation has published three sets of standards for evaluations. *The Personnel Evaluation Standards* was published in 1988, *The Programme Evaluation Standards* (2nd edition) was published in 1994, and *The Student Evaluation Standards* was published in 2003.

Each publication presents and elaborates a set of standards for use in a variety of educational settings. The standards provide guidelines for designing, implementing, assessing and improving the identified form of evaluation. Each of the standards has been placed in one of four fundamental categories to promote educational evaluations that are proper, useful, feasible, and accurate. In these sets of standards, validity and reliability considerations are covered under the accuracy topic. For example, the student accuracy standards help ensure that student evaluations will provide sound, accurate, and credible information about student learning and performance.

Dry Bulb Temperature

The dry-bulb temperature (DBT) is the temperature of air measured by a thermometre freely exposed to the air but shielded from radiation and moisture. DBT is the temperature that is usually thought of as air temperature, and it is the true thermodynamic temperature. As a matter of fact, it indicates the amount of heat in the air and it is directly proportional to the mean kinetic energy of the air molecules. Temperature is usually measured in degrees Celsius (°C), Kelvin (K), or Fahrenheit (°F).

Unlike wet bulb temperature, dry bulb temperature does not indicate the amount of moisture in the air. In construction, it is an important consideration when designing a building for a certain climate. Nall called it one of "the most important climate variables for human comfort and building energy efficiency."

DBT is indeed an important variable in Psychrometrics, being the horizontal axis of the Psychrometric chart.

Wet-bulb Temperature

The wet-bulb temperature is the temperature a parcel of air would have if it were cooled to saturation (100% relative humidity) by the evaporation of water into it, with the latent heat being supplied by the parcel. An actual wet-bulb thermometre indicates a temperature close to the true (thermodynamic) wet-bulb temperature. The wet-bulb temperature is the lowest temperature that can be reached under current ambient conditions by the evaporation of water only; it is the temperature felt when the skin is wet and exposed to moving air. Wet-bulb temperature is largely determined by both actual air temperature (dry-bulb temperature) and humidity, the amount of moisture in the air.

The thermodynamic wet-bulb temperature is the minimum temperature which may be achieved by purely evaporative cooling of a water-wetted (or even ice-covered), ventilated surface.

For a given parcel of air at a known pressure and dry-bulb temperature, the thermodynamic wet-bulb temperature corresponds to unique values of relative humidity, dew point temperature, and other properties. The relationships between these values are illustrated in a psychrometric chart.

For "dry" air, air that is less than saturated (i.e., air with less than 100 percent relative humidity), the wet-bulb temperature is lower than the dry-bulb temperature due to evaporative cooling. The greater the difference between the wet and dry bulb temperatures, the drier the air and lower the relative humidity. The dew point temperature is the temperature at which the ambient air must cool to reach 100% relative humidity where condensate and rain form; and conversely, the wet-bulb temperature rises to converge on the dry bulb temperature.

Cooling of the human body through perspiration is inhibited as the wet-bulb temperature (and absolute humidity) of the surrounding air increases in summer. Other mechanisms may be at work in winter if there is validity to the notion of a "humid" or "damp cold."

Lower wet-bulb temperatures that correspond with drier air in summer can translate to energy savings in air-conditioned buildings due to:

1. Reduced dehumidification load for ventilation air
2. Increased efficiency of cooling towers

Thermodynamic wet-bulb Temperature (Adiabatic Saturation Temperature)

The thermodynamic wet-bulb temperature is the temperature a volume of air would have if cooled adiabatically to saturation by evaporation of water into it, all latent heat being supplied by the volume of air.

The temperature of an air sample that has passed over a large surface of liquid water in an insulated channel is the thermodynamic wet-bulb temperature—it has become saturated by passing through a constant-pressure, ideal, adiabatic saturation chamber.

Meteorologists and others may use the term "isobaric wet-bulb temperature" to refer to the "thermodynamic wet-bulb temperature". It is also called the "adiabatic saturation temperature", though it should be pointed out that meteorologists also use "adiabatic saturation temperature" to mean "temperature at the saturation level", i.e. the temperature the parcel would achieve if it expanded adiabatically until saturated.

It is the thermodynamic wet-bulb temperature that is plotted on a psychrometric chart.

The thermodynamic wet-bulb temperature is a thermodynamic property of a mixture of air and water vapour. The value indicated by a simple wet-bulb thermometre often provides an adequate approximation of the thermodynamic wet-bulb temperature.

For an accurate wet-bulb thermometre, "the wet-bulb temperature and the adiabatic saturation temperature are approximately equal for air-water vapour mixtures at atmospheric temperature and pressure. This is not necessarily true at temperatures and pressures that deviate significantly from ordinary atmospheric conditions, or for other gas–vapour mixtures."

Temperature Reading of wet-bulb Thermometre

Wet-bulb temperature is measured using a thermometre that has its bulb wrapped in cloth—called a *sock*—that is kept wet with distilled water via wicking action. Such an instrument is called a *wet-bulb thermometre.* A widely used device for measuring wet and dry bulb temperature is a *sling psychrometre*, which consists of a pair of mercury bulb thermometres, one with a wet "sock" to measure the wet-bulb temperature and the other with the bulb exposed and dry for the dry-bulb temperature. The thermometres are attached to a swivelling handle which allows them to be whirled around so that water evaporates from the sock and cools the wet bulb until it reaches thermal equilibrium.

An actual wet-bulb thermometre reads a slightly different temperature than the thermodynamic wet-bulb temperature, but they are very close in value. This is due to a coincidence: for a water-air system the psychrometric ratio happens to be ~1, although for systems other than air and water they might not be close.

To understand why this is, first consider the calculation of the thermodynamic wet-bulb temperature: in this case, a stream of air with less than 100% relative humidity is cooled. The heat from cooling that air is used to evaporate some water which increases the humidity of the air. At some point the water vapour in the air becomes saturated (and has cooled to the thermodynamic wet-bulb temperature). In this case we can write the following:

$$(H_{\text{sat}} - H_0)\cdot\lambda = (T_0 - T_{\text{sat}})\cdot c_s$$

where H_0 is the initial water content of the air on a mass basis, H_{sat} is the saturated water content of the air, λ is the latent heat of water, T_0 is the initial air temperature, T_{sat} is the saturated air temperature and c_s is the heat capacity of the air.

For the case of the wet-bulb thermometre, imagine a drop of water with air of less than 100% relative humidity blowing over it. As long as the vapour pressure of water in the drop is more than the partial pressure of water in the air stream, evaporation will take place. Initially the heat required for the evaporation will come from the drop itself since the fastest moving water molecules are most likely to escape the surface of the drop, so the remaining water molecules will have a lower average speed and therefore a lower temperature. If this were the only thing that happened, then the drop would cool until the following was true:

$$P_{sat}(T_{drop}) = P_{vapor}$$

where P_{sat} is the saturation pressure of the water in the drop and is a function of the drop temperature and P_{vapor} is the partial pressure of water in the vapour phase. If the air started bone dry and was blowing sufficiently fast then P_{vapor} would be 0 and the drop could get infinitely cold. Clearly this doesn't happen. It turns out that as the drop cools, convective heat transfer begins to occur between the warmer air and the colder water. In addition, the evaporation does not occur instantly, but instead depends on the rate of convective mass transfer between the water and the air. At a certain point the water cools to a point where the heat carried away in evaporation is equal to the heat gain through convective heat transfer. At this point the following is true:

$$(H_{sat} - H_0)\cdot\lambda\cdot k' = (T_0 - T_{wb})\cdot h_c$$

where $(H_{sat} - H_0)$ is now the driving force for mass transfer, k' is the mass transfer coefficient (with English units of lb/(h·ft2)), h_c is the heat transfer coefficient and $(T_0 - T_{wb})$ is the temperature driving force.

Now if this equation is compared to the thermodynamic wet-bulb equation, we can see that if the quantity $\frac{h_c}{k'c_s} = 1$ (known as the psychrometric ratio) then $T_{sat} = T_{wb}$

Due to a coincidence, for air this is the case and the ratio is very close to 1.

Experimentally, the wet-bulb thermometre reads closest to the thermodynamic wet-bulb temperature if:

- The sock is shielded from radiant heat exchange with its surroundings

- Air flows past the sock quickly enough to prevent evaporated moisture from affecting evaporation from the sock
- The water supplied to the sock is at the same temperature as the thermodynamic wet-bulb temperature of the air

In practice the value reported by a wet-bulb thermometre differs slightly from the thermodynamic wet-bulb temperature because:

- The sock is not perfectly shielded from radiant heat exchange
- Air flow rate past the sock may be less than optimum
- The temperature of the water supplied to the sock is not controlled

At relative humidities below 100 percent, water evaporates from the bulb which cools the bulb below ambient temperature. To determine relative humidity, ambient temperature is measured using an ordinary thermometre, better known in this context as a dry-bulb thermometre. At any given ambient temperature, less relative humidity results in a greater difference between the dry-bulb and wet-bulb temperatures; the wet-bulb is colder. The precise relative humidity is determined by reading from a psychrometric chart of wet-bulb versus dry-bulb temperatures, or by calculation.

Psychrometres are instruments with both a wet-bulb and a dry-bulb thermometre.

A wet-bulb thermometre can also be used outdoors in sunlight in combination with a *globe thermometre* (which measures the incident radiant temperature) to calculate the Wet Bulb Globe Temperature (WBGT).

Adiabatic wet-bulb Temperature

The adiabatic wet-bulb temperature is the temperature a volume of air would have if cooled adiabatically to saturation and then compressed adiabatically to the original pressure in a moist-adiabatic process (AMS Glossary). Such cooling may occur as air pressure reduces with altitude, as noted in the article on lifted condensation level.

This term, as defined in this article, may be most prevalent in meteorology.

As the value referred to as "thermodynamic wet-bulb temperature" is also achieved via an adiabatic process, some engineers and others may use the term "adiabatic wet-bulb temperature" to refer to the "thermodynamic wet-bulb temperature". As stated in another section, meteorologists and others may use the term "isobaric wet-bulb temperature" to refer to the "thermodynamic wet-bulb temperature".

"The relationship between the isobaric and adiabatic processes is quite obscure. Comparisons indicate, however, that the two temperatures are rarely different by more than a few tenths of a degree Celsius, and the adiabatic version is always the smaller of the two for unsaturated air. Since the difference is so small, it is usually neglected in practice."

Wet-bulb Depression

The wet-bulb depression is the difference between the dry-bulb temperature and the wet-bulb temperature. If there is 100% humidity, dry-bulb and wet-bulb temperatures are identical, making the wet-bulb depression equal to zero in such conditions.

Wet-bulb Temperature and Health

Living organisms can only survive within a certain temperature range. When the ambient temperature is excessive, humans and many animals cool themselves below ambient by evaporative cooling of sweat (or other aqueous liquid; saliva in dogs, for example); this helps to prevent potentially fatal hyperthermia due to heat stress. The effectiveness of evaporative cooling depends upon humidity; wet-bulb temperature, or more complex calculated quantities such as Wet Bulb Globe Temperature (WBGT) which also takes account of solar radiation, give a useful indication of the degree of heat stress, and are used by several agencies as the basis for heat stress prevention guidelines.

A sustained wet-bulb temperature exceeding 35 °C is likely to be fatal even to fit and healthy people, unclothed in the shade next to a fan; at this temperature we switch from cooling the skin (losing heat to the environment), to warming it. As of 2010 wet-bulb temperatures only very rarely exceeded 30 °C anywhere.

Dew Point Temperature

The dew point is the temperature below which the water vapour in a volume of humid air at a given constant barometric pressure will condense into liquid water at the same rate at which it evaporates. Condensed water is called dew when it forms on a solid surface.

The dew point is a water-to-air saturation temperature. The dew point is associated with relative humidity. A high relative humidity indicates that the dew point is closer to the current air temperature. Relative humidity of 100% indicates the dew point is equal to the current temperature and that the air is maximally saturated with water. When the dew point remains constant and temperature increases, relative humidity decreases.

General aviation pilots use dew-point data to calculate the likelihood of carburetor icing and fog, and to estimate the height of the cloud base.

At a given temperature but *independent* of barometric pressure, the dew point is a consequence of the absolute humidity, the mass of water per unit volume of air. If both the temperature and pressure rise, however, the dew point will rise and the relative humidity will lower accordingly. Reducing the absolute humidity without changing other variables will bring the dew point back down to its initial value. In the same way, increasing the absolute humidity after a temperature drop brings the dew point back down to its initial level. If the temperature rises in conditions of constant pressure, then the dew point will remain constant but the relative humidity will drop. For this reason, a constant relative humidity (%) with different temperatures implies that when it's hotter, a higher fraction of the air is water vapour than when it's cooler.

At a given barometric pressure but *independent* of temperature, the dew point indicates the mole fraction of water vapour in the air, or, put differently, determines the specific humidity of the air. If the pressure rises without changing this mole fraction, the dew point will rise accordingly; Reducing the mole fraction, i.e., making the air less humid, would bring the dew point back down to its initial value. In the same way, increasing the mole fraction after a pressure drop brings the relative humidity back up to its initial level. Considering New York (33 ft elevation) and Denver (5,280 ft elevation), for example, this means that if the dew point and temperature in both cities are the same, then the mass of water vapour per cubic metre of air will be the same, but the mole fraction of water vapour in the air will be greater in Denver.

Relationship to Human Comfort

When the air temperature is high, the body's *thermoregulation* uses evaporation of perspiration to cool down, with the cooling effect directly related to how fast the perspiration evaporates. The rate at which perspiration can evaporate depends on how much moisture is in the air and how much moisture the air can hold. If the air is already saturated with moisture, perspiration will not evaporate. The body's cooling system will produce perspiration in an effort to keep the body at its normal temperature even when the rate it is producing sweat exceeds the evaporation rate. So even without generating additional body heat by exercising, one can become coated with sweat on humid days. It is the unevaporated sweat that tends to make one feel uncomfortable in humid weather.

As the air surrounding one's body is warmed by body heat, it will rise and be replaced with other air. If air is moved away from one's body with a natural breeze or a fan, sweat will evaporate faster, making perspiration more effective at cooling the body. The more unevaporated perspiration, the greater the discomfort.

A wet bulb thermometre also uses evaporative cooling, so it provides a good analog for use in evaluating comfort level.

Discomfort also exists when the dew point is low (below around –30 °C (–22 °F)). The drier air can cause skin to crack and become irritated more easily. It will also dry out the respiratory paths. OSHA recommends indoor air be maintained at 20 to 24.5 °C (68 to 76 °F) with a 20-60% relative humidity (a dew point of -4.5 to 15.5 °C (24 to 60 °F)).

Lower dew points, less than 10 °C (50 °F), correlate with lower ambient temperatures and the body requires less cooling. A lower dew point can go along with a high temperature only at extremely low relative humidity, allowing for relative effective cooling.

Those accustomed to continental climates often begin to feel uncomfortable when the dew point reaches between 15 and 20 °C (59 and 68 °F). Most inhabitants of these areas will consider dew points above 21 °C (70 °F) oppressive.

Dew point		*Human perception*	*Relative humidity at 32 °C (90 °F)*
Over 26 °C	Over 80 °F	Severely high. Even deadly for asthma related illnesses	65% and higher
24–26 °C	75–80 °F	Extremely uncomfortable, fairly oppressive	62%
21–24 °C	70–74 °F	Very humid, quite uncomfortable	52–60%
18–21 °C	65–69 °F	Somewhat uncomfortable for most people at upper edge	44–52%
16–18 °C	60–64 °F	OK for most, but all perceive the humidity at upper edge	37–46%
13–16 °C	55–59 °F	Comfortable	38–41%
10–12 °C	50–54 °F	Very comfortable	31–37%
Under 10 °C	Under 50 °F	A bit dry for some	30%

Measurement

Devices called dew point metres are used to measure dew point over a wide range of temperatures. These devices consist of a polished metal mirror which is cooled as air is passed over it. The temperature at which dew forms is, by definition, the dew point. Manual devices of this sort can be used to calibrate other types of humidity sensors, and automatic sensors may be used in a control loop with a humidifier or dehumidifier to control the dew point of the air in a building or in a smaller space for a manufacturing process.

Calculating the Dew Point

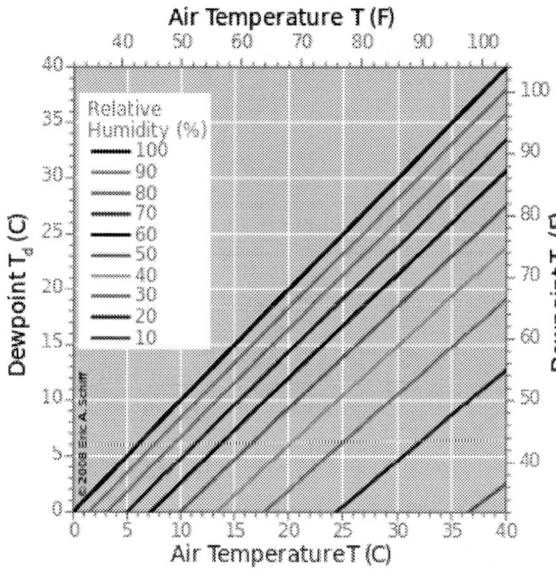

A well-known approximation used to calculate the dew point, T_{dp}, given just the actual ("dry bulb") air temperature, T and relative humidity (in percent), RH, is the *Magnus formula*:

$$\gamma(T,RH) = \ln\left(\frac{RH}{100}\exp\left(\frac{bT}{c+T}\right)\right) = \ln\left(\frac{RH}{100}\right) + \frac{bT}{c+T};$$

$$T_{dp} = \frac{c\gamma(T,RH)}{b-\gamma(T,RH)};$$

The more complete formulation and origin of this approximation involves the interrelated saturated water vapour pressure (in units of millibar, which is also hPa) at T, $P_s(T)$, and the actual water vapour pressure (also in units of millibar), $P_a(T)$, which can be either found with RH or approximated with the barometric pressure (in millibar units), BP_{mb}, and "wet-bulb" temperature, T_w is:

Note: unless declared otherwise, all temperatures are expressed and worked in degrees Celsius

$$P_s(T) = \frac{100}{RH} P_a(T) = a\exp\left(\frac{bT}{c+T}\right);$$

$$P_a(T) = \frac{RH}{100} P_s(T) = a\exp(\gamma(T,RH)),$$

$$\approx P_s(T_w) - BP_{mb}0.00066\left[1 + (0.00115T_w)\right](T - T_w);$$

$$T_{dp} = \frac{c\ln(P_a(T)/a)}{b - \ln(P_a(T)/a)};$$

For greater accuracy, $P_s(T)$ (and, therefore, $\gamma(T,RH)$) can be enhanced, using part of the *Bögel modification*, also known as the Arden Buck equation, which adds a fourth, d constant:

$$P_{s:m}(T) = a\exp\left(\left(b - \frac{T}{d}\right)\left(\frac{T}{c+T}\right)\right);$$

$$\gamma_m(T,RH) = \ln\left(\frac{RH}{100}\exp\left(\left(b - \frac{T}{d}\right)\left(\frac{T}{c+T}\right)\right)\right);$$

$$T_{dp} = \frac{c\gamma_m(T,RH)}{b - \gamma_m(T,RH)};$$

(where $a = 6.1121$ millibar; $b = 18.678$; $c = 257.14^\circ$C; $d = 234.5^\circ$C.*)*

There are several different constant sets in use. The ones used in NOAA's presentation are taken from a 1980 paper by David Bolton in the *Monthly Weather Review*:

$$a = 6.112 \text{ millibar}; \quad b = 17.67; \quad c = 243.5^\circ\text{C};$$

These valuations provide a minimum accuracy of 0.1%, for

-30°C ≤ T ≤ +35°C;

1% < RH < 100%;

Also noteworthy is the Sonntag1990,

$$a = 6.112 \text{ millibar}; \quad b = 17.62; \quad c = 243.12^\circ\text{C}: \quad -45^\circ\text{C} \le T \le +60^\circ\text{C} \quad (< -0.35^\circ\text{C})$$

Another common set of values originates from the 1974 *Psychrometry and Psychrometric Charts*, as presented by *Paroscientific*,

$$a = 6.105 \text{ millibar}; \quad b = 17.27; \quad c = 237.7^\circ\text{C}: \quad 0^\circ\text{C} \le T \le +60^\circ\text{C} \quad (\pm 0.4^\circ\text{C})$$

Also, in the *Journal of Applied Meteorology and Climatology*, Arden Buck presents several different valuation sets, with different minimum accuracies for different temperature ranges. Two particular sets provide a range of -40°C → +50°C between the two, with even greater minimum accuracy than all of the other, above sets (maximum error at given | C° | extreme):

$$a = 6.1121\text{ millibar};\quad b = 17.368;\quad c = 238.88^{\circ}\text{C}:\quad 0^{\circ}\text{C} \le T \le +50^{\circ}\text{C}\ (\le 0.05\%)$$

$$a = 6.1121\text{ millibar};\quad b = 17.966;\quad c = 247.15^{\circ}\text{C}:\quad -40^{\circ}\text{C} \le T \le 0^{\circ}\text{C}\ (\le 0.06\%)$$

Simple Approximation

There is also a very simple approximation that allows conversion between the dew point, temperature and relative humidity. This approach is accurate to within about ±1°C as long as the relative humidity is above 50%:

$$T_{dp} \approx T - \frac{100 - RH}{5};$$

and

$$RH \approx 100 - 5(T - T_{dp});$$

This can be expressed as a simple rule of thumb:

For every 1°C difference in the dew point and dry bulb temperatures, the relative humidity decreases by 5%, starting with RH = 100% when the dew point equals the dry bulb temperature.

The derivation of this approach, a discussion of its accuracy, comparisons to other approximations, and more information on the history and applications of the dew point are given in the Bulletin of the American Meteorological Society.

For temperatures in degrees Fahrenheit, these approximations work out to

$$T_{dp:f} \approx T_f - \frac{9}{25}(100 - RH);$$

and

$$RH \approx 100 - \frac{25}{9}(T_f - T_{dp:f});$$

For example, a relative humidity of 100% means dew point is the same as air temp. For 90% RH, dew point is 3 degrees Fahrenheit lower than air temp. For every 10 percent lower, dew point drops 3 °F.

Frost Point

The frost point is similar to the dew point, in that it is the temperature to which a given parcel of humid air must be cooled, at constant barometric pressure, for water vapour to be deposited on a surface as ice without going through the liquid phase. (Compare with sublimation.) The frost point for a given parcel of air is always higher than the dew point, as the stronger bonding between water molecules on the surface of ice requires higher temperature to break.

Specific Humidity

Specific humidity is the ratio of water vapour to dry air in a particular mass, and is sometimes referred to as humidity ratio. Specific humidity is expressed as a ratio of mass of water vapour, m_v, per unit mass of dry air m_a . This quantity is also known as the water vapour "mixing ratio".

That ratio is defined as:

$$SH = \frac{m_v}{m_a}.$$

Specific humidity can be expressed in other ways including:

$$SH = \frac{0.622 p_{(H_2O)}}{p_{(dry\,air)}} 0.622$$

$$0.622 = \frac{MM_{H_2O}}{MM_{dry\,air}}$$

or:

$$SH = \frac{0.622 p_{(H_2O)}}{p - p_{(H_2O)}}.$$

Using this definition of specific humidity, the relative humidity can be expressed as

$$\phi = \frac{SH * p}{(0.622 + SH) p^{*}_{(H_2O)}} \times 100$$

However, specific humidity is also defined as the ratio of water vapour to the total mass of the system (dry air plus water vapour). For example, the ASHRAE 2009 Handbook, Ch1,1.2, (9a) defines specific humidity as "the ratio of the mass of water vapour to total mass of the moist air sample".

Relative Humidity

Relative humidity is the ratio of the partial pressure of water vapour in an air-water mixture to the saturated vapour pressure of water at a prescribed temperature. The relative humidity of air depends on temperature and the pressure of the system of interest.

Definition

The relative humidity (ϕ) of an air-water mixture is defined as the ratio of the partial pressure of water vapour (H_2O) (e_w) in the mixture to the saturated vapour pressure of water $\left(e^*_w\right)$ at a given temperature.

Relative humidity is normally expressed as a percentage and is calculated by using the following equation:

$$\phi = \frac{e_w}{e^*_w} \times 100\%$$

Significance

Climate Control: Climate control refers to the control of temperature and relative humidity for human comfort, health, and safety; for the technical requirements of machines and processes; and in buildings, vehicles, and other enclosed spaces.

Comfort

Humans are sensitive to humid air because the human body uses evaporative cooling (perspiration) as the primary mechanism to rid itself of waste heat. The rate at which perspiration evaporates on the skin under humid conditions is lower than under arid conditions. Because humans perceive a low *rate of heat transfer* from the body the same as a higher air temperature the body experiences greater distress of waste heat burden at a lower temperature with high RH than at a higher temperature at lower RH.

If, for example, the air temperature is 24 °C (75 °F) and the relative humidity is zero percent then the air temperature feels like 21 °C (69 °F). If the relative humidity is 100 percent at the same air temperature then it feels like 27 °C (80 °F). If the air is 24 °C (75 °F) and contains saturated water vapour, in other words, then the human body cools itself at the same rate as it would if it were 27 °C (80 °F) and dry. The heat index and the humidex are indices that reflect the combined effect of temperature and humidity on the cooling effect of the atmosphere on the human body.

Humans can be comfortable within a wide range of humidities depending on the temperature — from thirty to seventy percent - but ideally between fifty and sixty percent.

Buildings

For climate control in buildings using HVAC systems, the key is to maintain the relative humidity at a comfortable range—low enough to be comfortable but high enough to avoid problems associated with very dry air. When the temperature is high and the relative humidity is low, evaporation of water is rapid; soil dries, wet clothes hung on a line or rack dry quickly, and perspiration readily evaporates from the skin. Wooden furniture can shrink, causing the paint that covers these surfaces to fracture.

When the temperature is high and the relative humidity is high, evaporation of water is slow. When relative humidity approaches 100 percent, condensation can occur on surfaces, leading to problems with mold, corrosion, decay, and other moisture-related deterioration. Certain production and technical processes and treatments in factories, laboratories, hospitals, and other facilities require specific relative humidity levels to be maintained using humidifiers, dehumidifiers and associated control systems.

Vehicles

The basic principles for buildings, above, also apply to vehicles. In addition, there may be safety considerations. For instance, high humidity inside a vehicle can lead to problems of condensation, such as misting of windshields and shorting of electrical components.

In sealed vehicles and pressure vessels such as pressurized airliners, submersibles and spacecraft, these considerations may be critical to safety, and complex environmental control systems including equipment to maintain pressure are needed. Airliners operate with low internal relative humidity, often under 10%, especially on long flights. The low humidity is a consequence of drawing in the very cold air with a low absolute humidity, which is found at airliner cruising altitudes. Subsequent warming of this air lowers its relative humidity. This causes discomfort such as sore eyes, dry skin, and drying out of mucosa, but humidifiers are not employed to raise it to comfortable mid-range levels because the volume of water required to be carried on board can be a significant weight penalty. As airliners descend from colder altitudes into warmer air (perhaps even flying through clouds a few thousand feet above the ground), the ambient relative humidity can increase dramatically. Some of this moist air is usually drawn into the pressurized aircraft cabin and into other non-pressurized areas of the aircraft and condenses on the cold aircraft skin. Liquid water can usually be seen running along the aircraft skin, both on the inside and outside of the cabin. Because of the drastic changes in relative humidity inside the vehicle, components must be qualified to operate in those environments. The recommended environmental qualifications for most commercial aircraft components is listed in RTCA DO-160.

Aviation

Cold humid air can provoke the formation of ice, which is a danger to aircraft as it affects the wing profile and increases weight. Carbureted engines have a further danger of ice forming inside the carburetor. Aviation weather reports (METARs) therefore include an indication of relative humidity, usually in the form of the dew point.

In addition, air with higher humidity is less dense, providing lower lift and lower propeller or turbine efficiency (this is somewhat offset by decreased drag). Pilots must take humidity into account when calculating takeoff distances because high humidity will require longer runways and will decrease the climb gradient.

Measurement

The humidity of an air-water vapour mixture is determined through the use of psychrometric charts if both the dry bulb temperature (T) and the wet bulb temperature (T_w) of the mixture are known. These quantities are readily estimated by using a sling psychrometre.

There are several empirical correlations that can be used to estimate the saturated vapour pressure of water vapour as a function of temperature. The Antoine equation is among the least complex of these formulas, having only three parametres (A, B, and C). Other correlations, such as those presented by Goff-Gratch and Magnus Tetens approximation, are more complicated but yield better accuracy. The correlation presented by Buck is commonly encountered in the literature and provides a reasonable balance between complexity and accuracy:

$$e^*_w = (1.0007 + 3.46 \times 10^{-6} P) \times (6.1121) e^{\left(\frac{17.502T}{240.97+T}\right)}$$

where T is the dry bulb temperature expressed in degrees Celsius (°C), P is the absolute pressure expressed in hectopascals (hPa), and e^*_w is the saturated vapour pressure expressed in hectopascals (hPa).

Buck has reported that the maximum relative error is less than 0.20% between -20°C and +50°C when this particular form of the generalized formula is used to estimate the saturated vapour pressure of water.

Misconception

The notion of air holding water vapour is sometimes used to describe the concept of relative humidity. This, however, is a misconception. Air is a mixture of gases (nitrogen, oxygen, argon, water vapour, and other gases) and as such the constituents of the mixture simply act as a transporter of water vapour but are not a holder of it.

Relative Humidity is defined in the physical properties of water and thus is unrelated to the notion of *air holding water*. An air-less volume can, in fact, contain water vapour and therefore the humidity of this volume can be readily determined.

Pressure Dependence

The relative humidity of an air-water system is dependent not only on the temperature but also on the absolute pressure of the system of interest. This dependence is demonstrated by considering the air-water system shown below. The system is closed (i.e., no matter enters or leaves the system).

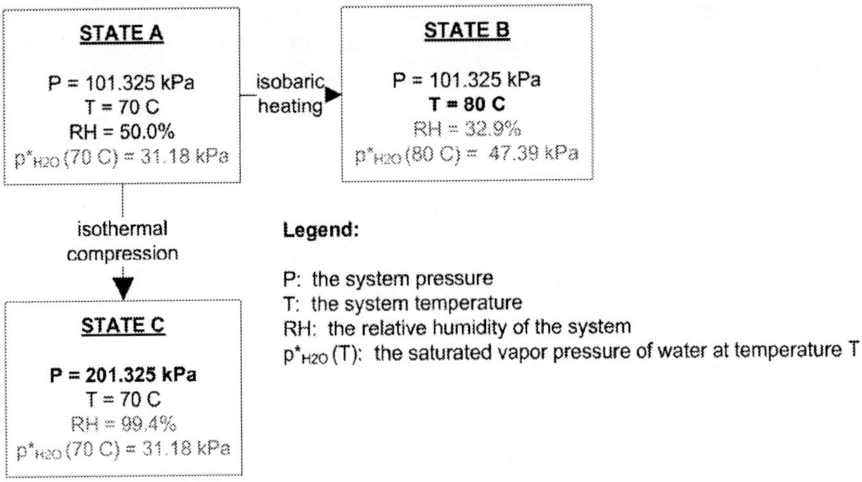

If the system at State A is isobarically heated (heating with no change in system pressure) then the relative humidity of the system decreases because the saturated vapour pressure of water increases with increasing temperature. This is shown in State B.

If the system at State A is isothermally compressed (compressed with no change in system temperature) then the relative humidity of the system increases because the partial pressure of water in the system increases with the volume reduction. This is shown in State

C. Above 202.64 kPa, the RH would exceed 100% and water may begin to condense.

If the pressure of State A was changed by simply adding more dry air, without changing the volume, the relative humidity would not change.

Therefore a change in relative humidity can be explained by a change in system temperature, a change in the volume of the system, or change in both of these system properties.

Enhancement Factor

The enhancement factor (f_w) is defined as the ratio of the saturated vapour pressure of water in moist air (e'_w) to the saturated vapour pressure of pure water.

$$f_W = \frac{e'_w}{e^*_w}$$

The enhancement factor is equal to unity for ideal gas systems. However, in real systems the interaction effects between gas molecules result in a small increase of the saturation vapour pressure of water in air relative to saturated vapour pressure of pure water vapour. Therefore, the enhancement factor is normally slightly greater than unity for real systems.

The enhancement factor is commonly used to correct the saturated vapour pressure of water vapour when empirical relationships, such as those developed by Wexler, Goff, and Gratch, are used to estimate the properties of psychrometric systems.

Buck has reported that, at sea level, the vapour pressure of water in saturated moist air amounts to an increase of approximately 0.5% over the saturated vapour pressure of pure water.

Related Concepts

The term relative humidity is reserved for systems of water vapour in air. The term *relative saturation* is used to describe the analogous property for systems consisting of a condensable phase other than water in a non-condensable phase other than air.

Other Important Facts

A gas in this context is referred to as saturated when the vapour pressure of water in the air is at the equilibrium vapour pressure for water vapour at the temperature of the gas and water vapour mixture; liquid water (and ice, at the appropriate temperature) will fail to lose mass through evaporation when exposed to saturated air. It may also correspond to the possibility of dew or fog forming, within a space that lacks temperature differences among its portions, for instance in response to decreasing temperature. Fog consists of very minute droplets of liquid, primarily held aloft by isostatic motion (in other words, the droplets fall through the air at terminal velocity, but as they are very small, this terminal velocity is very small too, so it doesn't look to us like they are falling, and they seem to be held aloft).

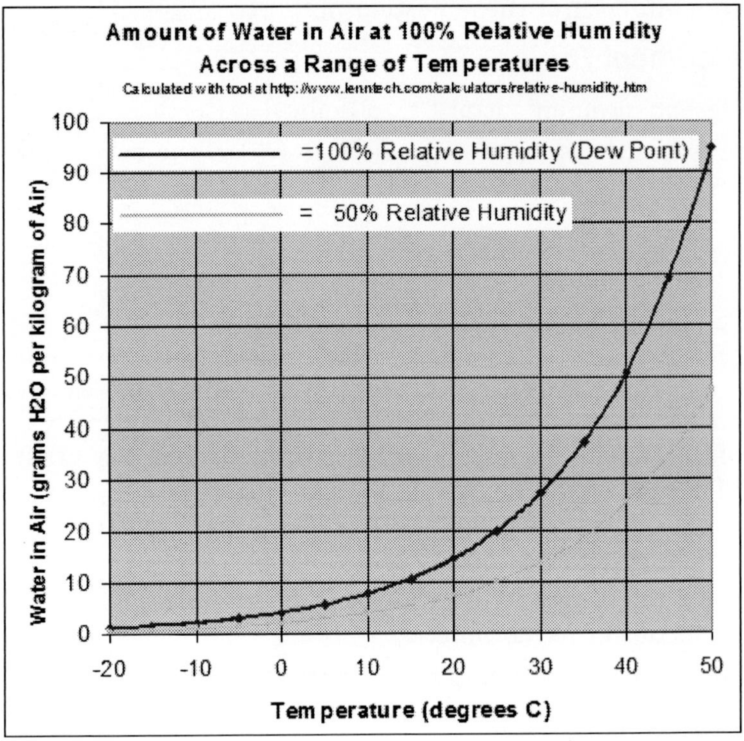

The statement that relative humidity (RH%) can never be above 100%, while a fairly good guide, is not absolutely accurate, without a more sophisticated definition of humidity than the one given here. An arguable exception is the Wilson cloud chamber, which uses, in nuclear physics experiments, an extremely brief state of "supersaturation" to accomplish its function.

For a given dewpoint and its corresponding absolute humidity, the relative humidity will change inversely, albeit nonlinearly, with the temperature. This is because the partial pressure of water increases with temperature – the operative principle behind everything from hair dryers to dehumidifiers.

Due to the increasing potential for a higher water vapour partial pressure at higher air temperatures, the water content of air at sea level can get as high as 3% by mass at 30 °C (86 °F) compared to no more than about 0.5% by mass at 0 °C (32 °F). This explains the low levels (in the absence of measures to add moisture) of humidity in heated structures during winter, resulting in dry skin, itchyeyes, and persistence of static electric charges. Even with saturation (100% relative humidity) outdoors, heating of infiltrated outside air that comes indoors raises its moisture capacity, which lowers relative humidity and increases evaporation rates from moist surfaces indoors (including human bodies and household plants.)

Similarly, during summer in humid climates a great deal of liquid water condenses from air cooled in air conditioners. Warmer air is cooled below its dewpoint, and the excess water vapour condenses. This phenomenon is the same as that which causes water droplets to form on the outside of a cup containing an ice-cold drink.

A useful rule of thumb is that the maximum absolute humidity doubles for every 20 °F or 10 °C increase in temperature. Thus, the relative humidity will drop by a factor of 2 for each 20 °F or 10 °C increase in temperature, assuming conservation of absolute moisture. For example, in the range of normal temperatures, air at 68 °F or 20 °C and 50% relative humidity will become saturated if cooled to 50°F or 10 °C, its dewpoint, and 41 °F or 5 °C air at 80% relative humidity warmed to 68 °F or 20 °C will have a relative humidity of only 29% and feel dry. By comparison, a relative humidity between 40% and 60% is considered healthy and comfortable in comfort-controlled environments.

Water vapour is a lighter gas than other gaseous components of air at the same temperature, so humid air will tend to rise by natural convection. This is a mechanism behind thunderstorms and other weather phenomena. Relative humidity is often mentioned in weather forecasts and reports, as it is an indicator of the likelihood of precipitation, dew, or fog. In hot summer weather, it also increases the apparent temperature to humans (and other animals) by hindering the evaporation of perspiration from the skin as the relative humidity rises. This effect is calculated as the heat index or humidex.

A device used to measure humidity is called a hygrometre; one used to regulate it is called a humidistat, or sometimes hygrostat. (These are analogous to a thermometre and thermostat for temperature, respectively.)

Saturated Air

The air consists of the water vapour or moisture at all times no matter what its temperature and pressure is. When we add more moisture to the air in an enclosed space at certain temperature, the air tends to keep on absorbing the moisture. However, after reaching certain stage, the air is no more able to absorb the moisture and all the extra moisture is converted into dew or fog. The air that contains maximum amount of moisture that it can hold at particular temperature is called as saturated air.

The quantity of the moisture that the air can hold depends on its temperature. The more is the temperature of the air, the more quantity of the moisture it can absorb.

Let us suppose that the partial pressure of water vapour when the air is saturated is P_{ws}, then as per the Dalton's law of partial pressure:

$$P = P_d + P_{ws}$$

Vapour Pressure

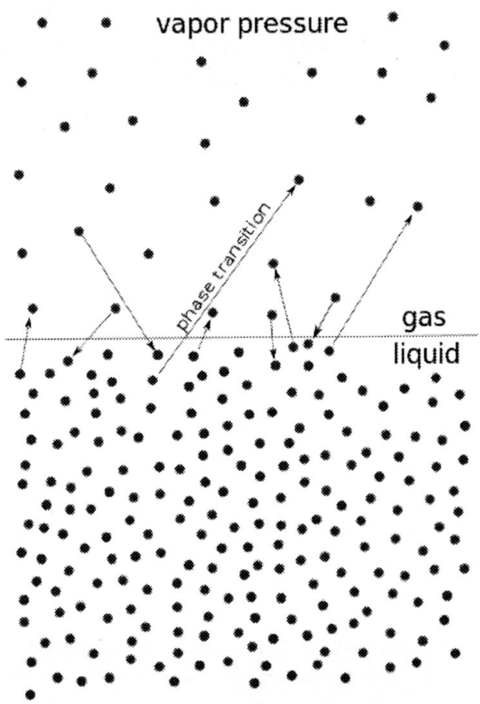

Figure: *The picture shows the particle transition, as a result of their vapour pressure, from the liquid phase to the gas phase and converse.*

Vapour pressure or equilibrium vapour pressure is the pressure exerted by a vapour in thermodynamic equilibrium with its condensed phases (solid or liquid) at a given temperature in a closed system. The equilibrium vapour pressure is an indication of a liquid's evaporation rate. It relates to the tendency of particles to escape from the liquid (or a solid). A substance with a high vapour pressure at normal temperatures is often referred to as *volatile.*

The vapour pressure of any substance increases non-linearly with temperature according to the Clausius–Clapeyron relation. The atmospheric pressureboiling point of a liquid (also known as the normal boiling point) is the temperature at which the vapour pressure equals the ambient atmospheric pressure. With any incremental increase in that temperature, the vapour pressure becomes sufficient to overcome atmospheric pressure and lift the liquid to form vapour bubbles inside the bulk of the substance. Bubble formation deeper in the liquid requires a higher pressure, and therefore higher temperature, because the fluid pressure increases above the atmospheric pressure as the depth increases.

The vapour pressure that a single component in a mixture contributes to the total pressure in the system is called partial pressure. For example, air at sea level, and saturated with water vapour at 20 °C, has partial pressures of about 23 mbar of water, 780 mbar of nitrogen, 210 mbar of oxygen and 9 mbar of argon.

Measurement and Units

Vapour pressure is measured in the standard units of pressure. The International System of Units (SI) recognizes pressure as a derived unit with the dimension of force per area and designates the pascal (Pa) as its standard unit. One pascal is one newton per square metre ($N \cdot m^{-2}$ or $kg \cdot m^{-1} \cdot s^{-2}$).

Experimental measurement of vapour pressure is a simple procedure for common pressures between 1 and 200 kPa. Most accurate results are obtained near the boiling point of substances and large errors result for measurements smaller than 1kPa. Procedures often consist of purifying the test substance, isolating it in a container, evacuating any foreign gas, then measuring the equilibrium pressure of the gaseous phase of the substance in the container at different temperatures. Better accuracy is achieved when care is taken to ensure that the entire substance and its vapour are at the prescribed temperature. This is often done, as with the use of an isoteniscope, by submerging the containment area in a liquid bath.

Estimating Vapour Pressures with Antoine Equation

The Antoine equation is a mathematical expression of the relation between the vapour pressure and the temperature of pure liquid or solid substances. The basic form of the equation is:

$$\log P = A - \frac{B}{C+T}$$

and it can be transformed into this temperature-explicit form:

$$T = \frac{B}{A - \log P} - C$$

where: P is the absolute vapour pressure of a substance

T is the temperature of the substance

A, $B$and C are substance-specific coefficients (i.e., constants or parametres)

log is typically either $\log_{10}$ or $\log_e$

A simpler form of the equation with only two coefficients is sometimes used:

$$\log P = A - \frac{B}{T}$$

which can be transformed to:

$$T = \frac{B}{A - \log P}$$

Sublimations and vaporizations of the same substance have separate sets of Antoine coefficients, as do components in mixtures. The Antoine equation is accurate to a few percent for most volatile substances (with vapour pressures over 10 Torr).

Relation to Boiling Point of Liquids

As a general trend, vapour pressures of liquids at ambient temperatures increase with decreasing boiling points. This is illustrated in the vapour pressure chart that shows graphs of the vapour pressures versus temperatures for a variety of liquids.

For example, at any given temperature, methyl chloride has the highest vapour pressure of any of the liquids in the chart. It also has the lowest normal boiling point (-24.2 °C), which is where the vapour pressure curve of methyl chloride (the blue line) intersects the horizontal pressure line of one atmosphere (atm) of absolute vapour pressure.

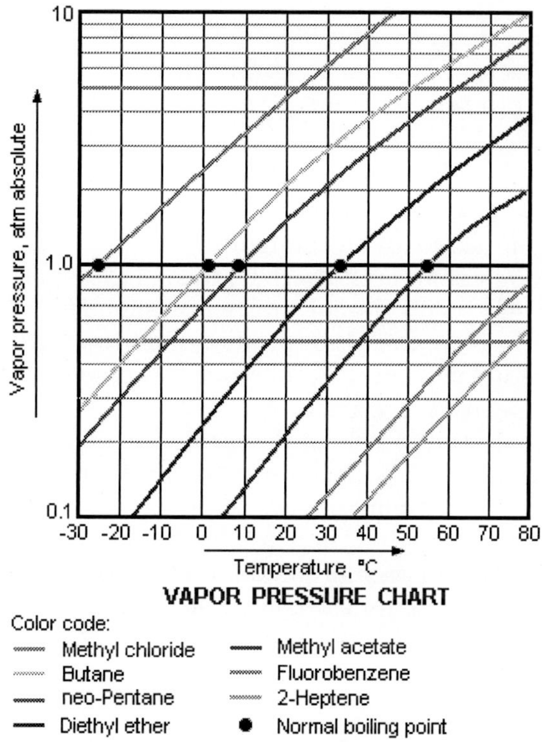

***Figure:** A typical vapour pressure chart for various liquids*

Although the relation between vapour pressure and temperature is non-linear, the chart uses a logarithmic vertical axis to produce slightly curved lines, so one chart can graph many liquids. A nearly straight line is obtained when the logarithm of the vapour pressure is plotted against 1/(T+230) where T is the temperature in degrees Celsius. The vapour pressure of a liquid at its boiling point equals the pressure of its surrounding environment.

Liquid Mixtures

Raoult's law gives an approximation to the vapour pressure of mixtures of liquids. It states that the activity (pressure or fugacity) of a single-phase mixture is equal to the mole-fraction-weighted sum of the components' vapour pressures:

$$p_{\text{tot}} = \sum_i p_i \chi_i$$

where p_{tot} is the mixture's vapour pressure, i is one of the components of the mixture and $\times_i$ is the mole fraction of that component in the liquid mixture. The term $p_i \times_i$ is the partial pressure of component i

in the mixture. Raoult's Law is applicable only to non-electrolytes (uncharged species); it is most appropriate for non-polar molecules with only weak intermolecular attractions (such as London forces).

Systems that have vapour pressures higher than indicated by the above formula are said to have positive deviations. Such a deviation suggests weaker intermolecular attraction than in the pure components, so that the molecules can be thought of as being "held in" the liquid phase less strongly than in the pure liquid. An example is the azeotrope of approximately 95% ethanol and water. Because the azeotrope's vapour pressure is higher than predicted by Raoult's law, it boils at a temperature below that of either pure component. There are also systems with negative deviations that have vapour pressures that are lower than expected. Such a deviation is evidence for stronger intermolecular attraction between the constituents of the mixture than exists in the pure components. Thus, the molecules are "held in" the liquid more strongly when a second molecule is present. An example is a mixture of trichloromethane (chloroform) and 2-propanone (acetone), which boils above the boiling point of either pure component.

Solids

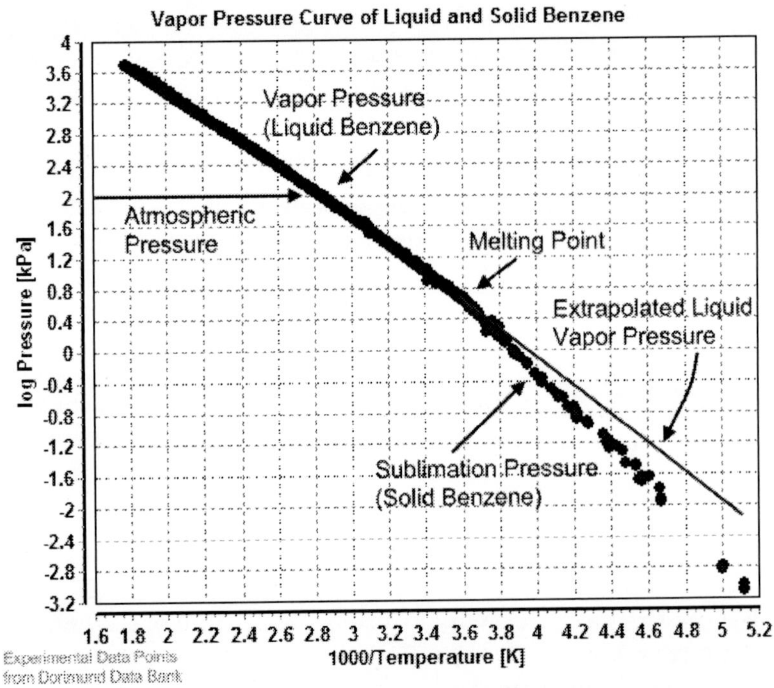

***Figure:** Vapour pressure of liquid and solid benzene*

Equilibrium vapour pressure can be defined as the pressure reached when a condensed phase is in equilibrium with its own vapour. In the case of an equilibrium solid, such as a crystal, this can be defined as the pressure when the rate of sublimation of a solid matches the rate of deposition of its vapour phase. For most solids this pressure is very low, but some notable exceptions are naphthalene, dry ice (the vapour pressure of dry ice is 5.73 MPa (831 psi, 56.5 atm) at 20 degrees Celsius, which causes most sealed containers to rupture), and ice. All solid materials have a vapour pressure. However, due to their often extremely low values, measurement can be rather difficult. Typical techniques include the use of thermogravimetry and gas transpiration.

There are a number of methods for calculating the sublimation pressure (i.e., the vapour pressure) of a solid. One method is to estimate the sublimation pressure from extrapolated liquid vapour pressures (of the supercooled liquid), if the heat of fusion is known, by using this particular form of the Clausius–Clapeyron relation:

$$\ln P^S_{solid} = \ln P^S_{liquid} - \frac{\Delta H_m}{R}\left(\frac{1}{T} - \frac{1}{T_m}\right)$$

with:

P^S_{solid}	=	Sublimation pressure of the solid component at the temperature $T < T_m$
P^S_{liquid}	=	Extrapolated vapour pressure of the liquid component at the temperature $T < T_m$
ΔH_m	=	Heat of fusion
R	=	Gas constant
T	=	Sublimation temperature
T_m	=	Melting point temperature

This method assumes that the heat of fusion is temperature-independent, ignores additional transition temperatures between different solid phases, and it gives a fair estimation for temperatures not too far from the melting point. It also shows that the sublimation pressure is lower than the extrapolated liquid vapour pressure (ΔH_m is positive) and the difference grows with increased distance from the melting point.

Boiling Point of Water in Nature

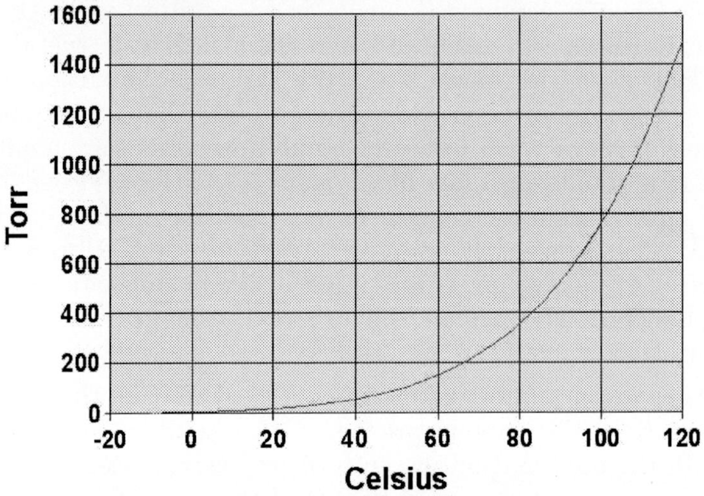

Figure: *Graph of water vapour pressure versus temperature. At the normal boiling point of 100°C, it equals the standard atmospheric pressure of 760 Torr or 101.325 kPa.*

Like all liquids, water boils when its vapour pressure reaches its surrounding pressure. In nature, the atmospheric pressure is lower at higher elevations and water boils at a lower temperature. The boiling temperature of water for atmospheric pressures can be approximated by the Antoine equation:

$$\log_{10} P = 8.07131 - \frac{1730.63}{233.426 + T_b}$$

or transformed into this temperature-explicit form:

$$T_b = \frac{1730.63}{8.07131 - \log_{10} P} - 233.426$$

where the temperature T_b is the boiling point in degrees Celsius and the pressure P is in Torr.

Dühring's Rule

Dühring's rule states that a linear relationship exists between the temperatures at which two solutions exert the same vapour pressure.

Examples

The following table is a list of a variety of substances ordered by increasing vapour pressure.

Substance	*Vapour Pressure (SI units)*	*Vapour Pressure (Bar);*	*Vapour Pressure (mmHg);*	*Temperature*
Tungsten	100 Pa	0.001	0.75	3203 °C
Ethylene glycol	500 Pa	0.005	3.75	20 °C
Xenon difluoride	600 Pa	0.006	4.50	25 °C
Water (H_2O)	2.3 kPa	0.023	17.5	20 °C
Propanol	2.4 kPa	0.024	18.0	20 °C
Ethanol	5.83 kPa	0.0583	43.7	20 °C
Methyl isobutyl ketone	2.66 kPa	0.0266	19.95	25 °C
Freon 113	37.9 kPa	0.379	284	20 °C
Acetaldehyde	98.7 kPa	0.987	740	20 °C
Butane	220 kPa	2.2	1650	20 °C
Formaldehyde	435.7 kPa	4.357	3268	20 °C
Propane	1.013 MPa	10.133	7600	25.6 °C
Carbonyl sulfide	1.255 MPa	12.55	9412	25 °C
Carbon dioxide	5.7 MPa	57	42753	20 °C

Estimating Vapour Pressure from Molecular Structure

Several empirical methods exist to estimate liquid vapour pressure from molecular structure for organic molecules. Some examples are SIMPOL, the method of Moller et al., and EVAPORATION.

Meaning in Meteorology

In meteorology, the term *vapour pressure* is used to mean the partial pressure of water vapour in the atmosphere, even if it is not in equilibrium, and the *equilibrium vapour pressure* is specified otherwise. Meteorologists also use the term *saturation vapour pressure* to refer to the equilibrium vapour pressure of water or brine above a flat surface, to distinguish it from equilibrium vapour pressure, which takes into account the shape and size of water droplets and particulates in the atmosphere.

Chapter

5

Power Cycles

Diesel

Diesel fuel /ÈdiĐzYl/ in general is any liquid fuel used in diesel engines. The most common is a specific fractional distillate of petroleum fuel oil, but alternatives that are not derived from petroleum, such as biodiesel, biomass to liquid (BTL) or gas to liquid (GTL) diesel, are increasingly being developed and adopted. To distinguish these types, petroleum-derived diesel is increasingly called petrodiesel. Ultra-low-sulfur diesel (ULSD) is a standard for defining diesel fuel with substantially lowered sulfur contents. As of 2007, almost all diesel fuel available in the United States of America, Canada and Europe is the ULSD type.

In the UK, diesel fuel for on-road use is commonly abbreviated DERV, standing for *Diesel Engined Road Vehicle*, which carries a tax premium over equivalent fuel for non-road use. In Australia Diesel fuel is often known as 'distillate'

History

Etymology: The word "diesel" is derived from the family name of German inventor Rudolf Diesel who in 1892 invented the compression-ignition engine.

Diesel Engine

Diesel engines are a type of internal combustion engine. Rudolf Diesel originally designed the diesel engine to use coal dust as a fuel.

He also experimented with various oils, including some vegetable oils, such as peanut oil, which was used to power the engines which he exhibited at the 1900 Paris Exposition and the 1911 World's Fair in Paris.

Sources

Diesel fuel is produced from petroleum and from various other sources.

Petroleum Diesel

Refining: Petroleum diesel, also called petrodiesel, or fossil diesel is produced from the fractional distillation of crude oil between 200 °C (392 °F) and 350 °C (662 °F) at atmospheric pressure, resulting in a mixture of carbon chains that typically contain between 8 and 21 carbon atoms per molecule.

Cetane Number

The principal measure of diesel fuel quality is its cetane number. A higher cetane number indicates that the fuel ignites more readily when sprayed into hot compressed air. European (EN 590 standard) road diesel has a minimum cetane number of 51. Fuels with higher cetane numbers, normally "premium" diesel fuels with additional cleaning agents and some synthetic content, are available in some markets.

Fuel Value and Price

As of 2010, the density of petroleum diesel is about 0.832 kg/l (6.943 lb/US gal), about 12% more than ethanol-free petrol (gasoline), which has a density of about 0.745 kg/l (6.217 lb/US gal). About 86.1% of the fuel mass is carbon, and when burned, it offers a net heating value of 43.1 MJ/kg as opposed to 43.2 MJ/kg for gasoline. However, due to the higher density, diesel offers a higher volumetric energy density at 35.86 MJ/L (128,700 BTU/US gal) vs. 32.18 MJ/L (115,500 BTU/US gal) for gasoline, some 11% higher, which should be considered when comparing the fuel efficiency by volume. The CO_2 emissions from diesel are 73.25 g/MJ, just slightly lower than for gasoline at 73.38 g/MJ. Diesel is generally simpler to refine from petroleum than gasoline, and contains hydrocarbons having a boiling point in the range of 180–360°C (360–680°F). The price of diesel traditionally rises during colder months as demand for heating oil rises, which is refined in much the same way. Because of recent changes in fuel quality regulations, additional refining is required to remove sulfur, which contributes to a sometimes higher cost. In many parts of the United States and throughout the United Kingdom and Australia, diesel may be priced higher than petrol. Reasons for higher-priced diesel include the shutdown of some refineries in the Gulf of Mexico, diversion of mass refining capacity to gasoline production, and a recent transfer to ultra-low-sulfur diesel (ULSD), which causes infrastructural complications. In Sweden, a diesel fuel designated as MK-1 (class 1 environmental diesel) is also being sold; this is a ULSD that also has a lower aromatics content, with a limit of 5%. This fuel is slightly more expensive to produce than regular ULSD.

Use as Vehicle Fuel

Unlike petroleum ether and liquefied petroleum gas engines, diesel engines do not use high-voltage spark ignition (spark plugs). An engine running on diesel compresses the air inside the cylinder to high pressures and temperatures (compression ratios from 14:1 to 18:1 are common in current diesel engines); the engine generally injects the diesel fuel directly into the cylinder, starting a few degrees before top dead centre (TDC) and continuing during the combustion event. The high temperatures inside the cylinder cause the diesel fuel to react with the oxygen in the mix (burn or oxidize), heating and expanding the burning mixture to convert the thermal/pressure difference into mechanical work, i.e., to move the piston. Engines have glow plugs to help start the engine by preheating the cylinders to a minimum operating temperature. Diesel engines are lean burn engines, burning the fuel in more air than is required for the chemical reaction. They thus use less fuel than rich burn spark ignition engines which use a Stoichiometric air-fuel ratio (just enough air to react with the fuel). Because they have high compression ratios and no throttle, diesel engines are more efficient than many spark-ignited engines.

Gas turbine internal combustion engines can also take diesel fuel, as can some other types of internal combustion. External combustion engines can easily use diesel fuel as well.

This efficiency and its lower flammability than gasoline are the two main reasons for military use of diesel in armored fighting vehicles. Engines running on diesel also provide more torque, and are less likely to stall, as they are controlled by a mechanical or electronic governor.

A disadvantage of diesel as a vehicle fuel in cold climates, compared to gasoline or other petroleum-derived fuels, is that its viscosity increases quickly as the fuel's temperature decreases, turning into a non-flowing gel at temperatures as high as –19 °C (–2.2 °F) or –15 °C

(5 °F), which cannot be pumped by regular fuel pumps. Special low-temperature diesel contains additives to keep it in a more liquid state at lower temperatures, but starting a diesel engine in very cold weather may still pose considerable difficulties.

Another disadvantage of diesel engines compared to petrol/gasoline engines is the possibility of runaway failure. Since diesel engines do not require spark ignition, they can sustain operation as long as diesel fuel is supplied. Fuel is typically supplied via a fuel pump. If the pump breaks down in an "open" position, the supply of fuel will be unrestricted, and the engine will runaway and risk terminal failure. (In vehicles or installations that use both diesel engines and bottled gas, a gas leak into the engine room could also provide fuel for a runaway, via the engine air intake.)

Use as Car Fuel

Diesel-powered cars generally have a better fuel economy than equivalent gasoline engines and produce less greenhouse gas emission. Their greater economy is due to the higher energy per-litre content of diesel fuel and the intrinsic efficiency of the diesel engine. While petrodiesel's higher density results in higher greenhouse gas emissions per litre compared to gasoline, the 20–40% better fuel economy achieved by modern diesel-engined automobiles offsets the higher per-litre emissions of greenhouse gases, and a diesel-powered vehicle emits 10–20 percent less greenhouse gas than comparable gasoline vehicles. Biodiesel-powered diesel engines offer substantially improved emission reductions compared to petrodiesel or gasoline-powered engines, while retaining most of the fuel economy advantages over conventional gasoline-powered automobiles. However, the increased compression ratios mean there are increased emissions of oxides of nitrogen (NO_x) from diesel engines. This is compounded by biological nitrogen in biodiesel to make NO_x emissions the main drawback of diesel versus gasoline engines.

Reduction of Sulfur Emissions

In the past, diesel fuel contained higher quantities of sulfur. European emission standards and preferential taxation have forced oil refineries to dramatically reduce the level of sulfur in diesel fuels. In the United States, more stringent emission standards have been adopted with the transition to ULSD starting in 2006, and becoming mandatory on June 1, 2010. U.S. diesel fuel typically also has a lower cetane number (a measure of ignition quality) than European diesel, resulting in worse cold weather performance and some increase in emissions.

Environment Hazards of Sulphur

High levels of sulfur in diesel are harmful for the environment because they prevent the use of catalytic diesel particulate filters to control diesel particulate emissions, as well as more advanced technologies, such as nitrogen oxide (NO_x) adsorbers (still under development), to reduce emissions. Moreover, sulfur in the fuel is oxidized during combustion, producing sulfur dioxide and sulfur trioxide, that in presence of water rapidly convert to sulfuric acid, one of the chemical processes that results in acid rain. However, the process for lowering sulfur also reduces the lubricity of the fuel, meaning that additives must be put into the fuel to help lubricate engines. Biodiesel and biodiesel/petrodiesel blends, with their higher lubricity levels, are increasingly being utilized as an alternative. The U.S. annual consumption of diesel fuel in 2006 was about 190 billion litres (42 billion imperial gallons or 50 billion US gallons).

Chemical Composition

Petroleum-derived diesel is composed of about 75% saturated hydrocarbons (primarily paraffins including *n*, *iso*, and cycloparaffins), and 25% aromatic hydrocarbons (including naphthalenes and alkylbenzenes). The average chemical formula for common diesel fuel is $C_{12}H_{23}$, ranging approximately from $C_{10}H_{20}$ to $C_{15}H_{28}$.

Algae, Microbes, and Water Contamination

There has been much discussion and misunderstanding of algae in diesel fuel. Algae need light to live and grow. As there is no sunlight in a closed fuel tank, no algae can survive, but some microbes can survive and feed on the diesel fuel.

These microbes form a colony that lives at the interface of fuel and water. They grow quite fast in warmer temperatures. They can even grow in cold weather when fuel tank heaters are installed. Parts of the colony can break off and clog the fuel lines and fuel filters.

Water in fuel can damage a fuel injection pump, some diesel fuel filters also trap water.

Road Hazard

Petrodiesel spilled on a road will stay there until washed away by sufficiently heavy rain, whereas gasoline will quickly evaporate. After the light fractions have evaporated, a greasy slick is left on the road which can destabilize moving vehicles. Diesel spills severely reduce tire grip and traction, and have been implicated in many accidents. The loss of traction is similar to that encountered on black ice. Diesel slicks are especially dangerous for two-wheeled vehicles such as motorcycles.

Synthetic Diesel

Synthetic diesel can be produced from any carbonaceous material, including biomass, biogas, natural gas, coal and many others. The raw material is gasified into synthesis gas, which after purification is converted by the Fischer–Tropsch process to a synthetic diesel.

The process is typically referred to as biomass-to-liquid (BTL), gas-to-liquid (GTL) or coal-to-liquid (CTL), depending on the raw material used.

Paraffinic synthetic diesel generally has a near-zero content of sulfur and very low aromatics content, reducing unregulated emissions of toxic hydrocarbons, nitrous oxides and PM.

Fame

Figure: *Biodiesel made from soybean oil*

Fatty-acid methyl ester (FAME), perhaps more widely known as biodiesel, is obtained from vegetable oil or animal fats (biolipids) which have been transesterified with methanol. It can be produced from many types of oils, the most common being rapeseed oil (rapeseed methyl ester, RME) in Europe and soybean oil (soy methyl ester, SME) in the USA. Methanol can also be replaced with ethanol for the transesterification process, which results in the production of ethyl esters. The transesterification processes use catalysts, such as sodium or potassium hydroxide, to convert vegetable oil and methanol into FAME and the undesirable byproducts glycerine and water, which will need to be removed from the fuel along with methanol traces. FAME can be used pure (B100) in engines where the manufacturer approves such use, but it is more often used as a mix with diesel, BXX where XX is the biodiesel content in percent.

FAME as a fuel is regulated under DIN EN 14214 and ASTM D6751.

FAME has a lower energy content than diesel due to its oxygen content, and as a result, performance and fuel consumption can be affected. It also can have higher levels of NOx emissions, possibly even exceeding the legal limit. FAME also has lower oxidation stability than diesel, and it offers favourable conditions for bacterial growth, so applications which have a low fuel turnover should not use FAME. The loss in power when using pure biodiesel is 5 to 7%.

Fuel equipment manufacturers (FIE) have raised several concerns regarding FAME fuels: free methanol, dissolved and free water, free glycerin, mono and diglycerides, free fatty acids, total solid impurity levels, alkaline metal compounds in solution and oxidation and thermal stability. They have also identified FAME as being the cause of the following problems: corrosion of fuel injection components, low-pressure fuel system blockage, increased dilution and polymerization of engine sump oil, pump seizures due to high fuel viscosity at low temperature, increased injection pressure, elastomeric seal failures and fuel injector spray blockage.

Unsaturated fatty acids are the source for the lower oxidation stability; they react with oxygen and form peroxides and result in degradation byproducts, which can cause sludge and lacquer in the fuel system.

As FAME contains low levels of sulfur, the emissions of sulfur oxides and sulfates, major components of acid rain, are low. Use of biodiesel also results in reductions of unburned hydrocarbons, carbon monoxide (CO), and particulate matter. CO emissions using biodiesel are substantially reduced, on the order of 50% compared to most petrodiesel fuels. The exhaust emissions of particulate matter from biodiesel have been found to be 30 percent lower than overall particulate matter emissions from petrodiesel. The exhaust emissions of total hydrocarbons (a contributing factor in the localized formation of smog and ozone) are up to 93 percent lower for biodiesel than diesel fuel.

Biodiesel also may reduce health risks associated with petroleum diesel. Biodiesel emissions showed decreased levels of polycyclic aromatic hydrocarbon (PAH) and nitrited PAH compounds, which have been identified as potential cancer-causing compounds. In recent testing, PAH compounds were reduced by 75 to 85 percent, except for benz(a)anthracene, which was reduced by roughly 50 percent. Targeted nPAH compounds were also reduced dramatically with biodiesel fuel, with 2-nitrofluorene and 1-nitropyrene reduced by 90 percent, and the rest of the nPAH compounds reduced to only trace levels.

Hydrogenated Oils and Fats

This category of diesel fuels involves converting the triglycerides in vegetable oil and animal fats into alkanes by refining and hydrogenation. The produced fuel has many properties that are similar to synthetic diesel, and are free from the many disadvantages of FAME.

DME

Dimethyl ether, DME, is a synthetic, gaseous diesel fuel that results in clean combustion with very little soot and reduced NOx emissions.

Transportation and Storage

Diesel fuel is widely used in most types of transportation. The gasoline-powered passenger automobile is the major exception.

Railroad

Diesel displaced coal and fuel oil for steam-powered vehicles in the latter half of the 20th century, and is now used almost exclusively for the combustion engines of self-powered rail vehicles (locomotives and railcars).

Aircraft

The first diesel-powered flight of a fixed-wing aircraft took place on the evening of September 18, 1928, at the Packard Motor Company proving grounds at Utica, USA, with Captain Lionel M. Woolson and Walter Lees at the controls (the first "official" test flight was taken the next morning). The engine was designed for Packard by Woolson, and the aircraft was a Stinson SM1B, X7654. Later that year, Charles Lindbergh flew the same aircraft. In 1929, it was flown 621 miles

(999 km) nonstop from Detroit to Langley Field, near Norfolk, Virginia. This aircraft is now owned by Greg Herrick, and is at the Golden Wings Flying Museum near Minneapolis, Minnesota. In 1931, Walter Lees and Fredrick Brossy set the nonstop flight record flying a Bellanca powered by a Packard diesel for 84 hours and 32 minutes. The Hindenburg rigid airship was powered by four 16-cylinder diesel engines, each with approximately 1,200 horsepower (890 kW) available in bursts, and 850 horsepower (630 kW) available for cruising.

The most-produced aviation diesel engine in history has been the Junkers Jumo 205, which, along with its similar developments from the Junkers Motorenwerke, had approximately 1000 examples of the unique opposed piston, two-stroke design power plant built in the 1930s leading into World War II in Germany.

Storage

In the US, diesel is recommended to be stored in a yellow container to differentiate it from kerosene and gasoline, which are typically kept in blue and red containers, respectively. In the UK, diesel is normally stored in a black container, to differentiate it from unleaded petrol (which is commonly stored in a green container) or, in the past, leaded petrol (which was stored in a red container).

Other Uses

Poor quality (high sulfur) diesel fuel has been used as an extraction agent for liquid–liquid extraction of palladium from nitric acid mixtures. Such use has been proposed as a means of separating the fission product palladium from PUREX raffinate which comes from used nuclear fuel. In this system of solvent extraction, the hydrocarbons of the diesel act as the diluent while the dialkyl sulfides act as the extractant. This extraction operates by a solvation mechanism. So far, neither a pilot plant nor full scale plant has been constructed to recover palladium, rhodium or ruthenium from nuclear wastes created by the use of nuclear fuel.

Diesel fuel is also often used as the main ingredient in oil-base mud drilling fluid. The advantage of using diesel is its low cost and that it delivers excellent results when drilling a wide variety of difficult strata including shale, salt and gypsum formations. Diesel-oil mud is typically mixed with up to 40% brine water. Due to health, safety and environmental concerns, Diesel-oil mud is often replaced with vegetable, mineral, or synthetic food-grade oil-base drilling fluids, although diesel-oil mud is still in widespread use in certain regions.

Emissions

Taxation: Diesel fuel is very similar to heating oil, which is used in central heating. In Europe, the United States, and Canada, taxes on diesel fuel are higher than on heating oil due to the fuel tax, and in those areas, heating oil is marked with fuel dyes and trace chemicals to prevent and detect tax fraud. Similarly, "untaxed" diesel (sometimes called "off-road diesel") is available in some countries for use primarily in agricultural applications, such as fuel for tractors, recreational and utility vehicles or other noncommercial vehicles that do not use public roads. Additionally, this fuel may have sulphur levels that exceed the limits for road use in some countries (e.g. USA).

This untaxed diesel is dyed red for identification, and should a person be found to be using this untaxed diesel fuel for a typically taxed purpose (such as "over-the-road", or driving use), the user can be fined (e.g. US$10,000 in the USA). In the United Kingdom, Belgium and the Netherlands, it is known as red diesel (or gas oil), and is also used in agricultural vehicles, home heating tanks, refrigeration units on vans/trucks which contain perishable items such as food and medicine and for marine craft. Diesel fuel, or marked gas oil is dyed green in the Republic of Ireland and Norway. The term "diesel-engined road vehicle" (DERV) is used in the UK as a synonym for unmarked road diesel fuel. In India, taxes on diesel fuel are lower than on petrol, as the majority of the transportation for grains and other essential commodities across the country runs on diesel.

In some countries, such as Germany and Belgium, diesel fuel is taxed lower than petrol (gasoline) (typically around 20% lower), but the annual vehicle tax is higher for diesel vehicles than for petrol vehicles. This gives an advantage to vehicles that travel longer distances (which is the case for trucks and utility vehicles) because the annual vehicle tax depends only on engine displacement, not on distance driven. The point at which a diesel vehicle becomes less expensive than a comparable petro vehicle is around 20,000 km a year (12,500 miles per year) for an average car. However, due to a rise in oil prices from about 2009, the advantage point started to drop, causing more people opting to buy a diesel car where they would have opted for a gasoline car a few years ago. Such an increased interest in diesel has resulted in slow but steady "dieseling" of the automobile fleet in the countries affected, sparking concerns in certain authorities about the harmful effects of diesel.

Taxes on biodiesel in the U.S. vary between states; some states (Texas, for example) have no tax on biodiesel and a reduced tax on biodiesel blends equivalent to the amount of biodiesel in the blend, so that B20 fuel is taxed 20% less than pure petrodiesel. Other states, such as North Carolina, tax biodiesel (in any blended configuration) the same as petrodiesel, although they have introduced new incentives to producers and users of all biofuels.

Dual Combustion Cycles

The dual combustion cycle (also known as the limited pressure or mixed cycle, Trinkler cycle, Seiliger cycle or Sabathe cycle) is a thermal cycle that is a combination of the Otto cycle and the Diesel cycle, first introduced by Russian-German engineer Gustav Trinkler. Heat is added partly at constant volume and partly at constant pressure, the advantage of which is that more time is available for the fuel to completely combust. Because of lagging characteristics of fuel this cycle is invariably used for diesel and hot spot ignition engines. The P-V diagram is given below:

The dual cycle consists of following operations:

1. 1-2 Adiabatic compression
2. 2-3 Addition of heat at constant volume.
3. 3-4 Addition of heat at constant pressure.
4. 4-5 Adiabatic expansion.
5. 5-1 Rejection of heat at constant volume.

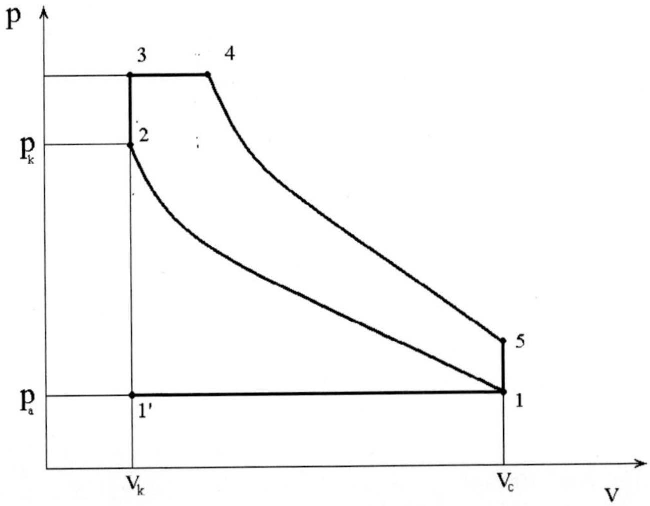

The dual combustion cycle (also known as the limited pressure or mixed cycle, Trinkler cycle, Seiliger cycle or Sabathe cycle) is a thermal cycle that is a combination of the Otto cycle and the Diesel cycle, first introduced by Russian-German engineer Gustav Trinkler. Heat is added partly at constant volume and partly at constant pressure, the advantage of which is that more time is available for the fuel to completely combust. Because of lagging characteristics of fuel this cycle is invariably used for diesel and hot spot ignition engines. The P-V diagram is given below:

The dual cycle consists of following operations:

1. 1-2 Adiabatic compression
2. 2-3 Addition of heat at constant volume.
3. 3-4 Addition of heat at constant pressure.
4. 4-5 Adiabatic expansion.
5. 5-1 Rejection of heat at constant volume.

Sterling Cycle

The Stirling cycle is a thermodynamic cycle that describes the general class of Stirling devices. This includes the original Stirling engine that was invented, developed and patented in 1816 by Reverend Dr. Robert Stirling with help from his brother, an engineer.

The cycle is reversible, meaning that if supplied with mechanical power, it can function as a heat pump for heating or cooling, and even for cryogenic cooling. The cycle is defined as a closed regenerative cycle with a gaseous working fluid. "Closed cycle" means the working fluid is permanently contained within the thermodynamic system. This also categorizes the engine device as an external heat engine. "Regenerative" refers to the use of an internal heat exchanger called a regenerator which increases the device's thermal efficiency.

The cycle is the same as most other heat cycles in that there are four main processes: compression, heat addition, expansion, and heat removal. However, these processes are not discrete, but rather the transitions overlap.

The Stirling cycle is a highly advanced subject that has defied analysis by many experts for over 190 years. Highly advanced thermodynamics is required to describe the cycle. Professor Israel Urieli writes: "...the various 'ideal' cycles (such as the Schmidt cycle) are neither physically realizable nor representative of the Stirling cycle".

The analytical problem of the regenerator (the central heat exchanger in the Stirling cycle) is judged by Jakob to rank "among the most difficult and involved that are encountered in engineering".

Idealized Stirling Cycle Thermodynamics

The idealized Stirling cycle consists of four thermodynamic processes acting on the working fluid:

1. Isothermal expansion. The expansion space is heated externally, and the gas undergoes near-isothermal expansion.
2. Constant-volume (known as isovolumetric or isochoric) heat removal. The gas is passed through the regenerator, thus cooling the gas, and transferring heat to the regenerator for use in the next cycle.
3. Isothermal compression. The compression space is intercooled, so the gas undergoes near-isothermal compression.
4. Constant-volume heat addition. The compressed air flows back through the regenerator and picks up heat on the way to the heated expansion space.

Piston Motion Variations

Most thermodynamics textbooks describe a highly simplified form of Stirling cycle consisting of four processes. This is known as an "ideal Stirling cycle", because it is an "idealized" model, and not necessarily an optimized cycle. Theoretically, the "ideal cycle" does have high net work output, but it is rarely used in practical applications, in part because other cycles are simpler or reduce peak stresses on bearings and other components. For convenience, the designer may elect to use piston motions dictated by system dynamics, such as mechanical linkage mechanisms. At any rate, the efficiency and cycle *power* are nearly as good as an actual implementation of the idealized case. A typical piston crank or linkage in a so named "kinematic" design often results in a near-sinusoidal piston motion. Some designs will cause the piston to "dwell" at either extreme of travel.

Many kinematic linkages, such as the well known "Ross yoke", will exhibit near-sinusoidal motion. However, other linkages, such as the "rhombic drive", will exhibit more non-sinusoidal motion. To a lesser extent, the ideal cycle introduces complications, since it would require somewhat higher piston acceleration and higher viscous pumping losses of the working fluid. The material stresses and pumping losses in an optimized engine, however, would only be intolerable when approaching the "ideal cycle" and/or at high cycle rates. Other issues include the time required for heat transfer, particularly for the isothermal processes. In an engine with a cycle approaching the "ideal cycle", the cycle rate might have to be reduced to address these issues.

In the most basic model of a free piston device, the kinematics will result in simple harmonic motion.

Volume Variations

In beta and gamma engines, generally the phase angle difference between the piston motions is *not* the same as the phase angle of the volume variations. However, in the alpha Stirling, they are the same. The rest of the article assumes sinusoidal volume variations, as in an alpha Stirling with co-linear pistons, so named an "opposed piston" alpha device.

Pressure-versus-volume Graph

This type of plot is used to characterize almost all thermodynamic cycles. The result of sinusoidal volume variations is the quasi-elliptical shaped cycle shown in figure. Compared to the idealized cycle, this cycle is a more realistic representation of most real Stirling engines. The four points in the graph indicate the crank angle in degrees.

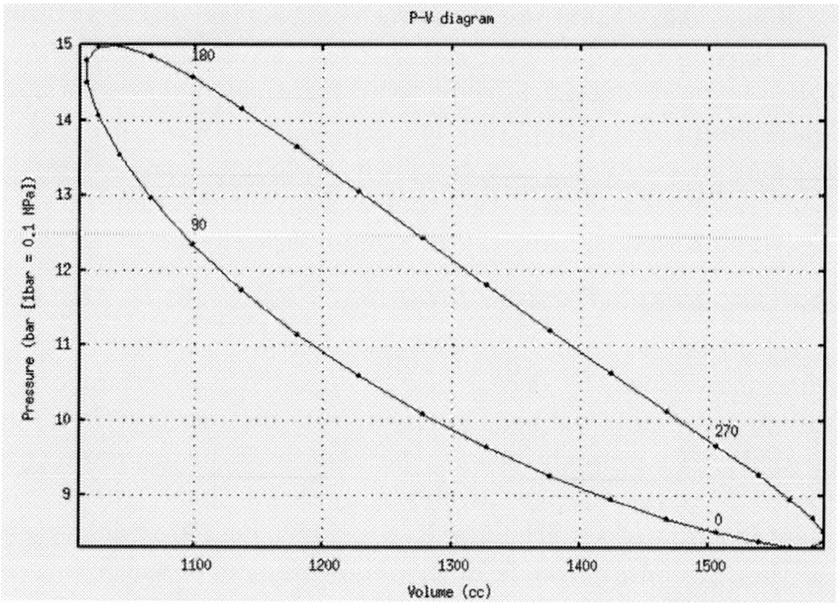

The *adiabatic* Stirling cycle is similar to the idealized Stirling cycle; however, the four thermodynamic processes are slightly different:

- 180° to 270°, pseudo-isothermal expansion. The expansion space is heated externally, and the gas undergoes near-isothermal expansion.
- 270° to 0°, near-constant-volume (or near-isometric or isochoric) heat removal. The gas is passed through the regenerator, thus

cooling the gas, and transferring heat to the regenerator for use in the next cycle.

- 0° to 90°, pseudo-isothermal compression. The compression space is intercooled, so the gas undergoes near-isothermal compression.
- 90° to 180°, near-constant-volume (near-isometric or isochoric) heat addition. The compressed air flows back through the regenerator and picks up heat on the way to the heated expansion space.

With the exception of a Stirling thermoacoustic engine, none of the gas particles actually flow through the complete cycle. So this approach is not amenable to further analysis of the cycle. However, it provides an overview and indicates the cycle work.

Particle/mass Motion

Figure shows the streaklines which indicate how gas flows through a real Stirling engine. The vertical coloured lines delineate the volumes of the engine. From left to right, they are: the volume swept by the expansion (power) piston, the clearance volume (which prevents the piston from contacting the hot heat exchanger), the heater, the regenerator, the cooler, the cooler clearance volume, and the compression volume swept by the compression piston.

Heat-exchanger Pressure Drop

Also referred to as "pumping losses", the pressure drops shown in figure are caused by viscous flow through the heat exchangers. The red line represents the heater, green is the regenerator, and blue is the cooler. To properly design the heat exchangers, multivariate optimization is required to obtain sufficient heat transfer with acceptable flow losses. The flow losses shown here are relatively low, and they are barely visible in the following image, which will show the overall pressure variations in the cycle.

Pressure Versus Crank Angle

Figure shows results from an "adiabatic simulation" with non-ideal heat exchangers. Note that the pressure drop across the regenerator is very low compared to the overall pressure variation in the cycle.

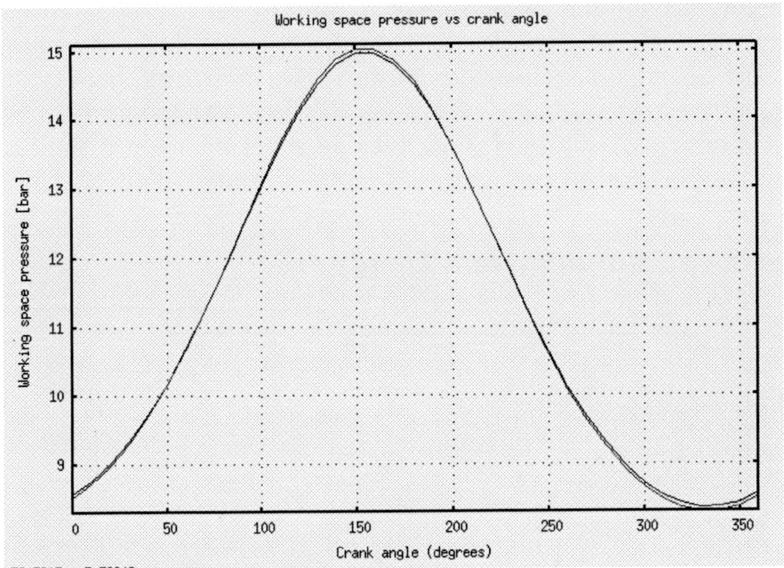

Temperature Versus Crank Angle

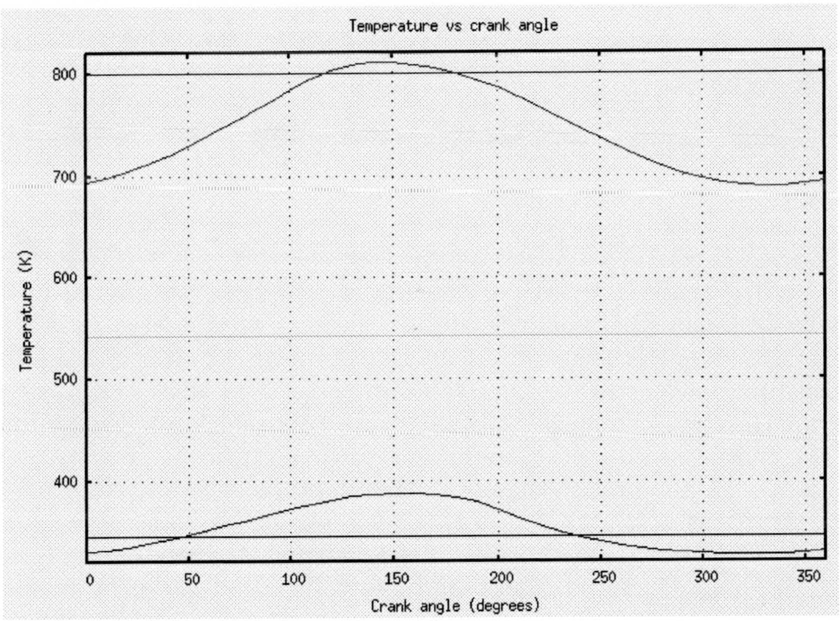

Figure illustrates the adiabatic properties of a real heat exchanger. The straight lines represent the temperatures of the solid portion of the heat exchanger, and the curves are the gas temperatures of the respective spaces. The gas temperature fluctuations are caused by the effects of compression and expansion in the engine, together with non-ideal heat exchangers which have a limited rate of heat transfer. When the gas temperature deviates above and below the heat exchanger temperature, it causes thermodynamic losses known as "heat transfer losses" or "hysteresis losses". However, the heat exchangers still work well enough to allow the real cycle to be effective, even if the actual thermal efficiency of the overall system is only about half of the theoretical limit.

Cumulative Heat and Work Energy

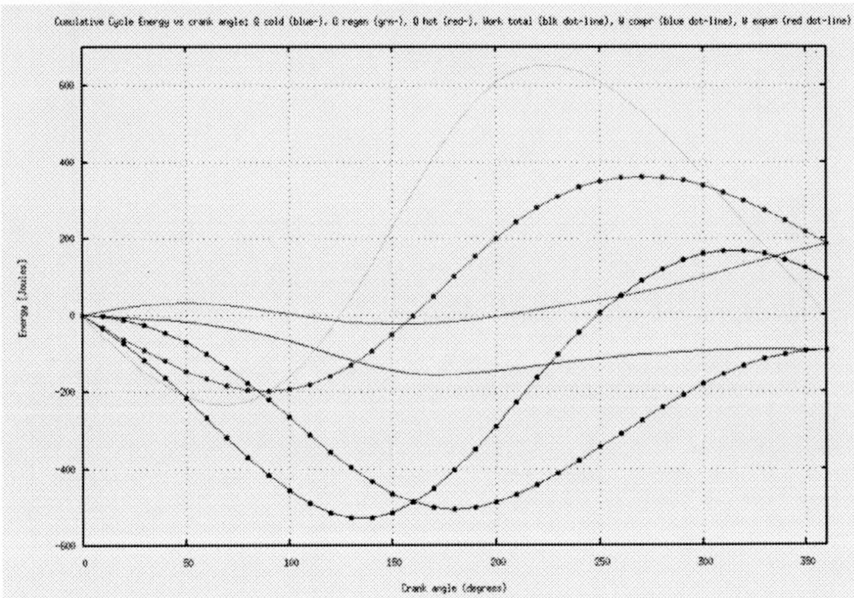

Figure shows a graph of the alpha-type Stirling engine data, where 'Q' denotes heat energy, and 'W' denotes work energy. The blue dotted line shows the work output of the compression space. As the trace dips down, and work is done on the gas as it is compressed. During the expansion process of the cycle, some work is actually done *on* the compression piston, as reflected by the upward movement of the trace. At the end of the cycle, this value is negative, indicating that compression piston requires a net input of work. The blue solid line shows the heat flowing out of the cooler heat exchanger. Notice that the heat from the cooler and the work from the compression piston have the same cycle energy! This is consistent with the zero-net heat transfer of the regenerator (solid green line). As would be expected, the heater and the expansion space both have positive energy flow. The black dotted line shows the net work output of the cycle. On this trace, the cycle ends higher that it started, indicating that the heat engine converts energy from heat into work.

Atkinson Cycle

The Atkinson cycle engine is a type of internal combustion engine invented by James Atkinson in 1882. The Atkinson cycle is designed to provide efficiency at the expense of power density, and is used in some modern hybrid electric applications.

Design

The original Atkinson cycle piston engine allowed the intake, compression, power, and exhaust strokes of the four-stroke cycle to occur in a single turn of the crankshaft and was designed to avoid infringing certain patents covering Otto cycle engines.

Due to the unique crankshaft design of the Atkinson, its expansion ratio can differ from its compression ratio and, with a power stroke longer than its compression stroke, the engine can achieve greater thermal efficiency than a traditional piston engine. While Atkinson's original design is no more than a historical curiosity, many modern engines use unconventional valve timing to produce the effect of a shorter compression stroke/longer power stroke, thus realizing the fuel economy improvements the Atkinson cycle can provide.

Ideal Thermodynamic Cycle

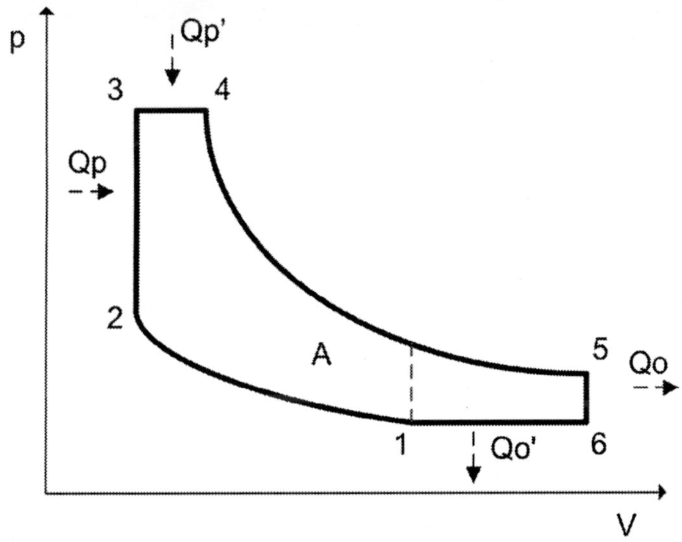

Figure: *Atkinson Gas Cycle*

The ideal Atkinson cycle consists of following operations:

- 1-2 Isentropic or reversible adiabatic compression

- 2-3 Isochoric heating (Qp)
- 3-4 Isobaric heating (Qp')
- 4-5 Isentropic expansion
- 5-6 Isochoric cooling (Qo)
- 6-1 Isobaric cooling (Qo')

Modern Atkinson Cycle Engines

Figure: *A small engine with Atkinson-style linkages between the piston and flywheel. Modern Atkinson cycle engines do away with this complex energy path.*

Recently, Atkinson cycle has been used to describe a modified Otto cycle engine in which the intake valve is held open longer than normal to allow a reverse flow of intake air into the intake manifold. The effective compression ratio is reduced (for a time the air is escaping the cylinder freely rather than being compressed) but the expansion ratio is unchanged. This means the compression ratio is smaller than the expansion ratio. Heat gained from burning fuel increases the pressure, thereby forcing the piston to move, expanding the air volume beyond the volume when compression began. The goal of the modern Atkinson cycle is to allow the pressure in the combustion chamber at the end of the power stroke to be equal to atmospheric pressure; when this occurs, all the available energy has been obtained from the combustion process. For any given portion of air, the greater expansion ratio allows more energy to be converted from heat to useful mechanical energy meaning the engine is more efficient.

The disadvantage of the four-stroke Atkinson cycle engine versus the more common Otto cycle engine is reduced power density. Due to a smaller portion of the compression stroke being devoted to compressing the intake air, an Atkinson cycle engine does not take in as much air as would a similarly designed and sized Otto cycle engine.

Four-stroke engines of this type with this same type of intake valve motion but with a supercharger to make up for the loss of power density are known as Miller cycle engines.

Rotary Atkinson Cycle Engine

The Atkinson cycle can be used in a rotary engine. In this configuration an increase in both power and efficiency can be achieved when compared to the Otto cycle. This type of engine retains the one power phase per revolution, together with the different compression and expansion volumes of the original Atkinson cycle. Exhaust gases are expelled from the engine by compressed-air scavenging. This modification of the Atkinson cycle allows the use of alternative fuels like diesel and hydrogen. Disadvantages of this design include the requirement that rotor tips seal very tightly on the outer housing wall and the mechanical losses suffered through friction between rapidly oscillating parts of irregular shape.

Vehicles Using Atkinson Cycle Engines

While a modified Otto cycle engine using the Atkinson cycle provides good fuel economy, it is at the expense of a lower power-per-displacement as compared to a traditional four-stroke engine. If demand for more power is intermittent, the power of the engine can be supplemented by an electric motor during times when more power is needed. This forms the basis of an Atkinson cycle-based hybrid electric drivetrain. These electric motors can be used independently of, or in combination with, the Atkinson cycle engine, to provide the most efficient means of producing the desired power. This drive train first entered production in late 1997 in the Japanese-market Toyota Prius.

At this writing, most production full hybrid-electric vehicles use Atkinson cycle engines:

- Chevrolet Tahoe Hybrid electric (four-wheel drive) with a compression ratio of 10.8:1
- Ford C-Max (front wheel drive / US market) hybrid & plug-in hybrid models
- Ford Escape/Mercury Mariner/Mazda Tribute electric (front- and four-wheel drive) with a compression ratio of 12.4:1

- Ford Fusion Hybrid/Mercury Milan Hybrid/Lincoln MKZ Hybrid electric (front-wheel drive) with a compression ratio of 12.3:1
- Honda Accord Plug-in Hybrid
- Honda Accord Hybrid (front-wheel drive)
- Hyundai Sonata Hybrid (front-wheel drive)
- Infiniti M35h Hybrid (rear-wheel drive)
- Kia Optima Hybrid (front-wheel drive)
- Lexus CT 200h (front-wheel drive)
- Lexus GS 450h hybrid electric (rear-wheel drive) with a compression ratio of 13:1
- Lexus HS 250h (front-wheel drive)
- Lexus RX 450h hybrid electric (four-wheel drive)
- Mazda 3 SkyActiv (front-wheel drive) with a 13:1 compression ratio (12:1 for North America)
- Mazda 6 SkyActiv-G 2.5L (front-wheel drive) with a 14:1 compression ratio (13:1 for North America)
- Mazda CX-5 (front- and all-wheel drive) with a 14:1 compression ratio (13:1 for North America)
- Mercedes ML450 Hybrid (four-wheel drive) electric
- Mercedes S400 Blue Hybrid (rear-wheel drive) electric
- Toyota Camry Hybrid electric (front-wheel drive) with a compression ratio of 12.5:1
- Toyota Highlander Hybrid (2011 and newer)
- Toyota Prius hybrid electric (front-wheel drive) with a (purely geometric) compression ratio of 13.0:1
- Toyota Yaris Hybrid (front-wheel drive) with a compression ratio of 13.4:1

Ericsson Cycle

The Ericsson cycle is named after inventor John Ericsson, who designed and built many unique heat engines based on various thermodynamic cycles. He is credited with inventing two unique heat engine cycles and developing practical engines based on these cycles. His *first* cycle is very similar to what is now called the "Brayton cycle", with the exception that it uses external combustion. His second cycle is now called the Ericsson cycle.

Ideal Ericsson Cycle

The following is a list of the four processes that occur between the four stages of the ideal Ericsson cycle:

- Process 1 -> 2: Isothermal compression. The compression space is assumed to be intercooled, so the gas undergoes isothermal compression. The compressed air flows into a storage tank at constant pressure. In the ideal cycle, there is no heat transfer across the tank walls.
- Process 2 -> 3: Isobaric heat addition. From the tank, the compressed air flows through the regenerator and picks up heat at a high constant-pressure on the way to the heated power-cylinder.
- Process 3 -> 4: Isothermal expansion. The power-cylinder expansion-space is heated externally, and the gas undergoes isothermal expansion.
- Process 4 -> 1: Isobaric heat removal. Before the air is released as exhaust, it is passed back through the regenerator, thus cooling the gas at a low constant pressure, and heating the regenerator for the next cycle.

Comparison with Stirling and Carnot Cycles

The Ericsson cycle is often compared to the Stirling cycle, since the engine designs based on these respective cycles are both external combustion engines with regenerators. The Ericsson is perhaps most similar to the so called "double-acting" type of Stirling engine, in which the displacer piston also acts as the power piston. Theoretically, both of these cycles have so called *ideal* efficiency, which is the highest allowed by the second law of thermodynamics. The most well known ideal cycle is the Carnot cycle, although a useful *Carnot engine* is not known to have been invented. The theoretical efficiencies for both, Ericsson and Stirling cycles acting in the same limits are equal to the Carnot Efficiency for same limits.

Comparison with the Brayton Cycle

The first cycle Ericsson developed is now called the "Brayton cycle", commonly applied to the rotary jet engines for airplanes.

The second Ericsson cycle is the cycle most commonly referred to as simply the "Ericsson cycle". The (second) Ericsson cycle is also the limit of an ideal gas-turbine Brayton cycle, operating with multistage intercooled compression, and multistage expansion with reheat and regeneration. Compared to the Brayton cycle which uses adiabatic compression and expansion, the second Ericsson cycle uses isothermal compression and expansion, thus producing more net work per stroke. Also the use of regeneration in the Ericsson cycle increases efficiency by reducing the required heat input. For further comparisons of thermodynamic cycles.

Cycle/Process	*Compression*	*Heat addition*	*Expansion*	*Heat rejection*
Ericsson (First, 1833)	adiabatic	isobaric	adiabatic	isobaric
Ericsson (Second, 1853)	isothermal	isobaric	isothermal	isobaric
Brayton (Turbine)	adiabatic	isobaric	adiabatic	isobaric

Ericsson Engine

The Ericsson engine is based on the Ericsson cycle, and is known as an "external combustion engine", because it is externally heated. To improve efficiency, the engine has a regenerator or recuperator between the compressor and the expander. The engine can be run open- or closed-cycle. Expansion occurs simultaneously with compression, on opposite sides of the piston.

Regenerator

Ericsson coined the term "regenerator" for his independent invention of the mixed-flow counter-current heat exchanger. However, Rev. Robert Stirling had invented the same device, prior to Ericsson, so the invention is credited to Stirling. Stirling called it an "economiser" or "economizer", because it increased the fuel economy of various types of heat processes. The invention was found to be useful, in many other devices and systems, where it became more widely used, since other types of engines became favoured over the Stirling engine. The term "regenerator" is now the name given to the component in the Stirling engine.

The term "recuperator" refers to a separated-flow, counter-current heat exchanger. As if this weren't confusing enough, a mixed-flow regenerator is sometimes used as a quasi-separated-flow recuperator. This can be done through the use of moving valves, or by a rotating regenerator with fixed baffles, or by the use of other moving parts. When heat is recovered from exhaust gases and used to preheat combustion air, typically the term recuperator is used, because the two flows are separate.

History

In 1791, before Ericsson, John Barber proposed a similar engine. The Barber engine used a bellows compressor and a turbine expander, but it lacked a regenerator/recuperator. There are no records of a working Barber engine. Ericsson invented and patented his first engine using an external version of the Brayton cycle in 1833 (number 6409/1833 British). This was 18 years before Joule and 43 years before Brayton. Brayton engines were all piston engines and for the most part, internal combustion versions of the un-recuperated Ericsson engine. The "Brayton cycle" is now known as the gas turbine cycle, which differs from the original "Brayton cycle" in the use of a turbine compressor and expander. The gas turbine cycle is used for all modern gas turbine and turbojet engines, however simple cycle turbines are often recuperated to improve efficiency and these recuperated turbines more closely resemble Ericsson's work.

Ericsson eventually abandoned the open cycle in favour of the traditional closed Stirling cycle.

Ericsson's engine can easily be modified to operate in a closed-cycle mode, using a second, lower-pressure, cooled container between the original exhaust and intake. In closed cycle, the lower pressure can be significantly above ambient pressure, and He or H_2 working gas can be used. Because of the higher pressure difference between the upward and downward movement of the work-piston, specific output can be greater than of a valveless Stirling engine. The added cost is the valve. Ericsson's engine also minimizes mechanical losses: the power necessary for compression does not go through crank-bearing frictional losses, but is applied directly from the expansion force. The piston-type Ericsson engine can potentially be the highest efficiency heat engine arrangement ever constructed. Admittedly, this has yet to be proven in practical applications.

The Ericsson-cycle engine (the second of the two discussed here) was used to power a 2,000-ton ship, the caloric ship *Ericsson*, and ran flawlessly for 73 hours. The combination engine produced about 300 horsepower (220 kW). It had a combination of four dual-piston engines; the larger expansion piston/cylinder, at 14 feet (4.3 m) in diametre, was perhaps the largest piston ever built. Rumor has it that tables were placed on top of those pistons (obviously in the cool compression chamber, not the hot power chamber) and dinner was served and eaten, while the engine was running at full power. At 6.5 RPM the pressure was limited to 8 psi (55 kPa). According to the official report it only consumed 4200 kg coal per 24 hours (original target was 8000 kg, which is still better than contemporary steam engines). The one sea trial proved that even though the engine ran well, the ship was underpowered. Sometime after the trials, the *Ericsson* sank. When it was raised, the Ericsson-cycle engine was removed and a steam engine took its place.

Ericsson designed and built a very great number of engines running on various cycles including steam, Stirling, Brayton, externally heated diesel air fluid cycle. He ran his engines on a great variety of fuels including coal and solar heat.

Ericsson was also responsible for an early use of the screw propeller for ship propulsion, in the USS *Princeton*, built in 1842–43.

Today's Potential

The Ericsson cycle (and the similar Brayton cycle) receives renewed interest today to extract power from the exhaust heat of gas (and producer gas) engines and solar concentrators. An important advantage of the Ericsson cycle over the widely known Stirling engine is often not recognized: the volume of the heat exchanger does not adversely affect the efficiency. For medium and large engines the cost of valves can be small compared to this advantage. Turbocompressor plus turbine implementations seem favourable in the MWe range, positive displacement compressor plus turbine for Nx100 kWe power, and positive displacement compressor+expander below 100 kW. With high temperature hydraulic fluid, both the compressor and the expander can be liquid ring pump even up to 400°C, with rotating casing for best efficiency.

Joule Cycle

The Brayton cycle or Joule Cycle is a thermodynamic cycle that describes the workings of a constant pressure heat engine. Gas turbine engines and airbreathing jet engines use the Brayton Cycle. Although the Brayton cycle is usually run as an open system (and indeed *must* be run as such if internal combustion is used), it is conventionally assumed for the purposes of thermodynamic analysis that the exhaust gases are reused in the intake, enabling analysis as a closed system.

The engine cycle is named after George Brayton (1830–1892), the American engineer who developed it, although it was originally proposed and patented by Englishman John Barber in 1791. It is also sometimes known as the Joule cycle. The Ericsson cycle is similar to the Brayton cycle but uses external heat and incorporates the use of a regenerator. There are two types of Brayton cycles, open to the atmosphere and using internal combustion chamber or closed and using a heat exchanger.

History

Figure: *Brayton Ready Motor*

In 1872, George Brayton applied for a patent for his Ready Motor. It is a reciprocating constant pressure engine. The engine used a separate piston compressor and expander. The compressed air was heated by internal fire as it entered the expander cylinder.

The first versions of the Brayton engine mixed vaporized fuel with air as it entered the compressor by means of a heated-surface carburetor., The fuel / air was contained in a reservoir / tank and then it was admitted to the expansion cylinder and burned. As the fuel / air mixture entered the expansion cylinder it was ignited by a pilot flame. A screen was used to prevent the fire from entering / returning to the reservoir.

This screen sometime failed and an explosion would occur. In 1874 Brayton solved the explosion problem by adding the fuel just prior to the expander cylinder. The engine now used heavier fuels such as kerosine and fuel oil. Ignition remained pilot flame. Brayton produced and sold "Ready Motors" to perform a variety of tasks like water pumping, mill operation, even marine propulsion. Critics of the day claimed the engines ran smoothly and had an efficiency of about 17%.

***Figure:** this Brayton cycle piston engine was made from historical drawings and pictures*

Brayton cycle engines were some of the first internal combustion engines used for motive power. In 1881 John Holland used a Brayton engine to power the worlds first successful self propelled submarine, the Fienian Ram John Philip Holland's submarine is preserved in the Paterson Museum in the Old Great Falls Historic District of Paterson, New Jersey.

In 1878, George B. Selden produced the first internal combustion automobile. Inspired by the internal combustion engine invented by George Brayton displayed at the Centennial Exposition in Philadelphia in 1876, Selden began working on a smaller lighter version, succeeding by 1878, some eight years before the public introduction of the Benz Patent Motorwagen in Europe.

The Selden auto was powered by a 3-cylinder, 400-pound version of the Brayton Cycle engine which featured an enclosed crankshaft.

Selden designed and constructed the engine with the help of Rochester machinist Frank H. Clement and his assistant William Gomm.

He filed for a patent on May 8, 1879 (in a historical cross of people, the witness Selden chose was a local bank-teller, George Eastman, later to become famous for the Kodak camera).

His application included not only the engine but its use in a 4 wheeled car. He then filed a series of amendments to his application which stretched out the legal process resulting in a delay of 16 years before the patent was granted on November 5, 1895.

Henry Ford fought the Selden patent. Ford argued his cars used the four-stroke Otto cycle and not the Brayton engine shown used in the Selden auto. Ford won the appeal of the original case.

Models

A Brayton-type engine consists of three components:

1. a compressor
2. a mixing chamber
3. an expander

In the original 19th-century Brayton engine, ambient air is drawn into a piston compressor, where it is compressed; ideally an isentropic process. The compressed air then runs through a mixing chamber where fuel is added, an isobaric process.

The heated (by compression), pressurized air and fuel mixture is then ignited in an expansion cylinder and energy is released, causing the heated air and combustion products to expand through a piston/ cylinder; another ideally isentropic process. Some of the work extracted by the piston/cylinder is used to drive the compressor through a crankshaft arrangement.

The term Brayton cycle has more recently been given to the gas turbine engine. This also has three components:

1. a gas compressor
2. a burner (or combustion chamber)
3. an expansion turbine

Ideal Brayton cycle:

- isentropic process - ambient air is drawn into the compressor, where it is pressurized.

- isobaric process - the compressed air then runs through a combustion chamber, where fuel is burned, heating that air—a constant-pressure process, since the chamber is open to flow in and out.
- isentropic process - the heated, pressurized air then gives up its energy, expanding through a turbine (or series of turbines). Some of the work extracted by the turbine is used to drive the compressor.
- isobaric process - heat rejection (in the atmosphere).

Actual Brayton cycle:

- adiabatic process - compression.
- isobaric process - heat addition.
- adiabatic process - expansion.
- isobaric process - heat rejection.

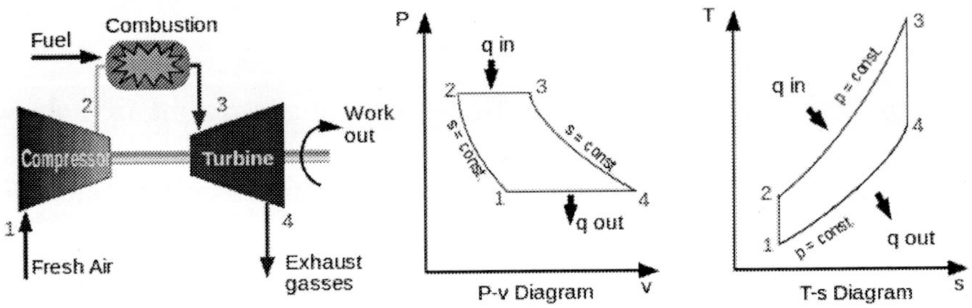

Figure: *Idealized Brayton cycle*

Since neither the compression nor the expansion can be truly isentropic, losses through the compressor and the expander represent sources of inescapable working inefficiencies. In general, increasing the compression ratio is the most direct way to increase the overall power output of a Brayton system.

The efficiency of the ideal Brayton cycle is $\eta = 1 - \frac{T_1}{T_2} = 1 - \left(\frac{P_1}{P_2}\right)^{(\gamma-1)/\gamma}$, where γ is the heat capacity ratio. Figure indicates how the cycle efficiency changes with an increase in pressure ratio. Figure indicates how the specific power output changes with an increase in the gas turbine inlet temperature for two different pressure ratio values.

The highest temperature in the cycle occurs at the end of the combustion process, and it is limited by the maximum temperature that the turbine blades can withstand.

This also limits the pressure ratios that can be used in the cycle. For a fixed turbine inlet temperature, the net work output per cycle increases with the pressure ratio (thus the thermal efficiency) and the net work output.

With less work output per cycle, a larger mass flow rate (thus a larger system) is needed to maintain the same power output, with may not be economical. In most common designs, the pressure ratio of a gas turbine ranges from about 11 to 16.

Methods to Increase Power

The power output of a Brayton engine can be improved in the following manners:

- *Reheat*, wherein the working fluid—in most cases air—expands through a series of turbines, then is passed through a second combustion chamber before expanding to ambient pressure through a final set of turbines. This has the advantage of increasing the power output possible for a given compression ratio without exceeding any metallurgical constraints (typically about 1000 °C). The use of an afterburner for jet aircraft engines can also be referred to as “reheat”; it is a different process in that the reheated air is expanded through a thrust nozzle rather than a turbine. The metallurgical constraints are somewhat alleviated, enabling much higher reheat temperatures (about 2000 °C). Reheat is most often used to improve the specific power (per throughput of air), and is usually associated with a drop in efficiency, this effect is especially pronounced in afterburners due to the extreme amounts of extra fuel used.
- *Overspray*, wherein, after a first compressor stage, water is injected into the compressor, thus increasing the mass-flow inside the compressor, increasing the turbine output power significantly and reducing compressor outlet temperatures. In a second compressor stage the water is completely converted to a gas form, offering some intercooling via its latent heat of vaporization.

Methods to Improve Efficiency

The efficiency of a Brayton engine can be improved in the following manners:

- *Increasing pressure ratio* - As Figure above shows, increasing the pressure ratio increases the efficiency of the Brayton cycle. This is analogous to the increase of efficiency seen in the Otto cycle when the compression ratio is increased. However, there are practical limits when it comes to increasing the pressure ratio. First of all, increasing the pressure ratio increases the compressor discharge temperature. This can cause the temperature of the gasses leaving the combustor to exceed the metallurgical limits of the turbine. Also, the diametre of the compressor blades becomes progressively smaller in higher pressure stages of the compressor. Because the gap between the blades and the engine casing increases in size as a percentage of the compressor blade height as the blades get smaller in diametre, a greater percentage of the compressed air can leak back past the blades in higher pressure stages. This causes a drop in compressor efficiency, and is most likely to occur in smaller gas turbines (since blades are inherently smaller to begin with). Finally, as can be seen in Figure, the efficiency levels off as pressure ratio increases. Hence, there is little to gain by increasing the pressure ratio further if it is already at a high level.
- *Regeneration*, wherein the still-warm post-turbine fluid is passed through a heat exchanger to preheat the fluid just entering the combustion chamber. This directly offsets fuel consumption for the same operating conditions, improving efficiency; it also results in less power lost as waste heat. However, at higher pressure ratios, the compressor discharge temperature can exceed the exhaust temperature. Under these conditions, regeneration would be counterproductive. Therefore, regeneration is only an option when the pressure ratio is sufficiently low that the exhaust temperature is higher than the compressor discharge temperature.
- *Recuperator* - A recuperator is a heat exchanger that acts as a counter-flow energy recovery device positioned within the

supply and exhaust air streams of an air handling system, in order to recover the waste heat. In many processes, heat is generated in the combustion chamber by combustion, and the recuperator helps in reclaiming this heat, for re-using or recycling. Adding a recuperator to the system can increase the overall efficiency of the Brayton cycle. For example, a gas-turbine engine works on the basic Brayton cycle. In this gas-turbine engine, gas is compressed, mixed with fuel which is then burned and then produces power in the turbine. The recuperator transfers some of the waste heat in the exhaust to the compressed air, thus preheating it before entering the fuel burner stage. Since the gases have been pre-heated, less fuel is needed to heat the gases up to the turbine inlet temperature. Because some of the energy (that would be usually lost as waste heat) is recovered, the recuperator increases the efficiency of this cycle.

This feature is only available if the exhaust heat is not used otherwise, as in cogeneration or combined cycle applications.

- A Brayton engine also forms half of the 'combined cycle' system, which combines with a Rankine engine to further increase overall efficiency. However, although this increases overall efficiency, it does not actually increase the efficiency of the Brayton cycle itself.
- *Cogeneration* systems make use of the waste heat from Brayton engines, typically for hot water production or space heating.

Variants

Closed Brayton Cycle: A closed Brayton cycle recirculates the working fluid, the air expelled from the turbine is reintroduced into the compressor, this cycle use a heat exchanger to heat the working fluid instead of an internal combustion chamber. The closed Brayton cycle is used for example in closed-cycle gas turbine and space power generation.

Solar Brayton Cycle

In 2002 a hybrid open solar Brayton cycle was operated for the first time consistently and effectively with relevant papers published, in the frame of the EU SOLGATE programme. The air was heated from 570 K to over 1000 K into the combustor chamber. Further hybridization was achieved during the EU Solhyco project running a hybridized Brayton cycle with solar energy and Biodiesel only.

Reverse Brayton Cycle

A Brayton cycle that is driven in reverse, via net work input, and when air is the working fluid, is the air refrigeration cycle or Bell Coleman cycle. Its purpose is to move heat, rather than produce work. This air cooling technique is used widely in jet aircraft.

Bibliography

Alberty, Robert, A.: *Biochemical Thermodynamics: Applications of Mathematica (Methods of Biochemical Analysis)*, Wiley-Interscience, 1968.

Beizer, Boris: *Software Testing Techniques,* International Thomson Computer Press, Boston, MA., 2004.

Bowker, R.: *The Nature of the Chemical Bond*, London: Free Press, 1995.

Callen, G.B.: *Thermodynamics and an Introduction to Thermostatistics*, Wiley, New York, 1985.

Cengel, Y. & Boles, M.: *Thermodynamics – an Engineering Approach,* New York: McGraw Hill, 2002.

De Groot, S.R., Mazur, P.: *Non-equilibrium Thermodynamics*, North-Holland, Amsterdam, 1962.

Defay, R.: *Chemical Thermodynamics*, Longmans, Green & Co., London, 1950.

Friedman, Ronald: *Molecular Quantum Mechanics*, Oxford University Press, New York, 2005.

George B. Hogaboom: *Principles of Electroplating and Electroforming (Electrotyping)*, McGraw-Hill, New York, 1949.

Guggenheim, E.A.: *Modern Thermodynamics by the Methods of J.W. Gibbs*, Methuen, London, 1933.

Harned, H. S., and B. B. Owen: *The Physical Chemistry of Electrolytic Solutions*, ACS Monograph 137, Reinhold, New York, 1958.

Haynie, Donald, T.: *Biological Thermodynamics*. Cambridge University Press, 2001.

Higuchi, T. : *Solubility Behavior ofOrganic Compounds*, Wiley-Interscience, New York, 1990.

Jackson, J.D.: *Classical Electrodynamics*, John Wiley and Sons, New York, 1975.

Jaynes, E.T.: *Probability Theory: The Logic of Science*, Cambridge University Press, Cambridge, 2003.

John A.: *Quantum Theory and Measurement*, Princeton: Princeton University Press, 2005.

Johnston, Jr., E.R.: *Mechanics of Materials*, McGraw-Hill, Inc., New York, 1981.

Keizer, J.: *Statistical Thermodynamics of Nonequilibrium Processes*, Springer-Verlag, New York, 1987.

Lavenda, B.H.: *Thermodynamics of Irreversible Processes*, Macmillan, London, 1978.

Lewis, Gilbert N.; Randall, Merle: *Thermodynamics and the Free Energy of Chemical Substances*. McGraw-Hill Book Co. Inc., 1923.

Marder, M. P.: *Condensed Matter Physics,* New York: John Wiley & Sons, 2006.

Mihalas, D., Weibel-Mihalas, B.: *Foundations of Radiation Hydrodynamics*, Oxford University Press, New York, 1984.

Mukherjee, P.N. : *A Textbook of Objective Organic Chemistry*, Wisdom Press, Delhi, 2011.

Nicholson, J. W.: *The Chemistry of Polymers,* Oxford: Blackwell Publishers, 2005.

Perrot, Pierre: *A to Z of Thermodynamics*. Oxford University Press, 1998.

Pooja Bhagwan: *A Handbook of Inorganic Chemistry : Nuclear, Atomic, Aqueous, Alkali and Alkaline*, ISPA, Delhi, 2005.

Rabek, J.: *Experimental Methods in Physical Chemistry and Photophysics*, New York: Oxford University Press, 1989.

Roland, O.: *Understanding Quantum Mechanics*, Princeton: Princeton University Press, 1998.

Sakurai: J. J.: *Modern Quantum Mechanics*, New York: John Wiley & Sons, 2006.

Sawyer, L. C.: *Polymer Microscopy*, London, Zed Books Publication, 2004.

Thompson, L. M.: *Theoretical Hydrodynamics*, Macmillan, India, 1968.

Walters, P.: *The New Quantum Universe*, Cambridge: University Press, 2004.

Wang, Y.: *Physical Chemistry and Photobiology of Nucleic Acids: Chemistry*, Athens: U of Georgia Press, 2005.

Weibel-Mihalas, B.: *Foundations of Radiation Hydrodynamics*, Oxford University Press, New York, 1984.

Index

J

L

M

N

O

P

Q

R

S

T

V

W

Z

❑❑❑